Preface

This book represents an attempt to present the core material for any first university course in the subject and upon which more advanced courses depend for their bases. It represents an evolution and up-dating of material contained in two previous books written by one of us (J. B.) with the addition of other new material by M. A. M. (Chapters 2, 8, 9 and 11 and the problems and their solutions for all the chapters). The book's contents are constrained by the advice of various anonymous and not-so-anonymous referees (some left their marks on their comments, Tip-Ex™ is no match for a chemist!) and the apparent necessity of producing a book of fewer than 320 pages. Thoughts of attempts at putting cubic metres into one litre flasks come to mind! Inorganic chemistry is an ever-expanding subject and university courses are becoming more and more diverse so the idea of producing one textbook which will be suitable for this diversity is impossible, with restraints of size and cost. The word *fundamentals* in the title needs some definition, in case it would appear to be somewhat over-optimistic. The OED provides two definitions of the word. One is *primary, from which others are derived.* The other is *pertaining to the foundation, basis or groundwork.* Of these definitions it is *groundwork* which is most relevant to the approach to the subject in this book. With this general approach to the subject, it is hoped that the book will be sufficient for the basic work to be covered in the first two years of a university course and that students will then be in a position to tackle both more specialist topics and to look further into the fundamentals of the subject as described by the first definition of the word.

In the theoretical treatment of bonding and molecular shape notice has been taken of R. S. Mulliken's opinion that *there are no electron deficient compounds, just deficient theories*! It is hoped that the reader will derive a view of these topics which extends beyond the ball-and-stick approach and will be tempted to undertake further and deeper studies.

Jack Barrett
Imperial College of Science, Technology and Medicine, University of London
Mounir A. Malati
Mid-Kent College of Higher and Further Education
1998

Acknowledgements

Both authors acknowledge the welcome comments of John Burgess of the University of Leicester and J. B. would like to thank David Goodgame of Imperial College for his very helpful and critical comments on the previous books upon which the theoretical chapters of this book are based. M. A. M. would like to thank Pamela Elsegood for word-processing his part of the manuscript (Chapters 2, 8, 9 and 11) and John-Paul Phillips for supplying the computer-based diagrams which form the basis of those that appear in those chapters. We are both grateful to the many students who have, by their many and varied questions, given us food for thought about the presentation of our subject.

Symbols and abbreviations used in the text

R.A.M. Relative atomic mass
R.M.M. Relative molar mass
ν frequency
λ wavelength
c velocity of light
Z atomic number, nuclear charge
A mass number
$t_{½}$ half-life
R Rydberg constant
n principal quantum number
l secondary quantum number
m_l magnetic quantum number
m_s spin quantum number
h Planck's constant
ψ atomic wavefunction
ϕ molecular wavefunction
ϵ_0 permittivity of a vacuum
a_0 Bohr radius, atomic unit of length
I ionization enthalpy
χ electronegativity coefficient
E_{ea} electron attachment enthalpy
L lattice enthalpy
D dissociation enthalpy
E(M–X) bond enthalpy term for the bond M–X
VSEPR valence shell electron pair repulsion
V.B. valence bond
M.O. molecular orbital
C.N. coordination number
$E^{\ominus}$ standard reduction potential
N_A Avogadro's number or constant
Δ_o octahedral splitting energy
ϵ molar absorption coefficient
Δ*G* change in Gibbs energy
Δ*H* change in enthalpy
Δ*S* change in entropy
The London Undergound symbol, $^{\ominus}$ is used to denote standard state, i.e. the state of an element, compound or ion at 1 atmosphere pressure and a particular temperature (usually 298 K unless otherwise specified).

FUNDAMENTALS OF INORGANIC CHEMISTRY
An introductory text for degree course studies

"Talking of education, people have now a-days" (said he) "got a strange opinion that every thing should be taught by lectures. Now, I cannot see that lectures can do so much good as reading the books from which the lectures are taken. I know nothing that can be best taught by lectures, except where experiments are to be shewn. You may teach chymestry by lectures — You might teach making of shoes by lectures!"

James Boswell: *Life of Samuel Johnson, 1766* (1709-1784)

"Every aspect of the world today - even politics and international relations - is affected by chemistry"

Linus Pauling, Nobel Prize winner for Chemistry, 1954, and
Nobel Peace Prize, 1962

"We chemists have not yet discovered how to make gold but, in contentment and satisfaction with our lot, we are the richest people on Earth."

Lord George Porter, OM, FRS
Nobel Prize winner for chemistry, 1967

ABOUT OUR AUTHORS

JACK BARRETT, BSc, PhD, CChem, MRSC

Jack Barrett's early obsession with this subject was so strong that during history lessons at the Grammar School in Chorley, Lancashire he dreamed of becoming a chemist, which might have damaged his broader education. Nevertheless, he obtained a State Scholarship to read chemistry at Manchester University and, after taking his first degree, stayed on to research photochemistry and liquid water, for which thesis he was awarded his PhD. After this followed periods of teaching and research at the University of Hertfordshire; and then Chelsea College, Queen Elizabeth College and King's College, all of London University. He is currently Personal Assistant to Professor Mike Mingos at the Imperial College of Science, Technology and Medicine, London University.

His elucidation of the mechanism of bacterial oxidation of gold-bearing minerals has been put to practical use in gold production plants world-wide, and recent work has argued strongly for a more scientific approach to evaluating likely climate change in the next century. He has written numerous papers and is the author of four other books: *An Introduction to Atomic and Molecular Structures* (John Wiley); *Understanding Inorganic Chemistry* and *Metal Extraction by Bacterial Oxidation of Minerals* (Ellis Horwood Limited); and in Ellis Horwood's new company *Chemistry in your Environment* (Horwood Publishing Limited)

MOUNIR MALATI, BSc, PhD, CChem, FRSC

Mounir Malati gained his BSc from the University of Cairo in 1943, and between 1943 and 1946 was Demonstrator in Chemistry, first in Alexandria and then in Cairo University. He taught science in Nile Independent School, Cairo from 1946 to 1953. He then came to England where he worked at Leeds University in the Physical Chemistry Department on full-time research and was awarded his PhD in 1957. Back in Egypt, he joined the National Research Centre in Giza, where he was a Research Associate Professor.

He worked in England at the Imperial College of Science, Technology and Medicine, London University for one year in 1963 and then became Senior Research Officer at Teddington Paint Research Station until 1965. That year, he joined Medway College of Technology (now Mid-Kent College of Higher and Further Education) where he became Senior Lecturer, and now continues as part-time lecturer. He has supervised numerous MSc and PhD degrees and has taught extensively on inorganic, physical and radiation chemistry over the years; and has read papers at international conferences in the UK, North America, Europe and Japan. His considerable contribution to science is reflected by an output of 150 published papers.

FUNDAMENTALS OF INORGANIC CHEMISTRY

An introductory text for degree course studies

Jack Barrett
Imperial College of Science, Technology and Medicine
University of London

and

Mounir A. Malati
Mid-Kent College of Higher and Further Education
Chatham

Horwood Publishing
Chichester

First published in 1998 by
HORWOOD PUBLISHING LIMITED
International Publishers
Coll House, Westergate, Chichester, West Sussex, PO20 6QL England

British Library Cataloguing in Publication Data
A catalogue record of this book is available from the British Library

ISBN 1-898563-38-1

Printed in Great Britain by Martins Printing Group, Bodmin, Cornwall

Contents

Contents

Contents

Contents

1

Introduction

This introductory chapter contains sections which are concerned with descriptions of chemically important fundamental particles, the nature of electromagnetic radiation and its relationship to changes of energy in nuclei, atoms, molecules and metals. A section is included which demonstrates the broad scope of modern inorganic chemistry and its applications. The last section of the chapter consists of a very brief outline of the remainder of the book.

1.1 FUNDAMENTAL PARTICLES

To describe adequately the chemical properties of atoms and molecules it is necessary only to consider three fundamental particles; protons and neutrons which are contained within atomic nuclei, and electrons which surround the nuclei. Protons and neutrons are composite particles, each consisting of three quarks, and are therefore not fundamental particles in the true sense of that term. They may, however, be regarded as being fundamental particles for all chemical purposes. The physical properties of electrons, protons and neutrons are given in Table 1.1, together with data for the hydrogen atom. The absolute masses are given, together with their values on the relative atomic mass (R.A.M.) scale which is based on the unit of mass being equal to that of one twelfth of the mass of the ^{12}C isotope, i.e. the R.A.M. of ^{12}C = 12.0000 exactly.

Table 1.1 Properties of some fundamental particles and the hydrogen atom. (*e* is the elementary unit of electronic charge)

Particle	Symbol	Mass/10^{-27} kg	R.A.M.	Charge
Proton	*p*	1.6726231	1.0072765	+*e*
Neutron	*n*	1.6749286	1.0086649	zero
Electron	*e*	0.0009109	0.0005486	−*e*
H atom	H	1.6735339	1.007825	zero

The three chemically important fundamental particles are the constituents of atoms, molecules and infinite arrays which include a variety of crystalline substances and metals. The interactions between constructional units are summarized in Table 1.2.

Table 1.2 A summary of material construction

Units	Cohesive force	Products
Protons, neutrons	Nuclear	Nuclei
Nuclei, electrons	Atomic	Atoms
Atoms	Valence	Molecules, infinite arrays
Molecules	Intermolecular	Liquid and solid aggregations

1.2 ELECTROMAGNETIC RADIATION

The electromagnetic spectrum ranges from γ-rays, through X-rays, ultra-violet radiation, visible light, infra-red radiation, microwave radiation to long wavelength radio waves. The various regions are distinguished by their different wavelength λ and frequency ν ranges as given in Table 1.3.

Table 1.3 The electromagnetic spectrum

Region	Wavelength λ/m	Frequency ν/Hz
	$> 1.0 \times 10^{4}$	$< 3.0 \times 10^{4}$
Radio		
	1.0	3.0×10^{8}
Microwave		
	1.0×10^{-3}	3.0×10^{11}
Infra-red		
	7.0×10^{-7}	4.3×10^{14}
Visible		
	4.0×10^{-7}	7.5×10^{14}
Ultra-violet		
	1.0×10^{-9}	3.0×10^{17}
X-ray		
	1.0×10^{-10}	3.0×10^{18}
γ-ray		
	$< 1.0 \times 10^{-10}$	$> 3.0 \times 10^{18}$

The relationship between frequency and wavelength is given by the equation:

$$\nu = c_0/\lambda_0 \tag{1.1}$$

where c_0 and λ_0 represent the speed of light and its wavelength respectively in a vacuum. The speed of light, c, and its wavelength, λ, in any medium are dependent upon the refractive index, n, of that medium according to the equations:

$$c = c_0/n \tag{1.2}$$

and

$$\lambda = \lambda_0/n \tag{1.3}$$

Under normal atmospheric conditions n has a value which differs little from unity, e.g. at 550 nm in dry air the value of n is 1.0002771, so the effect of measuring wavelengths in air may be neglected for most purposes. The frequency of any particular radiation is independent of the medium.

Electromagnetic radiation is a form of energy which is particulate in nature. The terms wavelength and frequency do not imply that radiation is to be regarded as a wave motion in a real physical sense. They refer to the form of the mathematical functions which are used to describe the behaviour of radiation. The fundamental nature of electromagnetic radiation is embodied in quantum theory which explains all the properties of radiation in terms of quanta or photons; packets of energy. A cavity, e.g. an oven, emits a broad spectrum of radiation which is independent of the material from which the cavity is constructed, but is entirely dependent upon the temperature of the cavity. Cavity radiation is the politically correct term for what was once called black-body radiation. Planck was able to explain the frequency distribution of broad spectrum cavity radiation only by postulating that the radiation consisted of quanta with energies given by the equation:

$$E = h\nu \tag{1.4}$$

where h is Planck's constant, the equation being known as the Planck-Einstein equation.

The quantum behaviour of radiation was demonstrated by Einstein in his explanation of the photoelectric effect. If radiation of sufficient energy strikes a clean metal surface, electrons (photoelectrons) are emitted, one electron per quantum. The energy of the photoelectron, E_{el}, is given by the difference between the energy of the incident quantum and the work function , W, (ionization energy) of the metal surface:

$$E_{el} = h\nu - W \tag{1.5}$$

Photons with energies lower than the work function do not have the capacity to cause the release of photoelectrons.

The other major application of the Planck-Einstein relationship is in the interpretation of transitions between energy states or energy levels in nuclei, atoms, molecules and in infinite arrays such as metals. If any two states i and j have energies E_i and E_j respectively (with energy $E_i > E_j$) and the difference in energy between the two states is represented by ΔE_{ij}, the appropriate frequency of radiation which will cause the transition between them in absorption (i.e. if state j absorbs energy to become state i or will be emitted (i.e. if state i releases some of its energy to become state j) is that given by the equation:

$$\Delta E_{ij} = E_i - E_j = h\nu \tag{1.6}$$

This modification of the Planck-Einstein relationship was suggested by Bohr and is sometimes referred to as the *Bohr frequency condition.* It applies to a large range of transitions between energy states of nuclei and electrons in atoms, and rotational, vibrational and electronic changes in ions and molecules, and those responsible for the optical properties of metals and semi-conducting materials.

1.3 DEVELOPMENTS IN MODERN INORGANIC CHEMISTRY

The scale of inorganic chemistry research may be indicated by an analysis of the 1993 Volume 90 Section A (Inorganic Chemistry) of Annual Reports of the Royal Society of Chemistry. The report consists of a series of twenty-six chapters written by acknowledged experts in the particular fields represented. The volume amounts to 583 pages (58 of these are devoted to the author index) covering the work of 10260 chemists and their 4614 published papers, each page of the report on average mentioning the work contained in 8-9 papers. Some of the references are to published reviews of specialized subjects which cover large numbers of original papers.

The senior reporters make their personal choices of the papers which they consider to illustrate the diversity of current work in inorganic chemistry. The following sections include some of their choices and some made by the authors of this book. It is hoped that the examples chosen illustrate the value of inorganic chemical research, and will encourage the further study of chemistry. Any terms which are unfamiliar to the reader are defined and discussed in the main text. Perhaps this section should be re-read after the main sections have been studied.

Small molecules and ions

There are still new small molecules and ions being prepared. The explosive compound iodine azide (IN_3) has been produced by the reaction of diiodine with silver azide (AgN_3) and shown to be polymeric. It is a compound in which the iodine has a formal oxidation state of +1. The reaction of Me_3SiN_3 with ICl_2^+ AsF_6^- gives the ion $I(N_3)_2^+$ in which the iodine has a formal oxidation state of +3.

Although the complex ions $[Cu(NH_3)_2]^+$ and $[Ag(NH_3)_2]^+$ have been known for some time, the comparable gold(I) complex $[Au(NH_3)_2]^+$ was synthesized for the first time in 1994. In the compound $[Au(NH_3)_2]Br$, the N–Au–N arrangement in the diamminegold(I) ion has been shown to be linear.

Clusters

Molecules containing clusters of metal atoms form the basis of much research concerning their potential as homogeneous catalysts. It is well-known that finely divided metals, e.g. nickel, palladium and platinum, act as heterogeneous catalysts in many reactions. Clusters with relatively small numbers of metal atoms compared to the constituents of finely divided metals might well be found to catalyse reactions in the solution phase. Clusters also have given rise to a new branch of theoretical chemistry and the formulation of rules (the Wade-Mingos rules) to rationalize their geometries.

Cluster molecules contain at least three central atoms in order that they can be arranged in a planar or three-dimensional manner. Chain polymers of metal atoms are excluded by such a definition. The cluster atoms form structures which can be understood in terms of the five regular Platonic solids; tetrahedra (four equilateral triangular faces and four vertices), cubes (six square faces and eight vertices), octahedra (eight equilateral triangular faces and six vertices), dodecahedra (twelve pentagonal faces and eight vertices), and icosahedra (twenty equilateral triangular faces and twelve vertices), and other structures which may be thought of as fragments of the regular solids or capped (i.e. with one or more extra atoms bonded to a face) regular solids.

The structure of the cationic gold cluster (its structure was predicted by Mingos five years before its synthesis, also by the same author) with the formula $[Au_{13}Cl_2(PMe_2Ph)_{10}]^{3+}$ is shown in Fig. 1.1 and has twelve gold atoms in an icosahedral formation which has a central gold atom. The two chlorine atoms are in *trans* positions to each other, the other ten ligand positions are filled by the dimethylphenylphosphine molecules.

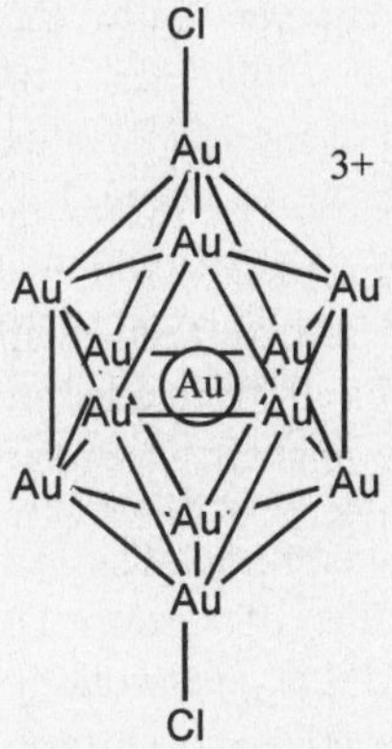

Fig. 1.1 The structure of the gold-atom-centred icosahedral +3 cation, $[Au_{13}Cl_2(PMe_2Ph)_{10}]$. The twelve peripheral gold atoms are each bonded to a dimethylphenylphosphine ligand.

The largest cluster known in 1981 was $Au_{55}(PPh_3)_{12}Cl_6$, but in 1993 the record belonged to the compound made by a reaction of copper(I) chloride with *bis*-trimethylsilylselenium, $Se(SiMe_3)_2$ and triphenylphosphine, PPh_3 (Ph = C_6H_5). The formula of the cluster is $[Cu_{146}Se_{73}(PPh_3)_{30}]$. Its structure is shown in Fig. 1.2.

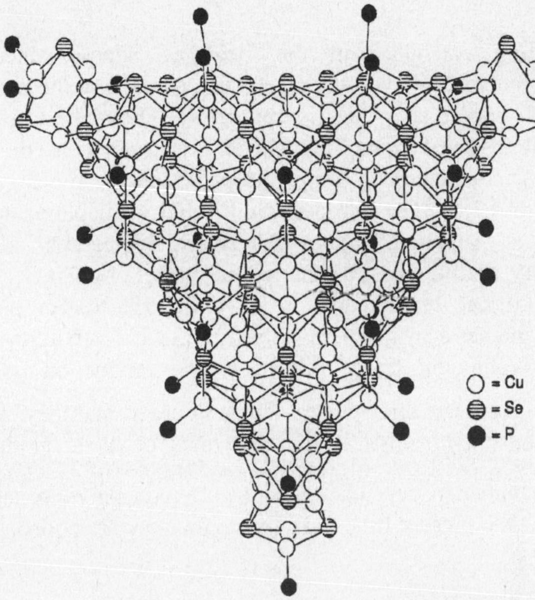

Fig 1.2 The structure of the cluster $[Cu_{146}Se_{73}(PPh_3)_{30}]$. The copper atoms are represented by open circles, the selenium atoms by hatched circles and the phosphorus atoms of the triphenylphosphine molecules by filled circles. (From Annual Reports on the Progress of Chemistry, Section A, 1993, Vol 90, p.75. Reproduced by permission of The Royal Society of Chemistry)

The scientific interest in this cluster arises from the observation that the one form of copper selenide, ß-Cu_2Se, has a selenium-atom layer structure and exhibits semi-metal electrical conductance. The cluster also shows electrical conductance, but not as high as that of the binary compound. This is thought to be probably because of the peripheral triphenylphosphine ligands attached to the outermost copper atoms and which effectively shield the Cu/Se core.

The sixty-atom carbon cluster, C_{60}, known as buckminsterfullerene (after the architect Buckminster Fuller who designed geodesic domes), fullerene or hexacontocarbon, was discovered in 1990 and has led to the award of the 1996 Nobel Prize for Chemistry to Sir Harry Kroto, Smalley and Curl. Its structure, shown in Fig. 1.3, is similar to the particular design of soccer ball which has a mixture of hexagonal and pentagonal panels, giving rise to another colloquial name for the cluster—soccerene. The compound K_3C_{60} consists of a closest packed arrangement (akin to that of oranges stacked up at the market place) of C_{60}^{3-} ions with all the tetrahedral and octahedral holes filled by K^+ ions. The compound shows some superconducting properties. Atoms of Group 18 elements have been incorporated into the C_{60} network by heating the two components at high gas pressures to form, for example, $C_{60}He$.

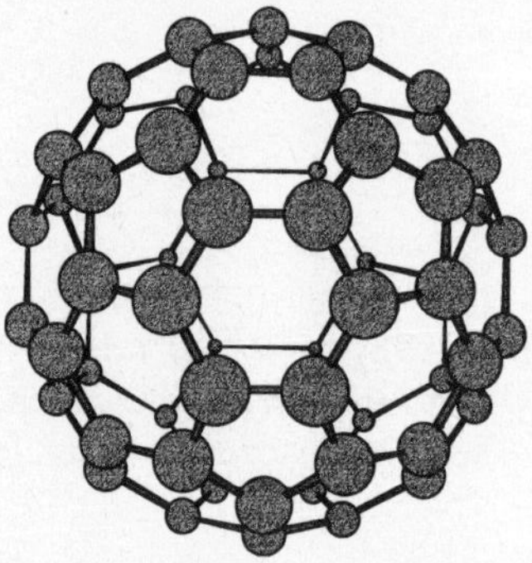

Fig. 1.3 The structure of the C_{60} molecule

Superconduction

Much research effort is being expended in the field of superconduction. Superconduction occurs in pure metals at very low temperatures (usually at the boiling point of helium, 4.3 K), below a particular critical temperature. The highest critical temperature for a pure metal is that observed for niobium, 9.5 K. The search is on for substances which will exhibit superconduction at temperatures which are more easily and economically attainable than 4.3 K. The cheapest cold liquid available is liquid dinitrogen which boils at 77 K and any superconductor with a critical temperature above 77 K would be commercially very useful. In 1987 the mixed oxide $YBa_2Cu_3O_7$ was prepared and found to have a critical temperature of 95 K. In 1993 the compound $HgBa_2Ca_2Cu_3O_{8+\delta}$ was prepared and has a critical temperature of 133K. The δ allows for some non-stoichiometry which is thought to be related to the production of the superconduction phenomenon.

The next stage is to find a substance capable of superconduction at temperatures in excess of 193 K, that of subliming solid carbon dioxide at atmospheric pressure. The commercial success of any superconductor depends upon the current density that can be sustained in the material. At the present time the superconductors are poor in this respect.

Polymers

There is considerable interest in the light-emitting properties of polysilanes. For example, a layer of polymethylphenylsilane doped with the compound 4-dicyanomethylene-2-methyl-6-(p-dimethylaminostyryl)-4H-pyran (DCM) forms the basis of a light emitting diode (LED). When subjected to a direct current with a voltage greater than 40V the polymer emits yellow light similar to that produced when the DCM compound photoluminesces. Polysilanes are generally better electrical conductors than carbon-based organic photoconductors.

MOCVD - Metal organic chemical vapour deposition

Extremely thin layers of metal oxides with particularly accurately controlled dimensions are of great importance and interest to the electronics industries. They can be produced by the thermal decomposition of a volatile organometallic. Decomposition under the correct conditions produces a layer of the oxide of the metal on the chosen surface.

Bioinorganic chemistry

Human and other forms of life depend not only upon the participatory chemistry of the bulk biological elements, C, H, O and N, which contribute to the 'organic' compounds and structures, but upon the contributions from many other elements, some that are essential in only very small concentrations. The elements already mentioned contribute about 95.9% of human body-weight. A second group of essential elements (Ca, P, K, S, Cl, Na, Mg and Fe) make up the majority of the remainder of 4.1%. A third group of elements (Cr, Mo, Mn, Co, Cu, Zn, Si, Se, F and I) represent only about 0.001% of body-weight, but are nevertheless essential for human well-being. A vast amount of research is continuing to establish the participation and mechanism of action of these elements in life-chemistry.

Self-assembling complexes

Much work is occurring in the area of molecular recognition and molecular self-assembly. Some of the research in this field is aimed at the self-construction of very large molecules, giving rise to the subject known as supramolecular chemistry. A very good example of this kind of chemistry is the production of a hexagonal ion with the formula, $Pt_{12}P_{24}N_{12}C_{570}H_{456}O_6^{12+}$. The structure of the ion and its synthesis are shown in Fig. 1.4.

The corner units are molecules of the compound *bis*-pyridyl ketone which has a ketonic carbon atom around which the bonds have a trigonally planar arrangement giving bond angles of 120°.

The linker units are molecules of *bis*-(platinum-*bis*-triphenylphosphine-triflate)-bipyridine bridged compound, in which the triflate ions are replaced by the nitrogen donor-atoms of the corner units.

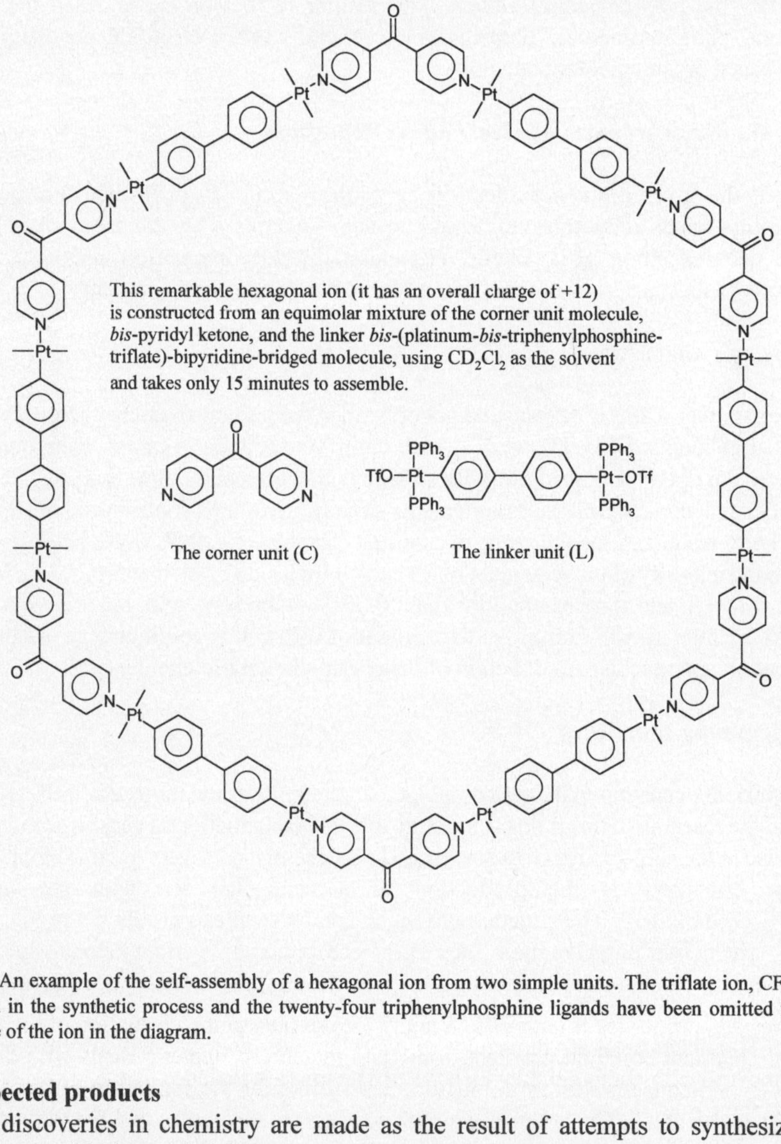

This remarkable hexagonal ion (it has an overall charge of +12)
is constructed from an equimolar mixture of the corner unit molecule,
bis-pyridyl ketone, and the linker bis-(platinum-bis-triphenylphosphine-
triflate)-bipyridine-bridged molecule, using CD2Cl2 as the solvent
and takes only 15 minutes to assemble.
The corner unit (C)
The linker unit (L)

Fig. 1.4 An example of the self-assembly of a hexagonal ion from two simple units. The triflate ion, $CF_3CO_2^-$, is removed in the synthetic process and the twenty-four triphenylphosphine ligands have been omitted from the structure of the ion in the diagram.

Unexpected products

Many discoveries in chemistry are made as the result of attempts to synthesize new compounds with the product being a complete surprise, rather than the one expected. One such example is the attempted preparation of a four-Pd cluster by mixing the palladium compound $[Pd_2(dba)_3].C_6H_6$ with 2 mole-equivalents of PBu^t_3. Chloroform was used as the solvent and crystals were obtained after recrystallization from acetone which had the formula, $[Pd_4(\mu_3\text{-}CH)\text{-}(\mu\text{-}Cl)_3(PBu^t_3)_4]$. The elements of the chloroform, $CHCl_3$, had been incorporated into the cluster formed - an unexpected result. The cluster has structure shown in Fig. 1.5, showing that the three chlorine atoms of the chloroform molecule have been separated from the CH group.

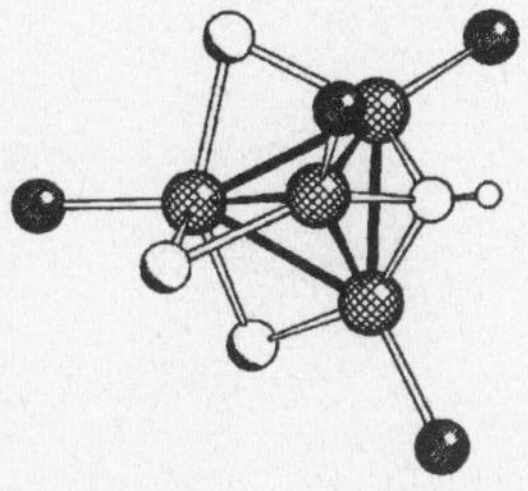

Fig. 1.5 The structure of the cluster $[Pd_4(\mu_3\text{-}CH)\text{-}(\mu\text{-}Cl)_3(PBu^t_3)_4]$. The palladium atoms are represented by hatched circles, the bridging chlorine atoms by the larger open circles with some shading, the phosphorus atoms of the tri-*t*-butylphoshine ligands by fully shaded circles, the triply-bridging carbon atom by the middle-sized open circle and the sole hydrogen atom by the small open circle. (From *Dalton Transactions* 1995, 2108 and reproduced by permission of The Royal Society of Chemistry)

The cluster is based on four palladium atoms which tetrahedrally disposed each having a phosphine ligand attached. The three chlorine atoms from the chloroform molecule act as bridging ligands between pairs of Pd atoms. The remainder of the chloroform molecule, CH, acts as a triply bridging ligand between three Pd atoms. The incorporation of a solvent molecule into a complex is not rare, but for it to break up and contribute to the bonding as in this case is almost unique and has led to a new aspect of cluster chemistry in which halogenated hydrocarbons can be degraded, a possible environmental advantage.

1.4 BOOK OUTLINE

The ten main chapters following are arranged to give a general discussion of inorganic chemistry at a level which is suitable for modern introductory undergraduate courses. Any terms in Chapter 1with which the reader is unfamiliar are fully explained in the following chapters. Chapter 2 consists of some basic information about the atomic nucleus and its stability. Although the details of nuclear stability belong in the realm of Physics, there is an overlap of interest with Chemistry in that nuclear decay processes almost always are associated with the disappearance and appearance of particular nuclei. One very important aspect of many nuclei is their *spin* which allows, in the form of nuclear magnetic resonance spectroscopy (NMR), structural analysis of many molecules. Chapter 3 deals with the arrangement of electrons in atoms and its relationship to the modern form of the periodic classification of the elements. Chapter 4 is a short introduction to group theory and its application to the symmetry properties of molecules and the orbitals used to describe their structures and properties. Chapters 5, 6 and 7 are devoted to the discussion of the various types of chemical bonding; covalent, ionic and metallic. Chapter 5 deals with diatomic molecules and concentrates on the factors responsible for the formation of bonds, their strengths and lengths. Chapter 6 deals with polyatomic molecules and the factors which govern the magnitudes of bond angles. Chapter 7 is concerned with the structures, properties and cohesion of crystalline solids. Chapters 8 and 9 are descriptions of some of the general chemistry of the s and p block elements of the periodic table, the so-called main group elements. Wherever possible, the chemistry is related to the principles outlined in Chapters 3–7. Chapter 10 deals with coordination compounds, those in which coordinate or dative bonds are present, i.e. bonds in which the electrons are supplied by one of the two bonded atoms. The theory of the coordinate bond is developed in this chapter and leads to the description of the typical chemistry of the d and f block elements (transition elements and the lanthanide and actinide elements respectively) which is the basis of the content of Chapter 11.

2

Nuclear and Radiochemistry

2.1 α, β AND γ 'RADIATION'

When radioactivity was discovered a hundred years ago, the 'radiation' emitted was labelled, α (positively charged as demonstrated by its deflection in a magnetic field), β^- (negatively charged) and the uncharged γ rays which were not deflected in the field. An α emitter was used by Rutherford to bombard thin metal foils, recording the fate of the α particles after penetrating the foil. From the results, he concluded that a heavy positively charged nucleus occupies a very small volume at the centre of the atom of the foil. By enclosing an α source in an evacuated thin glass vessel, helium gas was identified in the evacuated space around the vessel, showing that α particles are He nuclei. By studying the deflection of β 'radiation' in applied magnetic and electric fields, it was established that β particles are fast electrons generated in the radioactive nucleus. The uncharged γ rays were diffracted by crystals similarly to other electromagnetic radiation, their wavelengths being shorter than X-rays.

It was realised that radioactivity of an element is independent of its physical state, its chemical environment or temperature, suggesting that it is a property of the nucleus. When artificial radioactivity was produced, some particles emitted from their nuclei were found to be identical to the β particles but they were positively charged, hence the symbol β^+, using β^- for the negative particles.

2.2 THE GROUP DISPLACEMENT LAW

By careful study of the products of the decay of natural radioactive elements, Rutherford and Soddy realised that a product of decay was a nucleus of a different element, which was displaced in the Periodic Table to the right or to the left of the parent nucleus as shown in Fig. 2.1. This is a graphical representation of the Group Displacement Law.

During the investigation of the various decay products, some were found to be chemically identical although they had different radioactive properties. These were called **isotopes** (occupying the same position in the Periodic Table). It is now known that 20 of the naturally occurring elements are isotopically pure, i.e. their nuclei consist of only one isotope of the element, and the others are mixtures of isotopes, tin having as many as ten natural isotopes.

On bombarding some light elements, e.g. Be with α particles highly penetrating uncharged particles were emitted, which were called **neutrons** (*n*). These are slightly heavier than the nuclei of H, which carry the same charge as the β^+ particles, whose mass is about 1/1850 that of a H nucleus. The latter is a positive particle and is a constituent of all nuclei, responsible for the charge on the nucleus and is termed **proton** (*p*).

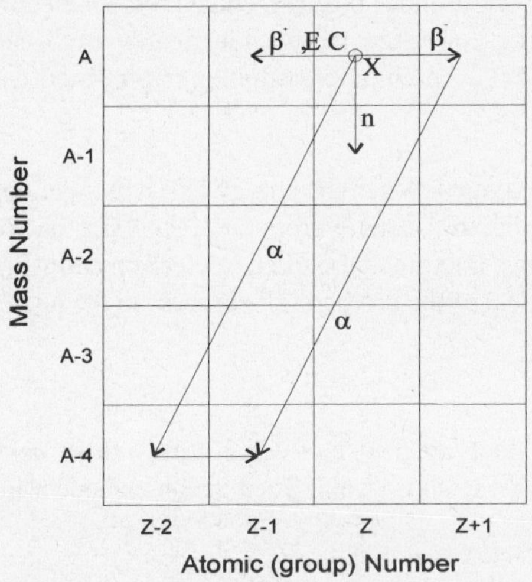

Fig. 2.1 The changes in *A* and *Z* of the nuclide ^A_ZX which result from nuclear decay processes

However, it is now known that a very small fraction of H_2 and its compounds contain a heavier isotope, named **deuterium** whose nucleus contains a neutron in addition to the proton. Another heavier isotope **tritium** is a β^- emitter. Its nucleus has 2 neutrons and a proton. Each isotopic nucleus is characterized by the number of protons *Z* (units of positive charge) and the number of **nucleons**, the term used for protons and neutrons. The number of nucleons is termed the mass number (*A*) which characterizes each isotope. The accepted notation for an isotope X is ^A_ZX, e.g., ^1_1H, ^2_1H and ^3_1H for the isotopes of H and ^4_2He, ^3_2He etc. The subscript could be dropped since it is implied from the chemical symbol. Because of the small masses of β^- or β^+, their emissions hardly change the mass of the nuclide (nucleus of a given isotope), but β emission increases *Z* or decreases it by one unit respectively. It is now believed that β^- emission involves the transformation within the nucleus of a neutron to a proton, according to:

$$n \rightarrow p^+ + \beta^- + \bar{\nu}$$

where $\bar{\nu}$ is an antineutrino (a particle with zero charge and very little mass). By contrast, a β^+ emission is ascribed to the process:

$$p^+ \rightarrow n + \beta^+ + \nu$$

where ν is a neutrino. The energy released by β emission is shared between the emitted β^- or β^+ and the emitted neutrino, $\bar{\nu}$ or ν, which is why the βs released have a range of energies known as a β-ray spectrum. The proton or neutron produced is retained by the nucleus.

Emission of an α reduces Z by 2 and A by 4 units. In the rare cases of neutron emission, A is reduced by one unit. These arguments account for the changes shown in Fig. 2.1.

Other types of decay processes which are observed are summarized below.

Annihilation radiation. When an emitted positron encounters an electron, they are both annihilated, their masses being converted to two identical γ ray quanta, each with a characteristic energy of 0.51MeV. Annihilation radiation is produced whenever β^+ decay occurs.

Electron capture (E.C.) or K-capture. When the energy of a radionuclide is insufficient to expel a β^+, it might capture an electron, usually from the K shell (i.e. one of the 1s electrons outside the nucleus). The vacancy thus created is filled by electrons from higher levels giving the characteristic X ray (K series) of the product. The process in the nucleus is represented by:

$$p + e^- \rightarrow n + \nu$$

Hence, the only particulate products are neutrinos which are not easily detected, the neutron produced remains in the nucleus. Both E.C. and β^+ emission can occur in parallel in some cases.

Internal conversion. When the γ rays emitted are not very fast they may expel an electron from the K or higher shells of the atom. Such electrons are known as conversion electrons. They are monoenergetic unlike emitted β^- particles, which have a wide range of energies.

Isomeric transition. The transitions between energy levels of the product nucleus are very rapid but when they take place at a measurable rate, the level from which this takes place is said to be metastable indicated by a superscript *m* attached to the nuclide symbol. The two states are said to be isomeric.

2.3 DECAY SERIES AND DECAY SCHEMES

Radioactive heavy nuclei decay by a series of α and/or β^- emissions finally resulting in the formation of a stable isotope of lead. There are four decay series distinguished by whether the mass numbers are exactly divisible by 4 or whether, when divided by four, there are remainders of 1, 2 or 3. The parent of the ($4n$) series is ^{232}Th and its end product is ^{208}Pb. The corresponding parents of the ($4n + 2$) and ($4n + 3$) series are ^{238}U and ^{235}U respectively, the final products being ^{206}Pb and ^{207}Pb respectively. An artificial series ($4n + 1$) starts with Neptunium ^{227}Np and ends in ^{209}Bi.

In some cases, an α or β^- particle emission is followed by a β^- or α emission respectively, both sequences ending with the same product. This is known as branched decay, also depicted in Fig. 2.1. Generally in a decay scheme, the parent and daughter are shown as well as the excited levels of the latter. Transitions between these levels or to the ground level is accompanied by γ emission, which takes place within 10^{-10} s of the α or β^- emission. Fig. 2.2 shows the decay scheme of ^{128}I. The half-life of the isotope is 25 min, but there are five modes of decay, some resulting in the production of excited states of the two product nuclei.

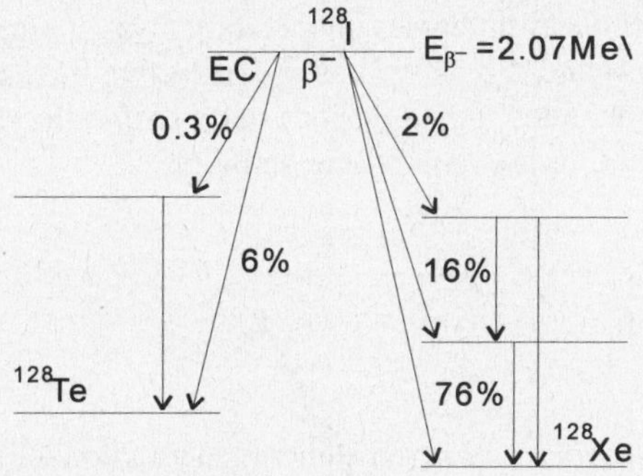

Fig. 2.2 The decay scheme for ^{128}I

An example of the decay of a metastable nucleus is shown in Fig. 2.3. The ^{110m}Ag nucleus has a half-life of 249.8 days and undergoes 99% β^- decay to give ^{110}Cd, the other 1% undergoes an isomeric transition to the ^{110}Ag nucleus. The latter also decays by β^- emission with a half-life of 24.6 s to give ^{110}Cd. Both routes to the product nucleus are associated with the emission of γ ray quanta as the initially produced ^{110}Cd excited state nuclei relax to give the ground state.

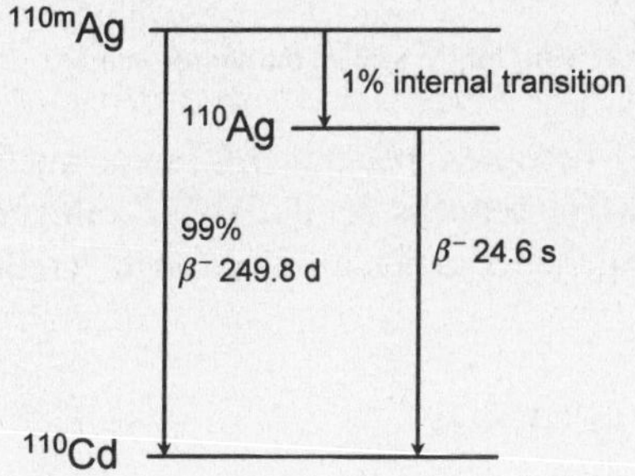

Fig. 2.3 A simplified decay scheme for ^{110m}Ag and ^{110}Ag

2.4 NUCLEAR STABILITY AND NUCLIDE CHARTS

A plot of N (the neutron number) against Z (atomic number or proton number) for stable nuclides, shown in Fig. 2.4, indicates that up to $Z = 20$, $N = 20$ (^{40}Ca), the relationship can be represented by a line with a slope of 45° i.e. maximum stability is attained when $N = Z$. At higher values of Z the graph becomes curved with the slope of the curve gradually increasing. To the right of the curve, where the N/Z ratio is lower than that required for stability (i.e. the nuclei are proton-rich), a radioactive nuclide can decay by β^+ emission or electron capture which produces a daughter nucleus with a ratio of $(N + 1)/(Z - 1)$. To the left of the curve a radioactive nuclide would be neutron-rich and have a value of the N/Z ratio which is larger than that appropriate for stability. Such a nuclide would decay by β^- emission to produce a daughter nucleus with a lower N/Z ratio of $(N - 1)/(Z + 1)$. In either of these cases the daughter nuclide might be stable (i.e. have an N/Z ratio within the stable range) or may undergo further decay processes until stability was attained. This behaviour is explained by the strong *n-n* and *n-p* as well as *p-p* attractive forces operative at short nuclear distances. For larger nuclides, *p-p* repulsions start to offset the attractive forces and an excess of neutrons over protons are required for stability.

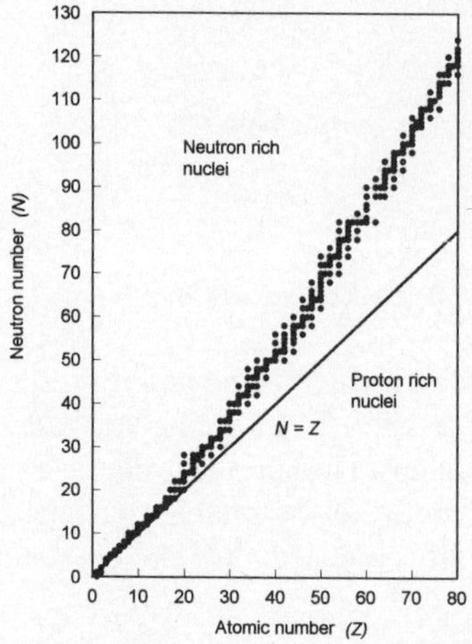

Fig. 2.4 A plot of the number of neutrons, *N*, against the atomic number, *Z*, for a range of stable nuclei

When the value of *Z* becomes greater than 82 some nuclides attain greater stability (i.e. decay) by α emission which reduces the initial N/Z value to $(N-2)/(Z-2)$, the more important consequence being the reduction in *Z* leading to a reduction in the extent of *p-p* repulsions.

2.5 NUCLEAR BINDING ENERGY

The mass (m) – energy (E) relationship postulated by Einstein is:

$$E = mc_0^2 \tag{2.1}$$

where c_0 is the speed of light in a vacuum. When a nucleus is formed from *Z* protons, each of mass m_p and $(A-Z)$ neutrons, each of mass m_n, the mass of the nucleus produced is less than the combined masses of its nucleons. The difference in mass, termed the mass defect, *MD*, is given by:

$$MD = Zm_p + (A-Z)\,m_n - M_{nuc} \tag{2.2}$$

where M_{nuc} is the mass of the bare nucleus ${}^A_Z\mathrm{X}$, or:

$$MD = Zm_H + (A-Z)m_n - M \tag{2.3}$$

where the m_H and M refer to the R.A.Ms of ${}^1\mathrm{H}$ and ${}^A_Z\mathrm{X}$ respectively (the latter has *Z* electrons equal to those contained by Z H atoms). The equivalent energy is called the nuclear binding energy, BE:

$$BE = MD.c_0^2 = [Zm_H + (A - Z)\, m_n - M]c_0^2 \tag{2.4}$$

It is more useful to express the nuclear binding energy in terms of the number of nucleons present, i.e. as the nuclear binding energy per nucleon:

$$BE/A = [Zm_H + (A - Z)\, m_n - M]c_0^2/A \tag{2.5}$$

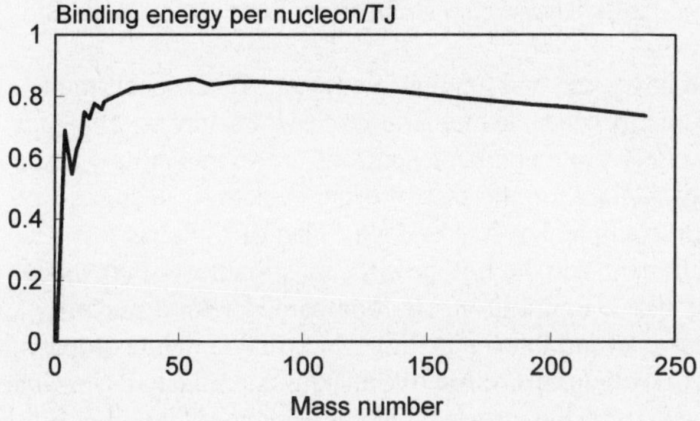

Fig. 2.5 A plot of the nuclear binding energy per nucleon, BE/A, against the mass number, A, for naturally occurring nuclides

The plot of BE/A versus A for the naturally occurring nuclides, shown in Fig. 2.5, exhibits a marked increase in BE/A as A increases until a broad maximum is reached at about $A \sim 56$ followed by a gradual decrease. The nuclei of the metals around the maximum (Fe and its neighbours) are the most stable and are thought to form the core of the earth. The exceptional stabilities of ^{4}He, ^{12}C, and ^{16}O, also shown in Fig. 2.5, suggest that the α particle is a basic constituent of their nuclei. When a heavy (high A) nuclide undergoes *fission* into two lighter fragments (of higher BE/A), the mass loss is transformed to nuclear fission energy, released explosively in the atom bomb or in a controlled manner in nuclear power stations. On the other hand, when 2 nuclides of very low A *fuse* to give a larger nucleus of higher BE/A energy is released as in the thermonuclear explosions of hydrogen bombs and in the processes occurring in stars.

2.6 NUCLEAR FISSION AND NUCLEAR ENERGY

The ^{235}U nucleus undergoes spontaneous fission, but fission can be induced in other U isotopes and in ^{232}Th by bombardment with energetic particles. When neutrons slowed down to thermal energy are used in the process:

$$^{235}\text{U} + {}^1n \rightarrow {}^{236}\text{U}$$

the energy released is higher than the activation energy of fission and hence fission is feasible. Uranium minerals are enriched in this isotope before fission. In a fission bomb, a chain fission reaction takes place leading to nuclear explosion, when 2 masses of fissile material are brought together to exceed the critical mass required. The chain reaction results from the release of more neutrons than those inducing it. Below the critical mass sufficient neutrons escape so that the chain reaction is not sustained. On the other hand, in a nuclear power station, control rods e.g. of Cd which is an efficient neutron absorber, are used to ensure that the neutron flux is under control. The energy released from the fission, through heat exchangers, is used for steam generation to produce electricity.

The fission products are on the high N side of the stability curve shown in Fig. 2.4 and hence decay by consecutive β^- emissions until stable isotopes are produced. In the thermal neutron fission of ^{235}U, a plot of the yield of fission products against A shows two maxima with two spikes at $A = 86$ and 136 (Kr and Xe very stable isotopes with 'magic' numbers $n = 50$ and 82 respectively). Unlike the *prompt* neutrons produced from the fission, excited ^{137}Xe emits a *delayed* neutron.

Nucleons are assumed to have spin and angular momenta. In the shell model, protons of opposite spin occupy an energy level and the same applies for neutrons. Nuclides which have a 'magic' number of either protons and/or neutrons are extra stable. These numbers represent filling of energy shells, as in the case of orbital electrons in atoms (see Chapter 3). The nucleus of ^{16}O, for example, has $N = 8$, $Z = 8$. That of ^{208}Pb has $N = 126$, $Z = 82$, both of which are 'magic numbers' and the isotope is the most stable. A bulk model assumes that the nucleus behaves like a liquid drop. If bombarded with a nucleon, a compound nucleus is formed, similar to an unstable liquid drop which may split into droplets, as in nuclear fission. The emission of particles from an active nucleus is similar to the escape of molecules from a liquid drop, overcoming the surface forces, also assumed for the nucleus. Although the liquid drop model can explain fission, the yield profile results from the shell model.

2.7 STELLAR CREATION OF ELEMENTS

About 73% of the mass of the universe is H whereas He makes up 25%, the remainder being all the other elements. It is believed that when the temperature of a star reaches 10^7 K, nuclear fusion of H to produce He can take place (in stages) in what is termed H burning. At higher temperatures, ^4He burning leads to the formation of ^{12}C and ^{16}O. In a similar fashion, heavier nuclei are produced until the Fe-Ni nuclei are formed (at the maximum of the binding energy versus A plot). Elements beyond these are believed to be formed by *n*-capture reactions.

2.8 NUCLEAR ENERGETICS

A nuclear process is accompanied by the release of energy if the products are lighter than the reactants i.e. if it is accompanied by mass loss. The maximum energy, E_{max}, for β^- decay:

$$ {}^{A}_{Z}X \rightarrow {}^{A}_{Z+1}Y + \beta^- \tag{2.6}$$

is given by:

$$E_{max} = (M_Z - M_{Z+1})\, c_0^2 \tag{2.7}$$

where the Ms are R.A.Ms. If the atomic mass units are expressed in kg and c_0 in m s^{-1}, E is obtained in J. Thus 1 a.m.u. = $1.49.10^{-10}$ J or usually 931.5 MeV (1 electron-volt, eV = 1.602×10^{-19} J). The mass of the β^- has not been ignored in (2.6) because the product has one more electron than the parent. Similar equations are applicable to other decay processes.

2.9 ARTIFICIAL RADIOACTIVITY AND SYNTHETIC NUCLIDES

In the early days of the study of radioactivity, artificial radioisotopes were obtained by the bombardment of stable nuclides with neutrons or positive particles, the latter usually producing β^+ emitters. The positive particles had to be accelerated to overcome the energy barrier posed by the positively charged target nucleus. However, nowadays, radionuclides are obtained routinely in large amounts by irradiating target elements in nuclear reactors or by separating the products of nuclear fission from the spent fuel from nuclear power stations.

The development of synthetic radionuclides began with the work of Irene Curie and Frederick Joliot who, in 1934, bombarded magnesium with alpha-particles to produce the ^{27}Si isotope then known as 'radio-silicon.' This was one of the first examples of an (α,*n*) reaction, in which the bombarding particle was an α particle and the particle emitted as a reaction product was a neutron. Curie (Marie's daughter) and Joliot were awarded the 1935 Nobel Prize for Chemistry. The symbolism generally used to describe nuclear reactions of this kind is:

target nucleus (bombarding particle, emitted particle or photon) product nucleus,

e.g. ^{24}Mg(α,*n*)^{27}Si.

In addition to the synthesis of new radioactive isotopes of the naturally occurring elements, the processes of nuclear transformation using neutron bombardment followed by β^- emission were used progressively to synthesize new elements. An (*n*,γ) process is used to build up mass and increase the N/Z ratio and the product isotope then emits a β^- particle to produce a daughter element with an extra unit of charge. Such reaction sequences applied to ^{238}U produces ^{239}Np (Z=93) which undergoes a further β^- emission to give ^{239}Pu (Z=94). Elements with higher atomic numbers than 94 have been prepared by bombarding target elements with suitably accelerated α-particles and ions of boron, carbon and other elements. To date the latest element to be synthesized is that with an atomic number of 112. The actinide and trans-actinide elements are discussed in Chapter 11.

2.10 THE RATE OF DECAY

The rate of decay of a nucleus follows the natural exponential law and if the number of parent nuclei present at any time is represented by N, the rate can be described by the equation:

$$-\mathrm{dN}/\mathrm{d}t = \lambda \mathrm{N} \qquad (2.8)$$

where λ is the decay constant, which characterizes the decaying species. The integrated form of equation (2.8) is:

$$\mathrm{N}_t = \mathrm{N}_0 e^{-\lambda t} \qquad (2.9)$$

where the subscripts refer to times t or zero (i.e. the beginning of the observations). The more useful logarithmic form is:

$$\ln \mathrm{N}_t = \ln \mathrm{N}_0 - \lambda t \qquad (2.10)$$

The time at which $N_t = \frac{1}{2}N_0$ is the half-life, $t_{1/2}$ which is related to λ by the relation:

$$t_{1/2} = \ln 2/\lambda \tag{2.11}$$

The measured activity (the rate of decay), A, is related to the number of active atoms N by the equation:

$$A = f.N \tag{2.12}$$

where f is the efficiency of the counter. Hence N in the above equations can be replaced by A and equation (2.10) in decadic logarithmic form gives:

$$\log_{10}A_t = \log_{10}A_0 - 0.301\lambda t \tag{2.13}$$

A plot of $\log A_t$ versus t is expected to be linear with A_0 and λ being obtained from the intercept and slope respectively. The half-life is then obtainable by using equation (2.11). When two or more radioisotopes of sufficiently different $t_{1/2}$ values are counted in the same sample its composite decay plot can be resolved into two or more straight lines.

Problem 2.1 Calculate $t_{1/2}$ for ^{241}Am in years given that it emits 1.2×10^{11} α particles per gram per second.

Answer. 1 gram of ^{241}Am contains $N_A/241$ nuclei. Using equation (2.8):

$$-dN/dt = \text{rate of decay} = 1.2 \times 10^{11} = \lambda \times N = \lambda \times N_A/241$$

$$\lambda = 1.2 \times 10^{11} \times 241/6.023 \times 10^{23};\ t_{1/2} = \ln 2/\lambda = (\ln 2 \times 6.023 \times 10^{23})/(1.2 \times 10^{11} \times 241)$$
$$= 1.44 \times 10^{10}\ \text{s} = 458\ \text{y}$$

Problem 2.2 A sample containing two radioisotopes I and II ($t_{1/2}$ of II is nearly 16 times that of I) was counted over a period at different times, t, giving the following count rates (in counts per second, c. p. s.):

t/s	60	130	245	370	660	910	1160	1440	1700
c. p. s	12.4	11.0	9.0	7.2	4.9	3.9	3.1	2.4	1.9

Plot the logarithm of the count rate against the time, t, and explain the shape of the resulting graph. This can be done conveniently by using semi-log graph paper or by using a spreadsheet. Estimate the initial count rate of each isotope.

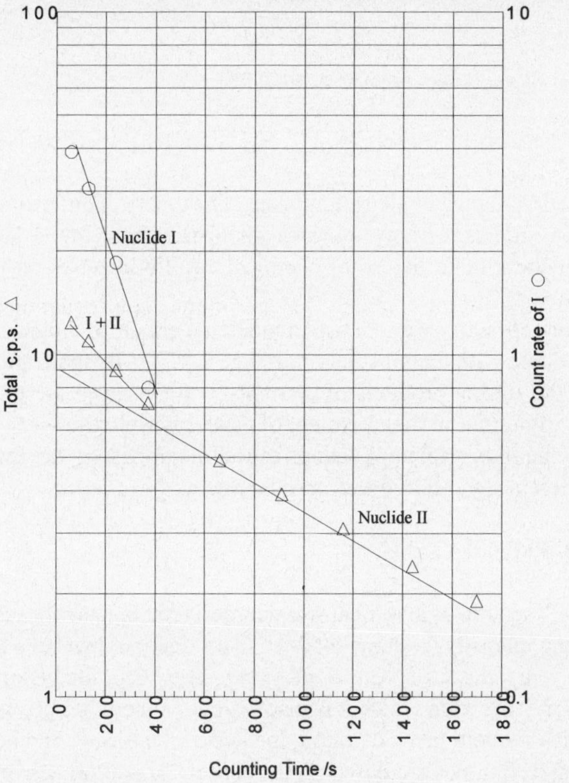

Answer. The composite decay curve, shown in the figure above, can be resolved into two straight lines. The earlier line is due to the decay of I and II, but the later line represents the decay of only II. By extrapolating the straight part of the line to $t = 0$, the initial activity of isotope II is found to be 9.0 c. p. s. The activity of isotope II at other times can be read from the straight-line graph and subtracted from the total activity to give the activities due to isotope I at the various times. These data were replotted (the dotted line in the Figure) and the line extrapolated to $t = 0$, giving the initial activity of isotope I as 5.25 c. p. s.

This method of separating the activities of two isotopes with different $t_{1/2}$ values is of general use in chemical kinetics and can be used, for example, in determining the rates of hydrolysis of two isomers.

2.11 CHEMICAL AND MEDICAL APPLICATIONS OF RADIOISOTOPES

A major application of radioisotopes is in the area of tracing the fate of labelled atoms during a reaction, e.g. the racemization of $[Cr(C_2O_4)_3]^{3-}$ (i.e. the production of an equilibrium mixture of the two optically active isomers of the complex ion) in presence of $^{14}C_2O_4^{2-}$ did not lead to any radioactivity in the complex ion, thus excluding an ionization mechanism. An intramolecular ring opening/closing mechanism is favoured. Some exchange reactions can be demonstrated by radiolabelling. Iodides exchange rapidly with I_2 through:

$$I^- + I_2 \rightleftharpoons I_3^-$$

The rates of electron exchange reactions, such as:

$$[{}^*Fe(CN)_6]^{4-} + [Fe(CN)_6]^{3-} \rightleftharpoons [{}^*Fe(CN)_6]^{3-} + [Fe(CN)_6]^{4-}$$

can be only studied by using labelled complexes. The theory of outer sphere redox reactions is based on such studies. Many other mechanistic applications have been reported. Establishment of the kinetic nature of chemical equilibria relies on the use of labelled compounds.

A number of isotopes are of use in medicine either for diagnosis or treatment. An example of an isotope with diagnostic importance is ^{99m}Tc. Its main source is the β^- decay of ^{99}Mo, one of the fission products of uranium. ^{131}I is used in the treatment of Graves' disease. The use of radium in the treatment of cancer is well known.

Among the industrial applications of radioisotopes are the measurement of bulk flow, mixing efficiency and liquid leak measurements.

2.12 RADIOACTIVE DATING

The most widely ^{14}C dating of historical wooden-derived objects is based on the knowledge that the cosmic ray intensity (responsible for ^{14}C production) has been practically constant for thousands of years, the $t_{1/2}$ of ^{14}C is 5568 years. The ^{14}C of the historical object starts to decay as soon as the tree from which it is made is cut. When the activity of such an object is compared with a modern piece of wood, the age of the former can be calculated. Other radiotracer dating techniques are based on the activity of natural active isotopes in certain rocks or minerals. The technique is used in the study of dendrochronology—the science of the study of tree-rings, to investigate past climates. The basis of the method is that the relative width of a tree-ring is proportional to the local climatic conditions so that the width is larger when the climate is warmer and/or wetter. The use of felled Bristlecone pines in the USA has allowed past climates to be studied up to around 2000 years ago.

2.13 NUCLEAR SPIN AND NUCLEAR MAGNETIC RESONANCE SPECTROSCOPY

A property of many nuclei which has found great application in chemistry is that of *nuclear spin* which relates to a nuclear spin quantum number, I. Every nucleus possessing spin has a magnetic moment which can be aligned in a quantized manner with respect to an applied magnetic field. In the absence of such a field the allowed alignments of the nuclear magnetic moment have equal energies; they are degenerate. When an external magnetic field is applied to a collection of nuclear magnets the permitted alignments are no longer degenerate and transitions are possible between the existing energy states, such changes of energy are known as nuclear magnetic resonance (NMR) transitions and are governed by the Bohr frequency condition of equation (1.6). The large magnetic fields used in modern spectrometers are designed to allow the NMR transitions to occur in the radiofrequency region of the electromagnetic spectrum between 100-600 MHz. The discussion which follows is restricted to some magnetic nuclei with spin quantum numbers, $I = ½$.

In an applied field such nuclei have two possible energies associated with the two possible alignments of their magnetic moments with the applied field. For any specific magnetic nucleus the frequency at which resonance absorption occurs is dependent upon (i) the chemical environment of the atom in its particular molecule and (ii) the magnetic nuclei it has as neighbours in the molecule which may be chemically identical or otherwise. For example, in the proton NMR spectrum of ethanol, CH_3CH_2OH, there are three absorptions corresponding to the three different environments experienced by the protons of the methyl group, those of the CH_2 group and the single proton of the OH group. The three different environments arise from the slightly different shielding of the protons from the applied field afforded by the electrons in the different regions of the molecule. The three different proton types absorb at three different frequencies which are known as *chemical shifts* from a standard frequency which is that at which the protons of tetramethylsilane (TMS, $(CH_3)_4Si$) absorb. The protons of the CH_2 group interact with those of the methyl group and that of the OH group, but there is no interaction between the proton of the OH group and the protons of the methyl group. The *coupling* between nuclei that are magnetically inequivalent is restricted usually to those attached to neighbouring atoms and gives information about the atoms attached to the neighbouring atoms. The coupling arises because the magnetic nuclei that couple do so by modifying the applied field experienced by each other. In the ethanol case the two protons of the CH_2 group can have spins which are both either parallel (↑↑) or anti-parallel (↓↓) to the applied field or they can have one spin parallel with the other spin anti-parallel (↑↓, ↓↑), the latter arrangement having twice the probability of occurrence than both of the other two. These arrangements of nuclear spins on the CH_2 group affect the applied field experienced by the protons on the CH_3 and OH groups whose signals are split into three with a 1:2:1 intensity in line with the relative probabilities of the spin arrangements on the CH_2 group. The three protons of the CH_3 group can have their spins arranged with all three spins parallel to the applied field (↑↑↑), or two parallel and one anti-parallel (↑↑↓), (↑↓↑) and (↓↑↑), or one parallel and two anti-parallel (↑↓↓), (↓↑↓) and (↓↓↑), or all three spins anti-parallel (↓↓↓). The effect of these four different arrangements of the three spins of the methyl group split the neighbouring CH_2 proton signals into a quartet with an intensity ratio 1:3:3:1. The single proton of the OH group can be parallel or anti-parallel to the applied field and splits the proton signals on the neighbouring CH_2 group into a 1:1 doublet. The result of this coupling is a proton signal from the CH_3 group split into a triplet, a proton signal from the CH_2 group split into a quartet of doublets and a proton signal from the OH group split into a triplet. This is the case in the spectrum of pure liquid ethanol, but the spectrum is considerably modified if a small amount of a aqueous protonic acid is added to the ethanol. Solvated protons catalyse the exchange of protons from the acidic OH group with the protons, $H^+(aq)$, in solution and the reactions are extremely rapid. The rate of the exchange process is such that the OH protons do not spend sufficient time bonded to the oxygen atom to exert the coupling effect, the protons adopting one or other of the two alignments of their nuclear magnets with and against the applied field so that the neighbouring protons experience an average effect and their splitting is not observed.

NMR spectroscopy can be used to decide the structure of a compound by analysis of the number of chemically shifted signals observed and the splitting pattern of the signals as neighbouring nuclei couple. It is also very useful in identification of exchange reactions and determining rates of exchange. Examples are dealt with in appropriate sections of the book.

2.14 PROBLEMS

1. Complete the equations for the following nuclear processes:

(a) $^{35}Cl + ^{1}n = ... + ^{4}He$
(b) $^{235}U + ^{1}n = ... + ^{137}Xe + 2^{1}n$
(c) $^{27}Al(\alpha,n)....$
(d) $....(n,p)^{35}S$;
(e) $^{55}Mn(n,\gamma)....$

2. Calculate the half life of ^{234m}Pa from the following corrected count rates (c.p.s.) at the time, *t*:

t/s	120	180	240	300	420
c.p.s.	105.7	56.3	30.5	16.7	4.9

3. Calculate the mass of ^{140}La in a sample whose activity is 3.7×10^{10} Bq (1 Bequerel, Bq = 1 disintegration per second) given that its $t_{1/2}$ is 40 hours.

4. Calculate the binding energy per nucleon for ^{12}C, ^{14}N, ^{16}O, ^{19}F and comment on their relative magnitudes.
(R. A. M: ^{12}C=12.000000, ^{14}N=14.003074, ^{16}O=15.994915, ^{19}F=18.998405)

5. The β^- activity of a sample of CO_2 prepared from a contemporary wood gave a count rate of 25.5 counts per minute (c. p. m.). The same mass of CO_2 from an ancient wooden statue gave a count rate of 20.5 c. p.m. in the same counter and conditions. Calculate its age to the nearest 50 years taking $t_{1/2}$ of ^{14}C as 5568 years. What would be the expected count rate of an identical mass of CO_2 from a sample which is 4000 years old?

6. Derive equations similar to that of (2.7) for (i) β^+ decay and (ii) α emission.

3

Electronic Configurations of Atoms

3.1 INTRODUCTION

The chemistry of an element is determined by the manner in which its electrons are arranged in the atom. Such arrangements and their chemical consequences are the subject of this chapter, leading to a general description of the structure of the modern periodic classification of the elements—the periodic table—and of the variations in some important atomic properties. The influence of relativistic effects upon some atomic properties, particularly those of the heavier elements, is described. The last section deals with the difference between electronic configurations and electronic states, the latter being important in the understanding of electronic spectra.

3.2 THE HYDROGEN ATOM

The hydrogen atom is the simplest atom with its nuclear charge of unity and its single orbital electron. Experimental evidence for the possible electronic arrangements in the hydrogen atom was provided by its emission spectrum. This consists of 'lines' of a particular frequency rather than being the continuous emission of all possible frequencies. Discrete 'line' emissions of characteristic wavelengths of light from electrical discharges through gases were observed in the early days of development of the subject. They are sometimes regarded as monochromatic (i.e. of a precise wavelength), but they have very small finite widths that depend upon the temperature and pressure of the system.

The discrimination of emission frequencies leads to the concept of there being discrete **energy levels** within the atom which are permitted for electron occupation. Detailed analysis of the frequencies of the lines in the emission spectrum of the hydrogen atom led to the formulation of the empirical Rydberg equation:

$$\nu_{ij} = cR\left[\frac{1}{n_i^2} - \frac{1}{n_j^2}\right] \tag{3.1}$$

where the frequency, ν_{ij}, relates to an electronic transition from the level j to the level i, j being larger than i, the terms n_i and n_j being particular values of the **principal quantum number**, n. The other terms in the equation are c—the velocity of light in a vacuum (the subscript zero is dropped from hereon), and R—the Rydberg constant, which has an experimentally observed value of 1.096776 x 10^7 m^{-1} for the hydrogen atom. The terms in brackets in equation (3.1) are dimensionless and if R is multiplied by c (m s^{-1}) the result is a frequency (s^{-1} or Hertz, Hz).

The line spectrum of the hydrogen atom supplies the basis of the concept of the quantization of electron energies— that the permitted energies for the electron in the atom of hydrogen are quantized. They have particular values so that it is not possible for the electron to possess any other values for its energy than those given by the Rydberg equation. As that equation implies, the electron energy is dependent upon the particular value of the quantum number, n. The Rydberg equation can be converted into one which relates directly to electron energies by multiplying both sides of equation (3.1) by Planck's constant, h, and by the Avogadro number, N_A to obtain the energy in units of J mol^{-1}. Such a procedure makes use of the Planck-Einstein equation which relates the frequency of electromagnetic radiation, ν, to its energy, E:

$$E = h\nu \tag{3.2}$$

The Rydberg equation in molar energy units is:

$$E_{ij} = N_A h\nu = N_A hcR\left[\frac{1}{n_i^2} - \frac{1}{n_j^2}\right] \tag{3.3}$$

Bohr interpreted spectral lines in terms of electronic transitions within the hydrogen atom. The Bohr equation expresses the idea that $E_i - E_j$ represents the difference in energy between the two levels, ΔE, and may be written in the form:

$$E_{ij} = \Delta E = E_j - E_i = N_A h\nu_{ij} \tag{3.4}$$

A combination of equations (3.3) and (3.4) gives:

$$E_j - E_i = N_A chR\left[\frac{1}{n_i^2} - \frac{1}{n_j^2}\right] \tag{3.5}$$

Equation (3.5) may be regarded as being the difference between the two equations:

$$E_j = -\frac{N_A chR}{n_j^2} \tag{3.6}$$

and

$$E_i = -\frac{N_A chR}{n_i^2} \tag{3.7}$$

so that a general relationship may be written as:

$$E_n = -\frac{N_A chR}{n^2} \tag{3.8}$$

Equation (3.8) describes the permitted quantized energy values for the electron in the atom of hydrogen. Some of these values are shown in the diagram of Fig. 3.1, together with the possible electronic transitions which form part of the emission spectrum of the atom. The reference zero for the diagram is that corresponding to the complete removal or ionization (to give the bare proton, $H^+(g)$), of the electron from the influence of the nucleus of the atom. The Lyman transitions are observed in the far ultra-violet region of the electromagnetic spectrum (wavelengths below 200 nm). The Balmer transitions are found mainly in the visible region, the Paschen transitions being in the infra-red. There are other series of transitions of lower energies still which are all characterized by their different final values of n.

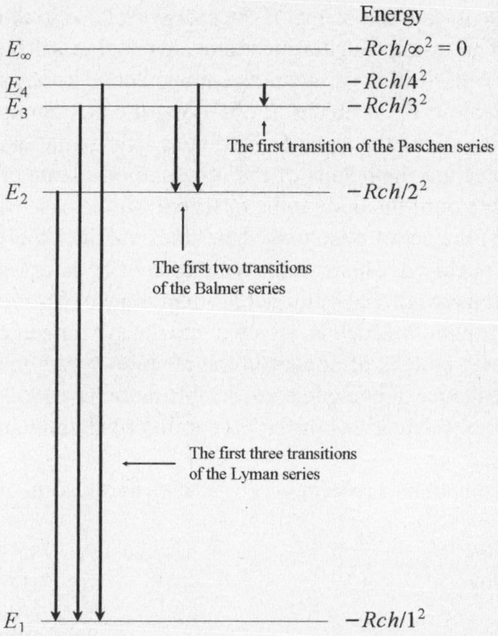

Fig. 3.1 The lower four energy levels of the hydrogen atom and some of the transitions which are observed in the emission spectrum of the gaseous atom

The ionization energy of the hydrogen atom from its ground state (the lowest energy electronic state) may be calculated from equation (3.5). The process of ionization is equivalent to the electronic transition between the energy levels corresponding to $n_i = 1$ and $n_j = \infty$ (since $E_j = 0$ when $n_j = \infty$). Using these values in equation (3.5) produces the equation describing the ionization energy of the hydrogen atom,

$$I_H = E_\infty - E_1 = N_A chR = 1312 \text{ kJ mol}^{-1} \tag{3.9}$$

The Rydberg equation refers to emissive transitions in which electrons are transferred from higher to lower energy levels. Equation (3.9) is also a calculation of the energy released when an electron with $n_j = \infty$, falls to the lowest energy level $n_i = 1$. The ionization energy is normally defined as the energy required to cause the reverse process. The values of the constants used in the above equations, N_A, c, h and R are known to a high degree of accuracy so the actual permitted energies for the electron in a hydrogen atom are also known very accurately. The consequence of such accurate knowledge is that the position of the electron within the atom is very uncertain. This situation is an example of the application of the Heisenberg uncertainty principle (or principle of indeterminacy) which may be stated as *It is impossible simultaneously to determine the position and momentum of an electron.*

Symbolically the principle may be written as:

$$\Delta p.\Delta q \approx h/2\pi \tag{3.10}$$

In equation (3.10) Δp represents the uncertainty (error in determining) in the momentum, p (p = mass times velocity, mv), and Δq the uncertainty in position of the electron. The two uncertainties bear an inverse relationship to each other so that if the energy (related to the momentum) is known with a high degree of precision then the uncertainty in position will be correspondingly large. That is the situation with electrons in atoms and molecules). The basis of the uncertainty principle is understandable if the problem of observing the electron is considered. Any process of observation involves the use of electromagnetic radiation. Eyesight depends upon the process of detecting the results of the reflexion of quanta or photons of electromagnetic radiation (light) from the body to be observed. In the case of massive bodies (such as everyday objects) the act of observing them does not alter their energies appreciably nor does it affect their positions. This is not so with elementary particles such as electrons whose energies and positions are altered by the act of attempting to observe them. In the attempt to 'see' an electron, the photon which is reflected into the eye (or some other detector) causes the electron to move, so it is no longer in the position where the observer might consider it to be. The consequence of being in considerable ignorance about the position of an electron in an atom is that calculations of the probability of finding an electron in a given position must be made.

The general mathematical expression of the problem may be written as one form of the Schrödinger equation:

$$H\psi = E\psi \tag{3.11}$$

The equation implies that if the operations represented by H are carried out on the function, ψ, the result will contain knowledge about ψ and its associated permitted energies. The term represented by ψ is the wave function which is such that its square, ψ^2 (strictly, this should be written as $\psi\psi^*$ where ψ^* is the complex conjugate of ψ in case there is an imaginery component), is the probability density. The value of $\psi^2 d\tau$ represents the probability of finding the electron in the volume element, $d\tau$ (which may be visualized as the product of three elements of the cartesian axes: $dx.dy.dz$). Although the solution of the Schrödinger equation for any system containing more than one electron requires the iterative techniques available to computers, it may be solved for the hydrogen atom (and for hydrogen-like atoms such as He^+ , Li^{2+},...) by analytical means, the energy solutions being represented by the equation:

$$E_n = -\frac{N_A \mu Z e^4}{8\epsilon_0^2 h^2}\left[\frac{1}{n^2}\right] \tag{3.12}$$

where Z is the atomic number, μ is the reduced mass of the system, defined by the equation:

$$\frac{1}{\mu} = \frac{1}{m_e} + \frac{1}{m_n} \tag{3.13}$$

in which m_e is the mass of the electron and m_n is the mass of the nucleus; e is the electronic charge and ϵ_0 is the permittivity of a vacuum. By comparing equations (3.8) and (3.12), with a Z value of 1, it may be concluded that the value of the Rydberg constant is given by:

$$R = \frac{\mu e^4}{8\epsilon_0^2 c h^3} \tag{3.14}$$

Equation (3.12) gives the permitted energies for the electron in the hydrogen atom.

The solution of the Schrödinger equation shows how ψ is distributed in the space around the nucleus of the hydrogen atom. The solutions for ψ are characterised by the values of three quantum numbers (essentially three because of the three spatial dimensions, x, y and z) and every allowed set of values for the quantum numbers describes what is termed an atomic orbital.

The quantum rules and atomic orbitals

The quantum rules are statements of the permitted values of the quantum numbers, n, l and m. They are outlined below.

(i) The princpal quantum number, n (the same n as in equation (3.12)), has values which are integral and non-zero:

$$n = 1, 2, 3, 4,...$$

It defines groups of orbitals which are distinguished, within each group, by the values of l and m.

(ii) The secondary or orbital angular momentum quantum number, l, as its name implies, describes the orbital angular momentum of the electron, and has values which are integral, including the value zero:

$$l = 0, 1, 2, 3,... (n - 1)$$

For a given value of n, the maximum permitted value of l is $(n - 1)$, so that for a value of $n = 3$, l is restricted to the values 0, 1 or 2.

(iii) The magnetic quantum number, m_l, so called because it is related to the behaviour of electronic energy levels when subjected to an external magnetic field, has values which are dependent upon the value of l. The permitted values are:

$$m_l = +l, +l-1, ...0, ...-(l-1), -l$$

For instance, a value of $l = 2$ would yield five different values of m_l : 2, 1, 0, −1 and −2. In general there are $2l + 1$ values of m_l for any given value of l. In the absence of an externally applied magnetic field the orbitals possessing a given l value would have identical energies. Orbitals of identical energy are described as being degenerate. In the presence of a magnetic field the degeneracy of the orbitals breaks down in that they then have different energies. The breakdown of orbital degeneracy (for a given l value) in a magnetic field is the explanation of the **Zeeman effect**. This is the observation that in the presence of a magnetic field the atomic spectrum of an element has more lines than in the absence of the field. In discussing such a phenomenon it is normal to refer to the values of m_l with their appropriate l values as subscripts; m_l. The values of l and m_l are dependent upon the value of n and so it can be concluded that the value of n is concerned with a particular set of atomic orbitals all characterised by the given value of n. For any one value of n there is the possibility of more than one permitted value of l (except in the case where $n = 1$, l can only be zero). The notation which is used to distinguish orbitals with different l values consists of a code letter associated with each value. The code letters are shown in Table 3.1.

Table 3.1 Code Letters for l Values

Value of l	Code letter
0	s
1	p
2	d
3	f

Table 3.1 gives only a portion of an infinite set of values of l. Those given are the only values of any interest for the majority of applications to known atoms. The selection of the code letters seems, and is, illogical in that the first four are the initial letters of the words: **s**harp, **p**rincipal, **d**iffuse and **f**undamental—words used historically to describe aspects of line spectra. The fifth letter, g, follows on alphabetically, the sixth being h and so on. The numerical value of n and the code letter for the value of l are sufficient for a general description of an atomic orbital. The main differences between atomic orbitals with different l values are concerned with their spatial orientations and those with different n values have different sizes. The main importance of the m_l values is that they indicate the number, $(2l + 1)$, of differently spatially oriented orbitals for the given l value. For instance, if $l = 2$ there are five different values of m_l corresponding to five differently spatially oriented d orbitals. The number of differently spatially oriented orbitals for particular values of l are given in

Table 3.2.

Application of all the above rules allows the compilation of the types of atomic orbital and the number of each type:

For $n = 1$ there is a single 1s orbital ($l = 0$, $m_l = 0$).

For $n = 2$ there is one 2s orbital ($l = 0$, $m_l = 0$) and three 2p orbitals ($l = 1$, $m_l = 1$, 0 or -1).

Table 3.2 Number of atomic orbitals for a given l value

Value of l	**Number of orbitals**
0	1
1	3
2	5
3	7

For $n = 3$ there is one 3s orbital, three 3p orbitals and five 3d orbitals ($l = 2$, $m_l =$ 2, 1, 0, -1 or -2).

For $n = 4$ there is one 4s orbital, three 4p orbitals, five 4d orbitals and seven 4f orbitals ($l = 3$, $m_l = 3, 2, 1, 0, -1, -2$ or -3) and so on up to and beyond the normally practical limit of $n = 7$.

For every increase of one in the value of n there is an extra type of orbital. The number of atomic orbitals associated with any value of n is given by n^2 so that for $n = 1$ there is one orbital (1s), for $n = 2$ there are four orbitals (one 2s and three 2p), for $n = 3$ there are nine orbitals (one 3s, three 3p and five 3d) and for $n = 4$ there are sixteen orbitals (one 4s, three 4p, five 4d and seven 4f). The sequence can be extended to infinity but in practice it is only necessary to consider values of n up to seven and not all the orbitals so described are needed for the electrons in known atoms in their ground electronic states. Higher values of the n quantum number are required to interpret the emission spectra of the heavier elements.

For the hydrogen atom, the orbitals having a particular value of n all have the same energy - they are **degenerate**. Level one ($n = 1$) is singly degenerate (i.e. non-degenerate), level two ($n - 2$) has a degeneracy of four and so on. The degeneracy of any level (i.e. the number of orbitals with identical energy) is given by thc value of n^2. Such widespread degeneracy of electronic levels in the hydrogen atom is the basis of the simplicity of the diagram of the levels shown in Fig. 3.1. The orbital energies are determined solely by the value of n in equation (3.8).

The spatial orientations of the atomic orbitals of the hydrogen atom are very important in the consideration of the interaction of such orbitals in the production of chemical bonds. The solution of the appropriate Schrödinger wave equation for the ground state of the hydrogen atom (i.e. one with its single electron in the 1s orbital) provides the equation:

$$\psi_{1s} = R_{1s}A_{1s} \tag{3.15}$$

where ψ_{1s} is the wave function for the 1s atomic orbital of the hydrogen atom, R_{1s} represents the radial function—the manner in which ψ_{1s} varies along any straight line from the nucleus, and A_{1s} is the angular function which takes into account variations of ψ_{1s} along any line radiating from the nucleus. The 1s atomic orbital of the hydrogen atom is spherically symmetrical and so the angular function is a constant term [$A_{1s} = 1/(2\pi^{1/2})$] chosen to normalize the wavefunction so that the integral of its square over all space has the value unity, as expressed by the general equation:

$$\int_0^\infty \psi^2 d\tau = 1 \quad (3.16)$$

This corresponds with the equating to unity of the mathematical certainty of finding the electron somewhere in the orbital. The radial function has the form:

$$R_{1s} = 2a_0^{-3/2}e^{-r/a_0} \quad (3.17)$$

where r is the distance from the nucleus and a_0 is the atomic unit of length consisting of a collection of universal constants and given by:

$$a_0 = h^2\epsilon_0/\pi\mu e^2 \quad (3.18)$$

where h is Planck's constant, ϵ_0 is the permittivity of a vacuum, μ is the reduced mass of the electron and e is the electronic charge.

The quantity, a_0, is referred to as the Bohr radius because it is identical to the radius of the orbit of the 1s electron in the 'Bohr atom'. The early Bohr theory of the atom invoked electron orbits of definite radii, but these are invalid since they violate the Heisenberg uncertainty principle (the position and momentum of an atomic particle are both subject to uncertainties, the product of which has a value of around $h/2\pi$, see above).

To make use of the wave function in a meaningful manner it is necessary to express it in terms of the radial distribution function, *RDF*. This is the variation with distance from the nucleus of the function $4\pi r^2\psi_{1s}^2$ which takes into account the probability of finding an electron between two spheres of respective radii r and $r + dr$, where dr is an infinitesimally small radius increment. The equation for this is:

$$RDF_{1s} = 4\pi r^2\psi_{1s}^2 = 4r^2a_0^{-3}e^{-2r/a_0} \quad (3.19)$$

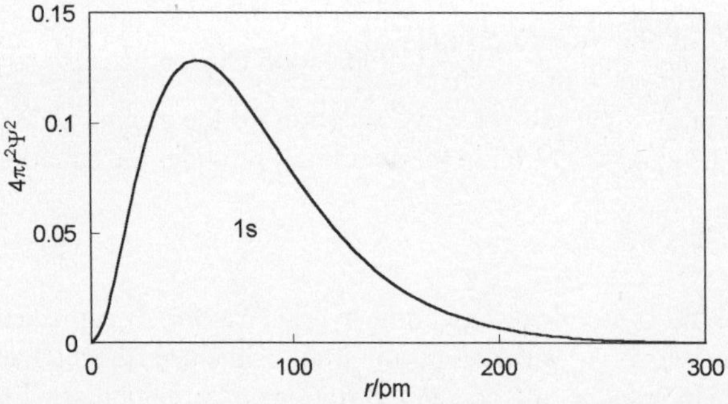

Fig. 3.2 The radial distribution function for the 1s atomic orbital of the H atom

A plot of RDF_{1s} against r is shown in Fig. 3.2 and indicates that the maximum probability of finding the electron is at a distance a_0 (= 52.9 pm) from the nucleus. The usual pictorial representation of atomic orbitals is the solid figure in which there is a 95% chance (a probability of 0.95) of finding the electron. The figure for the 1s atomic orbital of hydrogen is shown in Fig. 3.3.

Although the formal method of describing orbitals is to use mathematical expressions, much understanding of orbital properties may be gained by the use of pictorial representations. Pictorial representations are based upon contours of absolute values of ψ (i.e. with the sign of ψ being ignored), but with the sign of ψ being indicated in the various parts of the contour diagram (shaded for positive values, open for negative values of ψ). Those for the 2s, and 2p hydrogen orbitals are shown in Fig. 3.4.

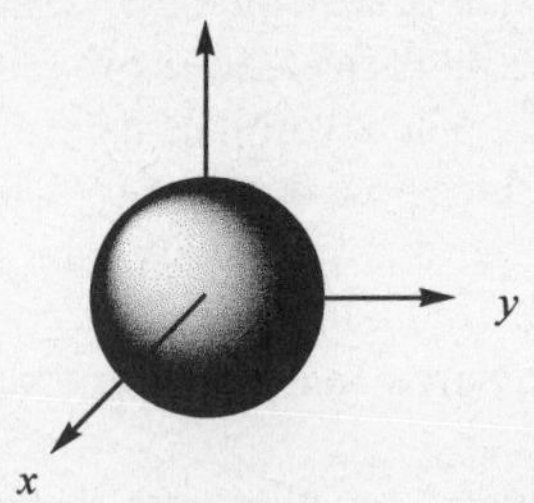

Fig. 3.3 The envelope of the 1s atomic orbital of the H atom

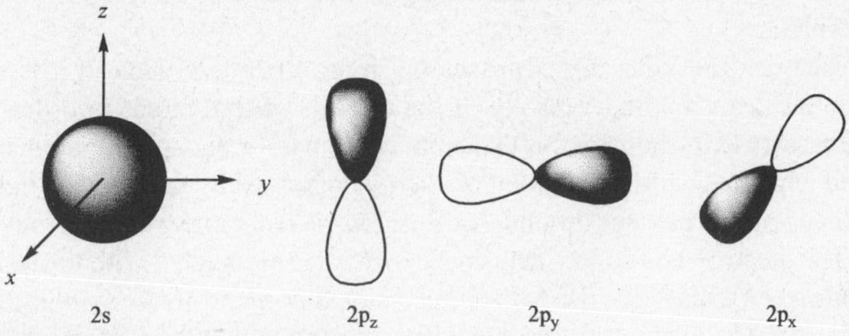

Fig. 3.4 The envelopes of the 2s and 2p atomic orbitals of the H atom

The s orbital is spherically symmetrical and the sign of ψ is everywhere positive. The three p orbitals are directed along the x, y and z axes and are described respectively as p_x, p_y and p_z. They each consist of two lobes, one of which has positive ψ values, the other having negative ψ values. The subscript descriptions of the d orbitals are chosen because such expressions are an important part of their respective mathematical formulations. It is generally accepted that the orbitals of polyelectronic atoms have similar spatial distributions as those of the hydrogen atom. The spatial orientations and the signs of ψ are important in the understanding of chemical bonding.

3.3 POLYELECTRONIC ATOMS

Deviations from the relative simplicity of one-electron atoms arise in atoms which contain more than one electron. Even an atom possessing two electrons is not treatable by the analytical mathematics which produced the solution of the wave equation for the hydrogen atom. Interelectronic repulsion must be considered. Electrons repel each other, the understanding of the effects of such repulsion being fundamental to the understanding of a great amount of chemistry. The wave equation (3.11) can only be solved for polyelectronic systems by using the iterative capacity of computers. The basis of the method is to guess the form of ψ for each orbital employed and to calculate the corresponding energy of the atomic system. The acceptable atomic orbitals as a solution of the wave equation are those which confer the minimum energy upon the system. The solutions, in general, mimic those for the hydrogen atom and the same nomenclature may be used to describe the orbitals of polyelectronic systems as have been described for the hydrogen atom. The shapes of their spatial distributions are like those for hydrogen. The major difference arises in the energies of the orbitals. The degeneracy of the levels with a given n value is lost. Orbitals with the same n value and with the samc l value are still degenerate. Those with the same n value but with different values of l are no longer degenerate. This does not affect the 1s orbital which is singly degenerate but it does affect the 2s and 2p orbitals. The 2p orbitals have a higher energy than do the 2s, the three 2p orbitals retaining their three-fold degeneracy. Likewise the sets of orbitals with higher values of n split into s, p, d,.... subsets which within themselves retain their degeneracy (given by the value of $2m_l + 1$). In general the energy of an atomic orbital decreases with the square of the nuclear charge, and superimposed upon this trend are the effects of interelectronic repulsion.

It is important to realise that the loss of degeneracy, together with the general decrease in energy as Z increases, causes changes in the order of energies of various sub-sets of atomic orbitals.

When dealing with atoms possessing more than one electron it is necessary to consider the electron-holding capacity of the orbitals of that atom. In order to do this it becomes necessary to introduce a fourth quantum number—s, the spin quantum number. This is concerned with the quantized amount of energy possessed by the electron independent of that concerned with its passage around the nucleus, the latter energy being controlled by the value of l. The electron has an intrinsic energy which is associated with the term spin. This is unfortunate since it may give rise to the impression that electrons are spinning on their own axes much as the Moon spins on its axis with a motion which is independent of its orbital motion around the Earth.

The uncertainty principle indicates that observation of the position of an electron is impossible so that the visualization of an electron spinning around its own axis must be left to the imagination! However, to take into account the intrinsic energy of an electron, the value of s is taken to be $\pm\frac{1}{2}$. Essentially the intrinsic energy of the electron may interact in a quantized manner with that associated with the angular momentum represented by l, such that the only permitted interactions are $l + s$ and $l - s$. For atoms possessing more than one electron it is necessary to specify the values of s with respect to an applied magnetic field which are expressed as values of m_s of $+\frac{1}{2}$ or $-\frac{1}{2}$.

The simple conclusion that the maximum number of electrons which may occupy any orbital is two arises from the Pauli *exclusion principle*. This is the cornerstone in the understanding the chemistry of the elements. It may be stated as: 'No two electrons in an atom may possess identical sets of values of the four quantum numbers, n, l, m_l and m_s'. The consequences are (i) to restrict the number of electrons per orbital to a maximum of two and (ii) to restrict the number of any particular orbital (defined by its set of n, l and m_l values) to one per atom. Consider the 1s orbital; $n = 1$, $l = 0$ and $m_l = 0$— there are no possibilities for changes in these values—any electron in a 1s orbital must be associated with them. One electron could have a value of m_s of ½, but the second electron must have the alternative value of m_s of −½. The two electrons occupying the same orbital must have opposite 'spins'. Since there are no other combinations of the values of the four quantum numbers it is concluded that only two electrons may occupy the 1s orbital and that there can only be one 1s orbital in any one atom. Similar conclusions are valid for all other orbitals. The application of the Pauli exclusion principle provides the necessary framework for the observed electronic configurations of the elements.

The orbital wavefunctions which have been discussed above do not represent the total wavefunctions of the electrons. The total wavefunction must include a spin wavefunction, ψ_{spin}, and can be written as:

$$\psi_{total} = \psi_{orbital}.\psi_{spin}$$

For two electrons, labelled 1 and 2 residing singly in two orbitals labelled *a* and *b*, taking into account the indistinguishability of electrons, there are two possible total wave functions which are written as:

$$\psi_m = [\psi_a(1).\psi_b(2) + \psi_a(2).\psi_b(1)].\psi_{spin}$$

$$\psi_n = [\psi_a(1).\psi_b(2) - \psi_a(2).\psi_b(1)].\psi_{spin}$$

The product $\psi_a(1).\psi_b(2)$ implies that electron 1 is resident in orbital *a* and electron 2 is in orbital *b*, etc. The linear combinations of the two ways of placing two electrons in the two orbitals represents the method of allowing for the indistinguishability of the two electrons. The orbital part of ψ_m is symmetric to electron exchange, i.e. it does not change sign if the two electrons are exchanged between their occupancies of the two orbitals. The orbital part of ψ_n does change sign if the two electrons are exchanged and is anti-symmetric to that operation.

If the two electrons possess identical spins, the two spin wavefunctions can be written as: $\alpha(1).\alpha(2)$ and $\beta(1).\beta(2)$, α and β being used to represent the two possible values of the electron spin. If the two electrons have opposed spins there are again two spin wavefunctions which can be written as the linear combinations:

$$\alpha(1).\beta(2) + \alpha(2).\beta(1) \text{ and } \alpha(1).\beta(2) - \alpha(2).\beta(1)$$

to incorporate the indistinguishability of the electrons. The former spin wavefunction is symmetric with respect to electron exchange and the latter is anti-symmetric to the same operation.

The relevance of the exchange properties of the orbital and spin wavefunctions is to an alternative manner of stating the Pauli exclusion principle which is that *the total wavefunction of a real system must be anti-symmetric*. This is a consistent property of all wavefunctions which are relevant to real systems. Applied to the two orbital − two electron case under consideration, the appropriate total anti-symmetric wavefunctions are produced by combining symmetric orbital wavefunctions with anti-symmetric spin wavefunctions or by combining anti-symmetric orbital wavefunctions with symmetric spin wavefunctions, since the products of two symmetric wavefunctions or that of two anti-symmetric wavefunctions are symmetric and would violate the Pauli principle. The appropriate total wavefunctions are thus:

$$\psi_p = [\psi_a(1).\psi_b(2) + \psi_a(2).\psi_b(1)].[\alpha(1).\beta(2) - \alpha(2).\beta(1)] \tag{3.20}$$

$$\psi_q = [\psi_a(1).\psi_b(2) - \psi_a(2).\psi_b(1)].[\alpha(1).\beta(2) + \alpha(2).\beta(1)] \tag{3.21}$$

$$\psi_r = [\psi_a(1).\psi_b(2) - \psi_a(2).\psi_b(1)].[\alpha(1).\alpha(2)] \tag{3.22}$$

$$\psi_s = [\psi_a(1).\psi_b(2) - \psi_a(2).\psi_b(1)].[\beta(1).\beta(2)] \tag{3.23}$$

The orbital state described by equation (3.20), two electrons in separate orbitals with opposed spins, is markedly different from that described by equations (3.21–3.23). The latter description, requiring three equations, is of two electrons occupying two separate orbitals and having parallel spins. This triplet state, represented by the three equations, is of lower energy than the singlet state represented by the single equation (3.20). The lower energy of the triplet state arises from the impossibility of the two electrons occupying the same space as would be the case if they occupied the same orbital. If ψ_a is made identical to ψ_b in equations (3.21–3.23) the wavefunctions become zero and there is zero probability of the occurrence of the double occupancy of one orbital by two electrons possessing the same spin. The interelectronic repulsion between two electrons in a triplet state is thus minimized by comparison to two electrons in a singlet state. In the latter case, if ψ_a is made identical to ψ_b in equation (3.20) the result is non-zero and indicates that two electrons can occupy the same orbital providing they have opposed spins.

3.4 THE ELECTRONIC CONFIGURATIONS OF THE ELEMENTS

The modern form of the periodic classification of the elements is shown in Fig. 3.5. The atomic number, Z, of each element is shown. There are eighteen groups according to modern convention. The quantum rules define the different types of atomic orbitals which may be used for electron occupation in atoms. The Pauli exclusion principle defines the number of each type of orbital and limits each orbital to a maximum electron occupancy of two. Experimental observation, together with some sophisticated calculations, indicates the energies of the available orbitals for any particular atom. The electronic configuration—the orbitals which are used to accommodate the appropriate number of electrons—may be decided by the application of what is known as the *aufbau* (German: building-up) principle which is that electrons in the ground state of an atom occupy the orbitals of lowest energy such that the total electronic energy is minimized.

In the ground state of the hydrogen atom, the electronic configuration is that in which the electron occupies the 1s orbital, written as: $1s^1$, the number of electrons occupying the orbital being indicated by the superscript.

For element number two (helium), there are two possible configurations which could be considered; $1s^2$ and $1s^12s^1$. There are two factors which decide which of the two configurations is of lower energy. These are the difference in energy between the 2s and 1s orbitals and the greater interelectronic repulsion energy in the $1s^2$ case. If the 2s–1s energy gap is larger than the interelectronic repulsion energy between the two electrons in the 1s orbital the two electrons will pair up in the lower orbital. The point may be made by the calculations which follow.

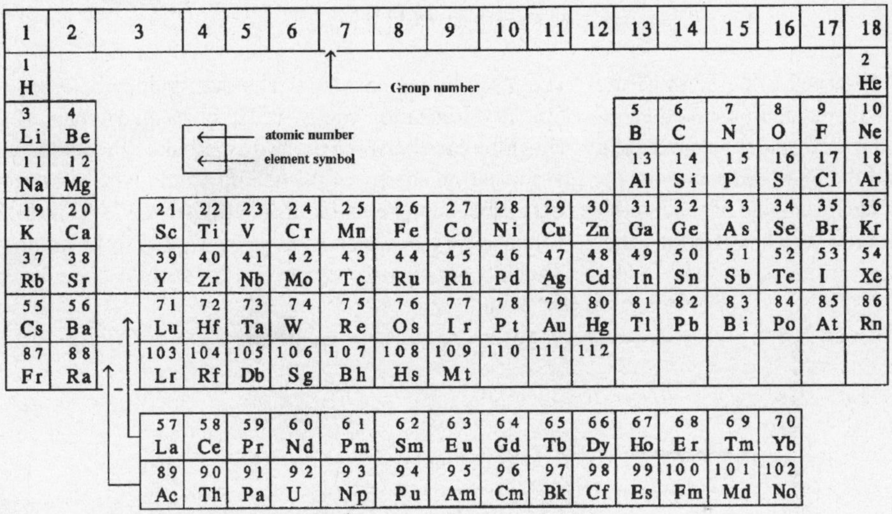
1 2 3 4 5 6 7 8 9 10 11 12 13 14 15 16 17 18
Group number
atomic number
element symbol
1 H
2 He
3 Li 4 Be
5 B 6 C 7 N 8 O 9 F 10 Ne
11 Na 12 Mg
13 Al 14 Si 15 P 16 S 17 Cl 18 Ar
19 K 20 Ca 21 Sc 22 Ti 23 V 24 Cr 25 Mn 26 Fe 27 Co 28 Ni 29 Cu 30 Zn 31 Ga 32 Ge 33 As 34 Se 35 Br 36 Kr
37 Rb 38 Sr 39 Y 40 Zr 41 Nb 42 Mo 43 Tc 44 Ru 45 Rh 46 Pd 47 Ag 48 Cd 49 In 50 Sn 51 Sb 52 Te 53 I 54 Xe
55 Cs 56 Ba 71 Lu 72 Hf 73 Ta 74 W 75 Re 76 Os 77 Ir 78 Pt 79 Au 80 Hg 81 Tl 82 Pb 83 Bi 84 Po 85 At 86 Rn
87 Fr 88 Ra 103 Lr 104 Rf 105 Db 106 Sg 107 Bh 108 Hs 109 Mt 110 111 112
57 La 58 Ce 59 Pr 60 Nd 61 Pm 62 Sm 63 Eu 64 Gd 65 Tb 66 Dy 67 Ho 68 Er 69 Tm 70 Yb
89 Ac 90 Th 91 Pa 92 U 93 Np 94 Pu 95 Am 96 Cm 97 Bk 98 Cf 99 Es 100 Fm 101 Md 102 No

Fig. 3.5 The modern form of the Periodic Classification of the Elements

Placing the value of the Rydberg constant from equation (3.14) into equation (3.12) and then substituting the terms representing the ionization energy of hydrogen from equation (3.9) gives the equation:

$$E = -I_H Z^2/n^2 \tag{3.24}$$

This equation is then applicable to hydrogen and hydrogen-like atoms (those possessing only one electron, e.g. He^+, Li^{2+},...). The energy of the 1s orbital in the helium atom is given by putting $Z = 2$, and $n = 1$, in equation (3.24) giving:

$$E(1s_{He}) = -4I_H \tag{3.25}$$

The energy of the 2s orbital of helium is given by putting $n = 2$, so that:

$$E(2s_{He}) = -I_H \tag{3.26}$$

Ignoring the effect of interelectronic repulsion, the energy of the $1s^2$ configuration is $-8I_H$ is thus lower than that of the $1s^1 2s^1$ configuration of $-5I_H$. An estimate of the interelectronic repulsion energy in the $1s^2$ configuration allows a conclusion to be made as to which of the two possible configurations is appropriate to the ground state of the helium atom. The first ionization energy of the helium atom - the energy required to cause the ionization:

$$He(g) \rightarrow He^+(g) + e^-(g)$$

is observed to be 2370 kJ mol^{-1} (rather than $4I_H$ which is 5248 kJ mol^{-1}) and the second ionization energy—that needed to cause the change:

$$He^{+}(g) \rightarrow He^{2+}(g) + e^{-}(g)$$

is observed to be 5248 kJ mol^{-1} (i.e. exactly equal to $4I_H$). The discrepancy between the calculated and observed values of the first ionization energy of He gives an estimate of the interelectronic repulsion energy. The difference between the hydrogen-like calculated value and the observed value for the first ionization energy of the helium atom gives an estimate of the magnitude of the interelectronic repulsion energy of 5248 − 2370 = 2878 kJ mol^{-1} an amount which is less than the 2s − 1s energy gap which is given by $3I_H$ (3936 kJ mol^{-1}).

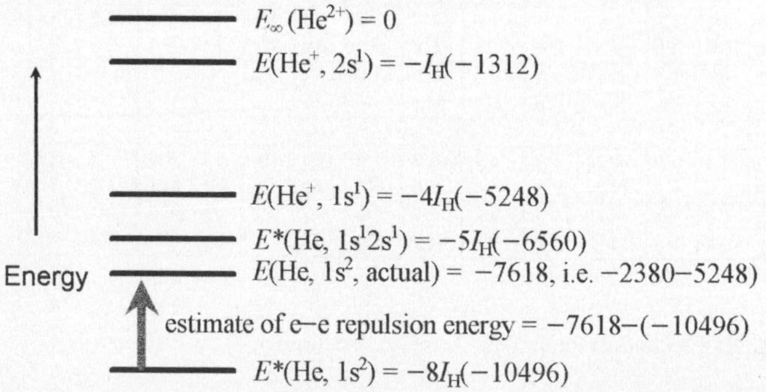

Fig. 3.6 Energies of some He and He$^+$ configurations. *The energy of the $1s^12s^1$ configuration of He does not include any contributions from Coulombic repulsion and exchange energy terms (exchange energy is explained below). If the two electrons were to have parallel spins, these terms would amount to a destabilization of the $-5I_H$ level to a higher enrgy of $-4.249I_H$ and would make it quite obvious that there is no question of the configuration representing the ground state of He. *The energy of this $1s^2$ configuration does not include the interelectronic repulsion energy.

Fig. 3.6 shows the calculated and experimental energies which are used in determining the ground state configuration of the helium atom. There is, therefore, no doubt that the ground state of the helium atom has the configuration, $1s^2$. Electrons only pair up in the same orbital when it is the lowest energy option.

The third element, lithium (Z = 3), has a full 1s orbital, with the third electron entering the 2s orbital giving the configuration: $1s^22s^1$. Beryllium (Z = 4) has the configuration, $1s^22s^2$, the interelectronic repulsion energy in $2s^2$ being lower than the 2p−2s energy gap.

The fifth electron in the boron (Z = 5) atom enters one of the three-fold degenerate 2p orbitals. Since these are truly degenerate it is not proper to specify which of the three orbitals is singly occupied, although some texts choose the $2p_x$ orbital for alphabetical reasons. Boron has the electronic configuration: $1s^22s^22p^1$. The carbon configuration is of considerable interest since it is the first example of electrons occupying degenerate orbitals. From previous considerations it is a simple matter to conclude that the two electrons occupying the 2p orbitals should occupy separate orbitals where interelectronic repulsion is less than if they doubly occupy a single orbital. What is not so obvious is that the two electrons, occupying two different 2p orbitals, should have identical values of the spin quantum number, m_s. The reason for this is that electrons with parallel spins (identical m_s values) have zero probability of occupying the same space—a kind of Pauli restriction, whereas two electrons with opposed spins (different m_s values) have a finite chance of occupying the same region of space within the atom and, in consequence, the interelectronic repulsion between them is greater than if their spins were parallel. This is the rationalization of *Hund's rules*. These may be stated in the following way: In filling a set of degenerate orbitals the number of unpaired electrons is maximized, and such unpaired electrons will possess parallel spins. An alternative, more exact, statement appears in section 3.7.

The electronic configuration of the carbon atom is $1s^22s^22p^2$ or, if the detailed content of the 2p orbitals is being discussed it may be written as: $1s^22s^22p_x^12p_y^1$, the choice of x and y being merely alphabetical. The 2p configuration is sometimes indicated diagrammatically by entering arrows (representing electrons with a particular spin) into two of three boxes (representing atomic orbitals):

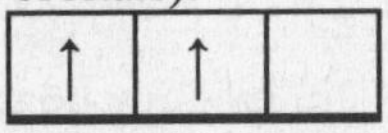

Such diagrams are only approximate descriptions of the electron arrangements in atoms, a full description is dealt with in Section 3.6. By similar arguments it is concluded that the nitrogen atom has a lowest energy electronic configuration of $1s^22s^22p_x^12p_y^12p_z^1$.

In the oxygen, fluorine and neon atoms the extra electrons doubly occupy the appropriate number of 2p orbitals since pairing is the lowest energy option - the 3s - 2p gap being greater than any interelectronic repulsion energy involved.

The atoms of the elements Li, Be, B, C, N, O, F and Ne form the second period of the Periodic Classification of the Elements. The third period contains the elements Na, Mg, Al, Si, P, S, Cl and Ar, which have **core** electronic configurations which are that of neon ($1s^22s^2\,2p^6$) plus those derived from the regular filling of the 3s and 3p orbitals as have been described for that of the 2s and 2p orbitals. Identical arguments apply and the elements of the third period are arranged under their counterparts in the second period with identical 'outer electronic configurations', except for the change in value of n from 2 to 3. The term 'outer' is a reference to the value of n and is related to the greater diffuseness of orbitals as n increases in value - the orbitals become larger. It is the nature of the outer electronic configuration which determines the chemistry of an element and it is for this reason that more emphasis is placed upon it rather than the arrangement of all the electrons of an atom.

The next elements are potassium and calcium in which the outer electrons occupy the next lowest energy orbital which is the 4s since it has a lower energy than the 3d orbitals. In the next ten elements (scandium to zinc) the five 3d orbitals are progressively occupied. The filling is in accordance with Hund's rules with two irregularities which are described below.

The outer electronic configurations of the elements Sc, Ti and V are: Sc $4s^23d^1$, Ti $4s^23d^2$, V $4s^23d^3$, but that of chromium is: Cr $4s^13d^5$. The explanation of this irregularity is that the 3d level is now of lower energy than the 4s and that the gap between the two sets of orbitals is small enough to prevent electron pairing in the 3d orbitals. An added factor in determining the stability of the Cr configuration is that of **exchange energy**. This stabilizing effect arises whenever two or more electrons with the same spin exist in a system. The interelectronic repulsion between two electrons with the same spin is less than that between two electrons with opposed spins. This is because, on average, electrons with parallel spins cannot approach each other as closely as can electrons with opposed spins. The effect is a kind of extension to the Pauli exclusion principle. The exchange stabilization of two electrons with the same spin is denoted by the amount *K*. The extent of exchange energy stabilization of a system depends upon the number of electrons with parallel spins. For a system with *n* parallel spins the stabilization is given by $^{n}C_2 \times K$— where $^{n}C_2$ represents the number of combinations of electron pairs— $n!/(2!(n-2)!)$. The $4s^1 3d^5$ configuration would have six electrons with parallel spins with an exchange energy of 15*K*. The alternative configuration for Cr of $4s^2 3d^4$ would be higher in energy by the energy difference between the 4s and 3d levels and would be less stable also by there being only five electrons with parallel spins leading to a loss of exchange energy stabilization of 5*K*.

Regularity returns with the next four elements: Mn $4s^2 3d^5$, Fe $4s^2 3d^6$, Co $4s^2 3d^7$, Ni $4s^2 3d^8$. With copper, the 3d level is sufficiently lower than the 4s to ensure complete pairing up in the 3d orbitals, leaving the sole unpaired electron in the 4s orbital: Cu $4s^1 3d^{10}$. The first series of transition elements is completed by zinc which has the configuration $4s^2 3d^{10}$.

The next orbitals to be used in building up the elements are the 4p set which are filled in a regular fashion in the elements gallium to argon, thus completing the fourth period or first long period of the periodic table. The first two elements (K and Ca) are arranged so that they come below Na and Mg and form parts of groups 1 and 2 respectively of the periodic table. The first set of transition elements form the first members respectively of groups 3 to 12 and the elements from Ga to Kr are placed under those from Al to Ar as members of groups 13 to 18.

The filling of the 5s, 4d and 5p orbitals accounts for the elements of the second long period Rb and Sr in the **s block**, Y to Cd in the second transition series or **d block**, and In to Xe in the **p block**. As was the case with the 3d orbitals, the filling of the 4d set is not regular due to the closeness of the energies of the 5s and 4d orbitals. The irregularities are not the same as in the first set of transition elements. The outer electronic configurations of the second series of transition elements are: Y $5s^2 4d^1$, Zr $5s^2 4d^2$, Nb $5s^1 4d^4$, Mo $5s^1 4d^5$, Tc $5s^1 4d^6$, Ru $5s^1 4d^7$, Rh $5s^1 4d^8$, Pd $4d^{10}$ (i.e. no 5s electrons), Ag $5s^1 4d^{10}$ and Cd $5s^2 4d^{10}$.

In Y and Zr the 5s orbital is sufficiently lower in energy than the 4d level to enforce the production of the $5s^2$ pairing. In the elements from Nb to Rh the 4d orbitals are marginally lower in energy than the 5s orbital and the two sets of orbitals behave as though they are degenerate with added stability arising from exchange energy effects. In the elements Pd, Ag and Cd the 4d energy is distinctly lower than that of the 5s orbital so that complete pairing occurs in Pd, and then the 5s filling follows in Ag and Cd. More irregularities are to follow in the formation of the third long period. The next orbital of lowest energy to be used is the 6s whose filling accounts for the outer electronic configurations of Cs and Ba and then come the 5d and 4f sets whose energies are very nearly identical, but vary with the nuclear charge. The electronic configurations of the next fifteen elements are shown in Table 3.3.

The 5d orbital is singly occupied at the beginning, in the middle and at the end of the above series of elements, which is where it has very similar energy to the 4f set of orbitals. For the other elements the 4f energy is sufficiently lower than that of the 5d for the latter not to be used. In gadolinium the $5d^1 4f^7$ arrangement is preferred to $4f^8$ as it maximizes the exchange energy. The elements La to Yb form the f block and are the fourteen elements concerned with the filling of the 4f orbitals with its completion giving lutetium ($6s^2 5d^{14}$) which is placed as the first member of the third transition series. Some versions of the periodic table have lanthanum as the first member of the third transition series, the lanthanide elements (cerium to lutetium) coming between lanthanum and the second member of the third transition series, Hf. Both lanthanum and lutetium have a single 5d electron, lutetium possessing fourteen 4f electrons ($4f^{14}$) as do the other members of the third transition series.

Table 3.3 Outer electronic configurations of elements 57 - 71

Element, symbol	6s	5d	4f
Lanthanum, La	2	1	0
Cerium, Ce	2	0	2
Praseodymium, Pr	2	0	3
Neodymium, Nd	2	0	4
Promethium, Pm,	2	0	5
Samarium, Sm	2	0	6
Europium, Eu	2	0	7
Gadolinium, Gd	2	1	7
Terbium, Tb	2	0	9
Dysprosium, Dy	2	0	10
Holmium, Ho	2	0	11
Erbium, Er	2	0	12
Thulium, Tm	2	0	13
Ytterbium, Yb	2	0	14
Lutetium, Lu	2	1	14

The next nine elements from hafnium to mercury complete the third transition series, although there are irregularities of filling the 5d orbitals since the 5d and 6s energies are close together. The outer electronic configurations of these elements are: Hf $6s^2 5d^2$, Ta $6s^2 5d^3$, W $6s^2 5d^4$, Re $6s^2 5d^5$, Os $6s^2 5d^6$ and Ir $6s^2 5d^7$—a regular filling of the 5d orbitals whose energies are higher than that of the 6s orbital, but after Ir the 5d energy becomes lower than that of the 6s and the small energy gap, with added exchange energy effects, causes the irregularities in the next two elements: Pt $6s^1 5d^9$ and Au $6s^1 5d^{10}$. The filling of the 6s orbital is completed in mercury: Hg $6s^2 5d^{10}$.

The regular filling of the 6p orbitals accounts for the outer electronic configurations of the elements from thallium to radon, thus completing the third long period with its integral **f block** elements.

The fourth long period is incomplete but, as far as it goes, mirrors the third long period. The first two elements, francium and radium have outer electronic configurations $7s^1$ and $7s^2$ respectively. The energies of the next orbitals to be filled (6d and 5f) are similar and give rise to irregularities dependent upon which of them is the lower in energy and by how much and exchange energies. The accepted outer electronic configurations are: Ac $7s^2 6d^1$, Th $7s^2 6d^2$ ($7s^2 6d^1 5f^1$ according to some sources), Pa $7s^2 6d^1 5f^2$, U $7s^2 6d^1 5f^3$, and Np $7s^2 6d^1 5f^4$,- the irregularities persist up to plutonium, $7s^2 5f^6$, and after that there is the presumption that there is a regular filling of the 5f orbitals, much like that of the 4f orbitals, with the exception of there being a reappearance of a single 6d electron in the case of curium (exchange energy is maximized as in the case of gadolinium). The trans-plutonium elements are intensely radioactive and have short lifetimes so that detailed studies of their electronic configurations have not been carried out and the published ones are speculative.

Similarly to the lanthanide elements, the actinides (Actinium to Nobelium) are placed as a set of fourteen elements as a separate f block, with element 104, Lr, being the first member of the fourth transition series. Alternative versions of the periodic table have actinium as the first member of the fourth transition series and have lawrencium as the last of the actinide elements in the series of fourteen from thorium to lawrencium. Lawrencium, as does lutetium, possesses a full f set of atomic orbitals. Some versions of the periodic table dodge the above considerations and have fifteen 'lanthanide' elements, La – Lu, and fifteen 'actinide' elements, Ac – Lr, placed in the group 3 column. The fourth transition series is unfinished and the element with the highest *Z* value, which has so far been synthesized, is number 112. The names of the translawrencium elements (i.e. those with atomic number values (*Z*) greater than 103) have only recently (1997) been settled after many years of argument. They have been ratified by the International Union of Pure and Applied Chemistry (IUPAC), and are as shown in Table 3.4.

The syntheses of isotopes of elements 110, 111 and 112 have been reported, but it is too early to have names for them, since it took over twenty years for the names in Table 3.4 to be accepted.

Table 3.4 Names and symbols of the translawrencium elements

Atomic number	IUPAC Name	Proposed symbol
104	Rutherfordium	Rf
105	Dubnium	Db
106	Seaborgium	Sg
107	Bohrium	Bh
108	Hassium	Hs
109	Meitnerium	Mt

The periodic table consists of eighteen groups corresponding to the filling of the *n*s, (*n* – 1)d and *n*p orbitals for the values of *n* up to seven, with the f block elements situated separately.

There are three short periods for $n = 1 - 3$, since there are no available d orbitals at that stage. Hydrogen and helium form the first short period with hydrogen placed as the first element in group one and helium as the first element in group eighteen. The elements from Li to Ne and Na to Ar form the second and third short periods respectively, although some older accounts describe them as the first and second short periods, discounting H and He. The fourth, fifth and sixth periods all contain 18 elements (associated with the filling of the *n*s, (*n* – 1)d and *n*p orbitals for $n = 4$, 5 and 6). The sixth period also contains 14 lanthanide elements associated with the filling of the seven 4f orbitals. The seventh period is similar to the sixth in having an extra 14 elements (associated with the filling of the seven 5f orbitals), but the later elements associated with the filling 6d sub-shell have yet to be synthesized.

The majority of elements in any particular group have identical outer electronic configurations (apart from having varying values of the principal quantum number, *n*). Because of the various irregularities in orbital filling it is not possible to state that all elements in a group have identical configurations. As already described, it is not possible for there to be a completely regular filling of orbitals which would allow any member of a group to have the same outer electronic configuration. This is usually not the case for any particular oxidation state of the elements of one group. These tend to have identical outer electronic configurations. The outer configurations of the elements, Ni, Pd and Pt, are s^2d^8, d^{10} and s^1d^9, respectively, but their +2 ions have the common d^8 configuration.

3.5 PERIODICITIES OF SOME ATOMIC PROPERTIES

This section deals with the variation, along the periods and down the groups of the periodic table, of the ionization energies, sizes and electronegativity coefficients of the elements.

Ionization Energies

The first ionization energy of an atom is the minimum energy required to convert one mole of the gaseous atom (in its ground—lowest energy—electronic state) into its unipositive ion:

$$A(g) \rightarrow A^{+}(g) + e^{-}$$

The first ionization energies of s and p block elements of periods two to four are plotted in Fig. 3.7.

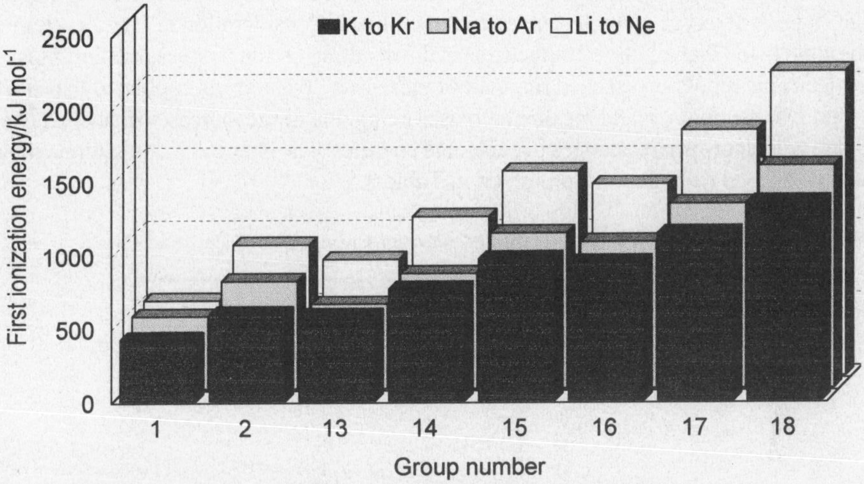

Fig. 3.7 The first ionization energies of the main group elements of the 2nd, 3rd and 4th periods

There is a pattern of the values for the elements Li to Ne which is repeated for the elements Na to Ar and which is repeated yet again for the elements K, Ca and Al to Kr (the s- and p-block elements of the fourth period). In addition to this periodicity it is obvious that the ionization energy decreases down any of the groups. Both observations are broadly explicable in terms of the electronic configurations of the elements.

Variation of Ionization Energy along a Period

Consider the variation of the first ionization energies of the elements of the second period. The electron which is removed in the ionization process is the one (or is one of those) with the highest energy - the one requiring the least energy to remove from the attractive influence of the atomic nucleus. The value of the first ionization energy of the lithium atom is 519 kJ mol^{-1} and corresponds to the change in electronic configuration $1s^2 2s^1$ to $1s^2$.

The increase to 900 kJ mol^{-1} in the case of the beryllium atom is due to the increase in effective nuclear charge. The electron most easily removed is one of the pair in the 2s orbital. In the case of the boron atom, in spite of an increase in nuclear charge, there is a decrease in the first ionization energy to 799 kJ mol^{-1}. This is because the electron removed is from a 2p orbital which is higher in energy than the 2s level.

There follows a general increase in the first ionization energies of the carbon (1090 kJ mol^{-1}) and nitrogen (1400 kJ mol^{-1}) atoms. The electrons are removed from the 2p orbitals and the increase is due to the increase in nuclear charge, there being one and two units of exchange energy stabilization to be overcome respectively. With oxygen there is a slight decrease to 1310 kJ mol^{-1} in spite of an increase in nuclear charge, there being no difference in exchange energy between the 2p^4 and 2p^3 configurations. In the cascs of B, C and N the electron removed in the ionization process is the sole resident of one of the 2p orbitals. In oxygen there are two 2p orbitals which are singly occupied and one which is doubly occupied. Consideration of the higher interelectronic repulsion associated with a pair of electrons in the same orbital leads to the conclusion that it is one of the paired-up electrons in the 2p level of oxygen which is most easily removed. Consideration of Hund's rules leads to the conclusion that all three 2p electrons in the resulting O^+ ion possess parallel spins. The interelectronic repulsion assisted removal of paired-up 2p electrons applies to the cases of the next two elements, F and Ne, the increases being due to the increases in nuclear charge coupled with appropriate changes in exchange energies. The changes in ionization energies along the second period are summarized in Table 3.5.

Table 3.5 Contributions to the first ionization energies of the elements Li – Ne

Atom	**First ionization energy**
Li	E(2s)
Be	E(2s) $-J$
B	E(2p)
C	E(2p) $+K$
N	E(2p) $+2K$
O	E(2p) $-J$
F	E(2p) $-J+K$
Ne	E(2p) $-J+2K$

The terms used in Table 3.5 are the orbital energies, E(2s) and E(2p), K which is a unit of exchange energy stabilization and J representing the extra coulombic repulsion energy possessed by a system containing two electrons paired up in the same orbital. It should be noted that the energies E(2s) and E(2p) are not constant along the period, but increase as the value of Z_{eff} increases. The combination of the three factors, varying orbital energies, exchange energy and coulombic energy differences along the period produce the observed pattern of changes in the first ionization energies.

The general pattern of the variation of the first ionization energies of the elements of the second short period is repeated for the respective elements of the s and p blocks of the subsequent periods of the periodic system except for those in period six (Cs, Ba, Tl, Pb, Bi, Po, At and Xe). The explanation of the different pattern is dealt with in Section 3.6.

In any of these series there are two variable quantities: the nuclear charge and the number of electrons of the atoms considered. It is possible to eliminate the effect of the nuclear charge (but not the effects of changes in the effectiveness of the nuclear charge) by considering the successive ionization energies of the neon atom, for example. Fig. 3.8 is a plot of the first eight successive ionization energies of the neon atom and, as would be expected from a reduction in the number of electrons being attracted by the constant nuclear charge, the values exhibit a general increase.

The first three ionizations arise from doubly occupied 2p orbitals by the same reasoning as is given above for the first ionization of the oxygen atom. All three ionizations are assisted by the large interelectronic repulsion associated with the double occupation of orbitals (J). As may be seen from the extrapolation of the line joining the first three points in Fig. 3.8 the next three electrons are considerably more difficult to remove than the first three. There is an increasing effectiveness of the nuclear charge with the additional difficulty of electron removal due to the absence of the high interelectronic repulsion assistance enjoyed by the first three electrons to be removed.

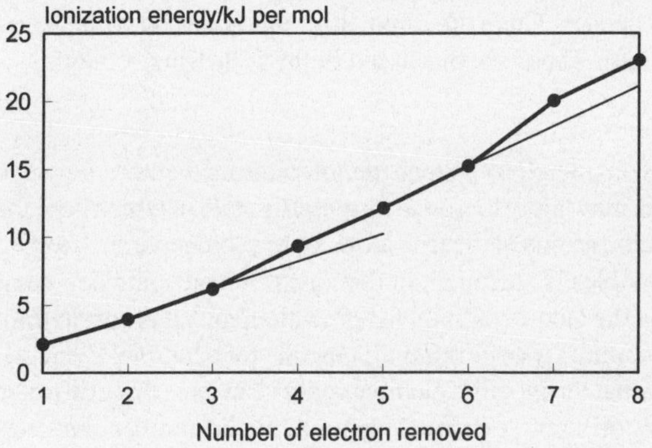

Fig. 3.8 The eight successive ionization energies of the Ne atom

The second discontinuity shown in Fig. 3.8 is associated with the ionization of the 7th and 8th electrons which are very much more difficult to remove because they originate in the 2s orbital of the neon atom.

The energy terms associated with the successive ionizations of the neon atom are given in Table 3.6. The values of E(2p) and E(2s) increase considerably as Z_{eff}^2 increases to produce the observed variations.

Table 3.6 Contributions to the successive ionization energies of the neon atom

Electron removed	Ionization energy
First	$E(2p) - J + 2K$
Second	$E(2p) - J + K$
Third	$E(2p) - J$
Fourth	$E(2p) + 2K$
Fifth	$E(2p) + K$
Sixth	$E(2p)$
Seventh	$E(2s) - J$
Eighth	$E(2s)$

Variation of Ionization Energy down a Group

In general there is a decrease in the first ionization energies of the atoms of any particular group of the periodic table as the nuclear charge increases. This may be seen from the data plotted in Fig. 3.7. Down any group the orbital from which the electron is ionized has a progressively larger value of n, and has the highest energy for each particular atom in the group. This essentially offsets the effect of the increasing nuclear charge. It is as though the outermost electrons are shielded from the effect of the nuclear charge by the full inner sets of orbitals. The exceptions to this general trend occur at the bottom of Groups 13 and 14, the first ionization energies of Tl and Pb being respectively higher than those of the corresponding elements in period 5, In and Sn. An explanation is given in Section 3.6.

Variations in Atomic Size across Periods and down Groups

The size of an atom is not a simple concept. There are three different modes by which sizes may be assigned to any particular atom. These are discussed in the following sections.

Atomic radius

The atomic radius of an element is considered to be half the interatomic distance between identical (singly bonded) atoms. This may apply to iron in its metallic state in which case the quantity may be regarded as the metallic radius of the iron atom, or to a molecule such as Cl_2. The difference between the two examples is sufficient to demonstrate that some degree of caution is necessary when comparing the atomic radii of different elements. It is best to limit such comparisons to elements with similar types of bonding - metals for example. Even that restriction is subject to the drawback that the metallic elements exhibit at least three different crystalline arrangements with possibly different coordination numbers (the number of nearest neighbours for any one atom).

Covalent radius

The covalent radius of an element is considered to be one half of the covalent bond distance of a molecule such as Cl_2 (equal to its atomic radius in this case) where the atoms concerned are participating in single bonding. Covalent radii for participation in multiple bonding are also quoted. In the case of a single bond between two different atoms the bond distance is divided up between the participants by subtracting the covalent radius of one of the atoms, whose radius is known, from it. A set of mutually consistent values is now generally accepted and, since the vast majority of the elements take part in some form of covalent bonding, the covalent radius is the best quantity to consider for the study of general trends. Only atoms of the Group 18 elements (except Kr and Xe) do not have covalent radii assigned to them because of their general inertness with respect to the formation of molecules. The use of covalent radii for comparing the sizes of atoms is subject to the reservation that its magnitude, for any given atom, is dependent upon the oxidation state of that element.

Van der Waals radius

The Van der Waals radius of an element is half the distance between two atoms of an element which are as close to each other as is possible without being formally bonded by anything except Van der Waals intermolecular forces. Such a quantity is used for the representation of the size of an atom with no chemical bonding tendencies—the Group 18 elements. That for krypton, for instance, is half of the distance between nearest neighbours in the solid crystalline state, and is equal to the atomic radius. Van der Waals radii of atoms and molecules are of importance in discussions of the liquid and solid states of molecular systems, and in the details of some molecular structures where two or more groups attached to the same atom may approach each other.

Fig. 3.9 shows how the covalent radii vary across periods and down groups of the periodic table.

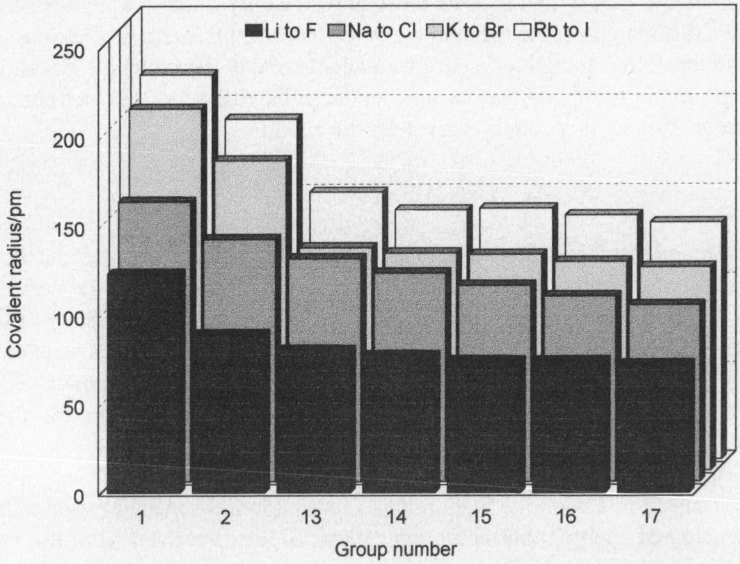

Fig. 3.9 The covalent radii of the main group elements of the 2nd, 3rd, 4th and 5th periods

Across periods there is a general reduction in atomic size, whilst down any group the atoms become larger. These trends are consistent with the understanding gained from the study of the variations of the first ionization energies of the elements. As the ionization energy is a measure of the effectiveness of the nuclear charge in attracting electrons it might be expected that an increase in nuclear effectiveness would lead to a reduction in atomic size. The trends in atomic size in the periodic system are almost the exact opposite to those in the first ionization energy.

Variations in Electronegativity across Periods and down Groups

The concept of electronegativity is derived from such experimental observations that the elements, fluorine and chlorine, are highly electronegative in that they exhibit a very strong tendency to become negative ions. The metallic elements of Group 1, on the other hand, are not electronegative and are better described as being electropositive - they have a strong tendency to form positive ions. A scale of electronegativity coefficients would be useful in allowing a number to represent the tendency of an element in a molecule of attracting electrons to itself. The establishment of such a scale has involved the powers of two Nobel prize winners (Pauling and Mulliken) and, after many other efforts, there are still doubts about the currently accepted values and about their usefulness.

Digression on the Pauling and Mulliken scales of electronegativity

Pauling's scale of electronegativity was based on the observation that the bond dissociation energy of a diatomic molecule, AB, was normally greater than the average values for the diatomic molecules, A_2 and B_2, for any pair of elements, A and B. It is important to note that the three molecules AB, A_2 and B_2 were those in which only single covalent bonds existed. He stipulated that the geometric mean of the bond dissociation energies of the molecules A_2 and B_2 represented the strength of a purely covalent bond in the molecule AB and that any extra strength of the A-B bond was because of the of the difference in electronegativity of the two atoms. This extra strength is given by the equation:

$$\Delta = D(\text{A-B}) - [D(\text{A}_2) \times D(\text{B}_2)]^{1/2}$$

and Pauling found that there was a correlation between the square root of Δ and the positive differences between values of the electronegativity coefficents of the two elements concerned. By assigning the value of 2.1 for the electronegativity coefficient of hydrogen (this was done to ensure that all values of the coefficients would be positive) Pauling was able to estimate values for a small number of elements for which the thermochemical data were available. The scale was limited by lack of data and by the perversity of the majority of elements not to form diatomic molecules.

Mulliken's scale was based on the logic that electronegativity is a property of an element that can be related to the magnitude of its first ionization energy and to that of its electron attachment energy, both being indications of the effectiveness of the nucleus in attracting electrons. The ionization energy is a measure of the effectiveness of the nucleus in retaining electrons and the electron attachment energy is a measure of the attraction of the neutral atom for an external electron. The Mulliken coefficients were calculated as the average of the first ionization energy and the energy released when the atom accepted an electron (i.e the value of the first electron attachment energy with its sign changed) with a suitable modifying multiplier to ensure the best correlation with the already established Pauling values

An electronegative atom is expected to have a high first ionization energy and to have a large negative value for its electron attachment energy and to arrange for the two energies to work in the same direction it is necessary to change the sign of the latter.

The scale was limited by the available accurate data, particularly with regard to values of the electron attachments energies. These latter are now well established, but the up-to-date Mulliken scale has not found general acceptance. The correlation with the Allred-Rochow scale (see below) is highly significant with a correlation coefficient of 0.9, the Mulliken values being, on average, 8% lower than the corresponding Allred-Rochow values.

This is not surprising since the Allred-Rochow values were parameterized to a smaller set of elements, those having either or both Pauling and/or Mulliken values.

The now generally accepted scale of electronegativity was derived by Allred and Rochow and is known by their names. It is based on the concept that the electronegativity of an element is related to the force of attraction experienced by an electron at a distance from the nucleus equal to the covalent radius of the particular atom. According to Coulomb's law this force is given by:

$$F = \frac{Z_{eff} e^2}{r_{cov}^2} \tag{3.27}$$

where Z_{eff} is the effective atomic number and e is the electronic charge. The effective atomic number is considered to be the difference between the actual atomic number, Z, and a shielding factor, S, which is estimated by the use of Slater's rules. These represent an approximate method of calculating the screening constant, S, such that the value of the effective atomic number, Z_{eff}, is given by $Z - S$. The value of S is obtained from the following rules.

1. The atomic orbitals are to be grouped in the order of their increasing distance from the nucleus:
(1s); (2s, 2p); (3s, 3p); (3d); (4s, 4p); (4d); (4f); etc.

2. For an electron (principal quantum number n) in a group of s and/or p electrons, the value of S is given by the sum of the following contributions.
(i) Zero from electrons in groups further away from the nucleus than the one considered.
(ii) 0.35 from each other electron in the same group, unless the group considered is the 1s when an amount 0.3 is used.
(iii) 0.85 from each electron with principal quantum number of $n - 1$.
(iv) 1.00 from all other inner electrons.

3. For an electron in a d or f group the parts (i) and (ii) apply, as in rule 2, but parts (iii) and (iv) are replaced with the rule that all the inner electrons contribute 1.00 to S.

Slater's rules are used to calculate the values of the Allred-Rochow electronegativity coefficients, the best fit with the older accepted Pauling/Mulliken values for electronegativity coefficients is given by the equation:

$$\chi = \frac{3590 Z_{eff}}{r_{cov}^2} + 0.744 \tag{3.28}$$

in which r_{cov} is expressed in picometres.

The values given in modern versions of the periodic table and in data books are the Allred-Rochow values rounded off (usually) to one decimal place. To attempt to represent χ more accurately would be imprudent after consideration of the difficulties in obtaining any values at all! The method of calculation offers the possibilities of assigning χ values to the different oxidation states of the same element - it would be expected, for instance, that the electronegativities of manganese in its oxidation states of 0, II, III, IV, V, VI and VII would be different and providing that the necessary covalent radii are known the χ values are easily calculated. As the element becomes smaller (as the oxidation state increases) the χ value rises. It would be expected that the smaller and more highly charged an atom becomes, the higher would be its attraction for electrons.

Fig. 3.10 shows the variation of Allred-Rochow electronegativity coefficients for singly bonded elements along periods and down groups of the periodic table. In general the value of the electronegativity coefficient increases across the periods and decreases down the groups. That is precisely the opposite of the trends in covalent radii but similar to the trends in first ionization energy (Fig. 3.7). This latter conclusion is no surprise as it is to be expected that elements with a high tendency to attract electrons possess high first ionization energies.

There have been attempts to relate differences in electronegativity coefficients to the percentage ionic character of bonds between different elements. These are reasonable on a qualitative level but quantitative equations are found only to apply roughly to particular series of compounds (e.g. H–F to H–I), no broader generalizations being reasonable.

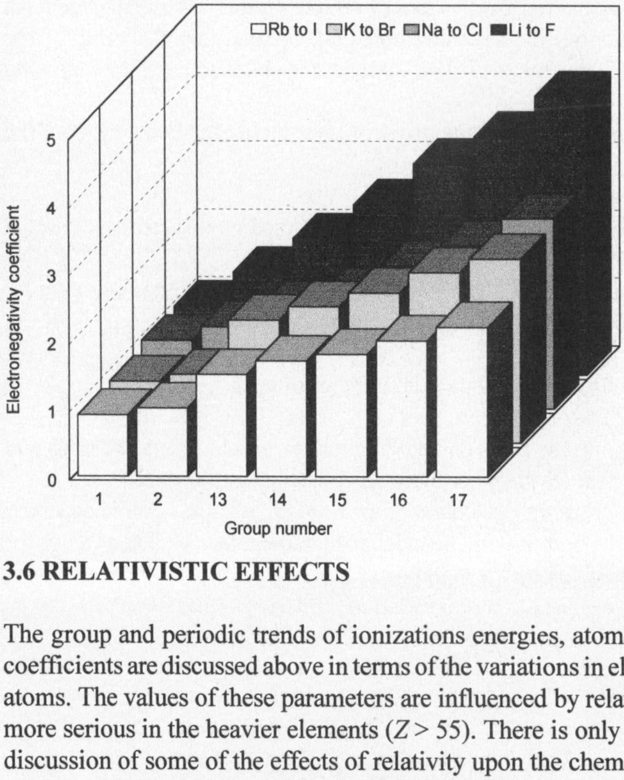

Fig. 3.10 The Allred-Rochow electronegativity coefficients of the main group elements of the 2nd, 3rd, 4th and 5th periods

3.6 RELATIVISTIC EFFECTS

The group and periodic trends of ionizations energies, atomic sizes and electronegativity coefficients are discussed above in terms of the variations in electronic configurations of the atoms. The values of these parameters are influenced by relativistic effects which become more serious in the heavier elements ($Z > 55$). There is only scope in this book for a brief discussion of some of the effects of relativity upon the chemistry of the elements.

The theory of relativity expresses the relationship between the mass m of a particle travelling at a velocity v and its rest mass m_0:

$$m = \frac{m_0}{\left(1 - \frac{v^2}{c^2}\right)^{1/2}} \tag{3.29}$$

where c_0 is the velocity of light. The average velocity of a 1s electron is proportional to the atomic number (Z) of the element. That of an atom of mercury ($Z = 80$) is 80/137 of the velocity of light. Application of equation (3.29) shows that the mass is 23% greater than the rest mass. Since the radius of the 1s orbital is inversely proportional to the mass of the electron, the radius of the orbital is reduced by 23% compared to that of the non-relativistic radius. This s orbital contraction affects the radii of all the other orbitals in the atom up to, and including, the outermost orbitals. The s orbitals contract, the p orbitals also contract, but the more diffuse d and f orbitals can become more diffuse as electrons in the contracted s and p orbitals offer a greater degree of shielding to any electrons in the d and f orbitals.

The effects are observable by a comparison of the metallic radii and the first two ionization energies of the elements of Group 12 (Zn, Cd and Hg) as shown in Table 3.7.

Table 3.7 The metallic radii and first two ionization energies of the elements zinc, cadmium and mercury I_1 /kJ mol^{-1}

Atom	r/pm	I_1 /kJ mol^{-1}	I_2 /kJ mol^{-1}
Zn	133	908	1730
Cd	149	866	1630
Hg	152	1010	1810

The mercury atom is smaller than expected from the zinc-cadmium trend and is more difficult to ionize than the lighter atoms. In consequence the metal-metal bonding in mercury is relatively poor resulting in the element being a liquid in its standard state. This almost 'inert gas' behaviour of mercury may be compared to that of the real inert gas xenon which has first and second ionization energies of 1170 and 2050 kJ mol^{-1}.

Relativistic effects are also observable in the properties of the Group 11 elements (Cu, Ag and Au). Their metallic radii, first three ionization energies and their electron attachment energies are given in Table 3.8.

Table 3.8 The metallic radii, first three ionization energies and first electron attachment energies of copper, silver and gold (energies in kJ mol^{-1})

Atom	r/pm	I_1	I_2	I_3	E
Cu	128	745	1960	3550	-119
Ag	144	732	2070	3360	-126
Au	144	891	1980	2940	-223

The relatively high value of the first ionization energy of gold is a major factor in the explanation of the nobility of the metal. The extraordinarily highly negative electron attachment energy of gold is demonstrated by the existence of the red crystalline ionic compound Cs^+Au^- —produced by heating an equimolar mixture of the two elements.

The general Group similarities exhibited by members of the second and third transition series are due to their almost identical atomic radii. The reason for the anomalously low sizes of atoms of the third transition series is usually given in terms of the lanthanide contraction. As the 4f set of orbitals is filled in the elements lanthanum–lutetium there is a contraction in atomic radius from 187 pm (La) to 172 pm (Lu). This contraction is regarded as having no more than a 14% relativistic contribution so this is not the major factor responsible for the apparent contraction of the transition metals of the third series. The main factor affecting the sizes of the elements of the third transition series is the increasing relativistic contraction due to their high Z values, the maximum effects being in gold.

The exceptions to the general trends of first ionization energies across periods and down groups, mentioned in above, are explicable when relativistic effects are considered. Thallium and lead have higher values of their first ionization energies than expected from the trends down their respective groups because their s and p orbitals are more compact. The relativistic effect upon the p orbitals of the elements from Tl to Rn is to produce a stabilization of one orbital with respect to the other two. Instead of the expected trend of the first ionization energies of Tl, Pb and Bi (589, 715 and 703 kJ mol^{-1}) to exhibit a general increase like those of In, Sn and Sb, (558, 709 and 834 kJ mol^{-1}) the value for Bi is less than that of Pb. This is because the highest energy electron in Bi is in a p orbital higher in energy than the other two p electrons.

Another notable difference in properties down groups is the 'inert pair effect' as exhibited by the chemical behaviour of Tl, Pb and Bi. The main oxidation states of these elements are I, II and III respectively which are lower by two units than those expected from the behaviour of the lighter members of each Group which exhibit oxidation states of III, IV and V respectively. This effect is largely explicable by the relativistic effects on the appropriate ionization energies which make the achievement of the higher oxidation states (the removal of the pair of s electrons) more difficult.

The irregularities in filling the 3d and 4s, 4d and 5s, and the 5d and 6s sets of atomic orbitals (described in Section 3.4) to produce the three series of transition elements are explicable by the relativistic effects upon the relative energies of the d and s orbitals concerned in each series.

3.7 ELECTRONIC CONFIGURATIONS AND ELECTRONIC STATES

The term *electronic state* is used when referring to a very precise method of describing the distribution of electrons within a given electronic configuration. It is necessary because the electronic configuration as such is not sufficiently precise for some purposes and because there are many instances when one configuration represents more than one state. The electronic state is representative of the precise manner in which atomic orbitals are occupied and takes into account the various spin possibilities for a given configuration.

The simplest example of a configuration leading to more than one state is the carbon atom. It also exemplifies the proper application of Hund's rules. The general expression of the carbon atom's electronic configuration is $1s^22s^22p^2$ and the 'ground state' has been decided to be that in which the two 2p electrons possess the same value of the spin quantum number, m_s.

Electronic states of the carbon atom

More quantum numbers are required to define the possible states arising from the $2p^2$ configuration of the carbon atom. This is because of the quantized nature of the interaction of the two electron's orbital and spin momenta to give a resultant total orbital angular momentum (L) and a resultant total spin angular momentum (S). The method of interaction described in this section is known as Russell-Saunders coupling. It depends upon there being strong interactions between the individual l values to give an L value and strong interactions between the individual s values to give an S value. This is followed by the quantized interaction between the L and S values. The rule for the total orbital angular momentum quantum number (L) is that it is made up from the individual l values in general as:

$$L = l_1 + l_2, l_1 + l_2 - 1, \ldots. |l_1 - l_2| \tag{3.30}$$

There is a similar rule for the total spin angular momentum quantum number (S):

$$S = s_1 + s_2, |s_1 - s_2| \tag{3.31}$$

Rather than use the rules as written in equations (3.30) and (3.31) it is better and more certain to use rules which refer to the permitted L and S values with respect to an externally applied magnetic field (M_L and M_S) and use their combinations to decide upon the permitted values of L and S. The rules for M_L and M_S values with respect to an external magnetic field are:

$$M_L = L, L-1, \ldots 0, \ldots -(L-1), -L \tag{3.32}$$

$$M_S = S, S-1, \ldots 0, \ldots -(S-1), -S \tag{3.33}$$

where the upper case Ms refer to totals of individual m_l and m_s values respectively as defined by:

$$M_L = \Sigma m_l \text{ and } M_S = \Sigma m_s$$

the l and s subscripts being present to distinguish the orbital and spin magnetic quantum number values.

By using these rules it is possible to derive the electronic states associated with the $2p^2$ electronic configuration with certainty. It is necessary to write down all the possible microstates associated with the two 2p electrons. This is done in Table 3.9.

This is done by writing down the permitted combinations of the possible values of m_l and m_s, each different combination representing a microstate. There are fifteen ways of distributing two indistinguishable electrons between three orbitals given that there is a choice of two values for the spin quantum number. With three orbitals and two spin values there are six ways of assigning the first electron. That leaves only five ways of assigning the second electron, the combinations of the two assignments being thirty in total. This has to be divided by two to take account of the indistinguishability of the two electrons giving a final total of fifteen microstates. These are specified in Table 3.9. Each microstate could be represented by a 'box' diagram, this section illustrating the need to take into account all possible microstates to produce an accurate description of an electronic state.

Table 3.9 The fifteen microstates of the $2p^2$ configuration

Microstate number	$m_l = 1$	$m_l = 0$	$m_l = -1$	Σm_l	Σm_s
1	↑↓			2	0
2		↑↓		0	0
3			↑↓	-2	0
4	↑	↑		1	1
5	↓	↓		1	-1
6	↑		↑	0	1
7	↓		↓	0	-1
8		↑	↑	-1	1
9		↓	↓	-1	-1
10	↑	↓		1	0
11	↓	↑		1	0
12	↑		↓	0	0
13	↓		↑	0	0
14		↑	↓	-1	0
15		↓	↑	-1	0

All of the microstates have equal probability of occurrence. The problem is to sort them out into members of proper electronic states. This is done by identifying sets of microstates which represent individual electronic states. This means finding sets of values of Σm_l which are consistent with there being a valid value of L in accordance with equation (3.28) and the best way of doing this is to look for the highest value of Σm_l in Table 3.9—this corresponds to microstate number 1, with a Σm_l value of 2 and its associated value of Σm_s of zero. If equation (3.32) is to be satisfied there should be additional values of Σm_l of 1, 0, -1 and -2, all with the same value of $\Sigma m_s = 0$. There are such values present in microstates numbers 10 and 11 ($\Sigma m_l = 1$), numbers 2, 12 and 13 ($\Sigma m_l = 0$), numbers 14 and 15 ($\Sigma m_l = -1$) and number 3 ($\Sigma m_l = -2$), all of which have total spin values of zero. Only one microstate is required for each value of Σm_l and the duplicate and triplicate microstates can and do contribute to other states present. What is certain is that the $M_L = \Sigma m_l$ values of 2, 1, 0, -1 and -2, all with an $M_S = \Sigma m_s$ value of zero, are present and satisfy the rules for there being a particular electronic state arising from the $2p^2$ configuration. Before isolating the other possible states it is necessary to deal with the symbolism used in the description of electronic states.

For any electronic state the value of L (the maximum Σm_l value) is used to assign a letter to the state, in the same way in which l is used to assign a descriptive letter to an orbital. The coding is just the same as that used for the values of l except that upper case letters are used for the L value codes. Some values of L and the associated code letters are given in Table 3.10.

Table 3.10 Coding letters for some L values

Value of L	**Code letter**
0	S
1	P
2	D
3	F
4	G

Table 3.10 is a portion of an infinite series of L values and the coding follows alphabetical order after F. The L value of the electronic state already identified for the $2p^2$ configuration is a D state. The five microstates forming the D state are associated with a single value of Σm_s and accordingly it is to be described as a singlet state. A more formal method of arriving at this conclusion is to calculate the **multiplicity** of the state by using the equation:

$$\text{Multiplicity} = 2S + 1 \tag{3.34}$$

in which S is given by the maximum value of Σm_s for any particular state. The multiplicity is indicated as a left-hand superscript to the code letter for the state. Thus the state under discussion is written as: ^{1}D.

There is a quantized interaction between the total angular orbital momentum and the total spin angular momentum which may be expressed as a rule for yet another quantum number, J, the resultant total angular momentum quantum number. The rule for J is:

$$J = L + S, L + S - 1, \ldots |L - S| \tag{3.35}$$

The multiplicity of a state indicates the number of different J values it possesses. For the ^{1}D state under discussion, since $S = 0$, there is only one J value which is equal to the value of L (2). The J value of a state is indicated by a right-hand subscript on the code letter, in the current example as: $^1D_2$. This symbolic description of the electronic state is called the **term symbol** for the particular state and is sometimes preceded by the value of the principal quantum number of the participating electrons, in this example $n = 2$ so the full description of the state is 2^1D_2.

Returning to the other microstates of the $2p^2$ configuration, the next highest value of Σm_l is 1 and the associated values of Σm_s are 1, 0 and -1, which are indicative of a triplet state. If S is equal to 1, then the multiplicity is three (by equation (3.34). The value of Σm_l indicates the presence of a P state, but to confirm this and that it is a triplet state it is essential to identify the **nine** microstates associated with such a state. There are nine microstates which form the ^{3}P state because the three values of Σm_l of 1, 0 and -1 (for a P state $L = 1$) are in turn associated with the three values of Σm_s of 1, 0 and -1 (for a triplet state $S = 1$). The appropriate microstates are identified in Table 3.11.

As may be seen from Table 3.11 there is an excess of microstates over the nine actually required, but that is also the situation with regard to the ^{1}D state. For instance the $\Sigma m_l = 1$, $\Sigma m_s = 0$ combination arises in microstates numbers 10 and 11 of Table 3.8. One is needed for the ^{1}D state and one for the ^{3}P state and since they are identical in value they

both contribute to both states in the form of the linear combinations, $\psi_{10} + \psi_{11}$, and $\psi_{10} - \psi_{11}$.

Table 3.11 The microstate numbers of Table 3.7 contributing to the ^{3}P state

	$\Sigma m_l = 1$	$\Sigma m_l = 0$	$\Sigma m_l = -1$
$\Sigma m_s = 1$	4	6	8
$\Sigma m_s = 0$	10 and 11	2, 12 and 13	14 and 15
$\Sigma m_s = -1$	5	7	9

The ^{3}P state has three permitted *J* values: 2, 1 and 0 by the operation of the *J* rule expressed by equation (3.35). Fully written down these are: $^3P_2$, $^3P_1$ and $^3P_0$. They have energies which differ because of *spin-orbit coupling*, but are considerably lower in energy than the two excited states. This statement is not true for heavier elements (see the next section on *j–j* coupling).

The ^{1}D and ^{3}P states require fourteen (5 + 9) microstates for their complete description so that there is one microstate from the fifteen of Table 3.8 which is unaccounted for. This is the one (of a possible three) which has $\Sigma m_l = \Sigma m_s = 0$ and which can only be a singlet-S state: ^{1}S. There is only one *J* value (zero) so that the full term symbol for the state is $^1S_0$.

Using this rather lengthy procedure it is possible to conclude that the $2p^2$ configuration has associated with it the ^{1}D, ^{3}P and ^{1}S states. It is now possible to re-state Hund's rules in a more exact and understandable form. They may be formulated as indicating that the electronic ground state is:

(1) that which has the highest multiplicity

and if there are more than one state in that category the one of lowest energy is:

(2) that with the highest value of *L*.

There is an additional rule concerning the values of *J*:

(3) for states of identical *L* values the lowest in energy is that with the lowest *J* value if the electron shell is less than half full, and is the one with the highest value of *J* if the electron shell is more than half full. The ground states of those atoms with a half full shell are S states ($L = 0$) with only one *J* value, whatever the multiplicity may be. In such a case this sub-rule is irrelevant. The ground state of the carbon atom is now revealed to be the ^{3}P state and its make-up from nine microstates—a more complex but more accurate description than by considering it as 'two singly occupied 2p orbitals with the two electrons possessing parallel spins' as expressed by the 'box' notation explained in section 3.4. The energies of the ^{1}D and ^{1}S states of the carbon atom are 95.7 kJ mol^{-1} and 251 kJ mol^{-1} higher than the ground state, respectively.

The formulation and accurate descriptions of electronic states and the assignment of term symbols is of great importance in the elucidation of the electronic spectra of atoms and molecules and is of particular importance in the interpretation of the spectra of complexes containing transition metal ions.

j-j Coupling and the heavier elements

Russell-Saunders coupling of the *L* and *S* quantum numbers described in the previous section is adequate for the description of the electronic states of most elements, but significant deviations are observed in the states of the heavier elements from $Z > 55$. These are caused by relativistic effects which, as discussed in Section 3.6, stabilize the s and p valence electrons. The effect is to produce what is known as *j-j* coupling in which the quantum numbers l and s for individual electrons combine to give a 'small j' value. The rule for this combination is that:

$$j = l + s, l + s - 1, \ldots |l - s| \tag{3.36}$$

With regard to an external magnetic field the corresponding m_j values are given by:

$$m_j = j, j-1, \ldots -(j-1), -j \tag{3.37}$$

For the lead atom the two 6s electrons occupy the orbital in which the two values of m_j are ½ and −½, the orbital being labelled by its j value of ½: $s_{1/2}$. The three 6p orbitals are labelled as $p_{1/2}$ and $p_{3/2}$, the former being singly degenerate and of lower energy than the doubly degenerate latter level. The two electrons in the lower $s_{1/2}$ orbital are relativistically stabilized and form the 'inert pair' typical of lead chemistry. The single occupation of the destabilized $p_{3/2}$ orbital in bismuth explains the observation that the first ionization energy of the element (703 kJ mol^{-1}) is lower than that of lead (715 kJ mol^{-1}).

3.8 PROBLEMS

1. The wavelength of the first line in the Balmer series is 656.3 nm. Calculate the wavelength of (a) the third line in the Balmer series and (b) the second line in the Paschen series.

2. Plot the following logarithms of the successive ionization energies of an element against the number of electrons removed. Assume that eventually all the electrons are removed. Explain the shape of the plot in terms of the electronic configuration of the element.

2.77	3.06	3.69	3.81	3.91	4.02	4.09	4.15	4.26	4.31
4.76	4.80	4.85	4.90	4.94	4.97	5.02	5.05	5.69	5.72

3. Allocate the elements, B, Be, Mg and Si to the following sets of successive ionization energies (given in kJ mol^{-1}) and justify your choice.

Set	I	II	III	IV
I_1	736	786	799	900
I_2	1450	1580	2429	1760
I_3	7740	3230	3660	14500

4. The following table contains the first three successive ionization energies of the Group 11 elements (given in kJ mol^{-1}).

	Cu	Ag	Au
I_1	745	732	891
I_2	1960	2070	1980
I_3	3550	3360	2940

Make plots of (a) I_1, (b) $I_1 + I_2$, and (c) $I_1 + I_2 + I_3$ against atomic number and correlate the plots with the oxidation states exhibited by the three elements.

5. Write down the electronic configuration of the fluorine atom and calculate the effective nuclear charge experienced by the valence electrons by using Slater's rules. Taking the covalent radius of F to be 72 pm, calculate the Allred-Rochow electronegativity coefficient for the element.

4

Molecular Symmetry and Group Theory

This chapter deals with group theory and its application to molecular symmetry. The applications to molecular orbital theory, molecular shapes and the spectroscopy of complex ions are demonstrated in the remainder of the book.

4.1 MOLECULAR SYMMETRY

A *group* consists of a set of *elements* which are related to each other according to certain rules. The particular kind of elements which are relevant to the symmetries of molecules are *symmetry elements*. With each symmetry element there is an associated *symmetry operation*. The necessary rules are referred to when appropriate.

Elements of symmetry and symmetry operations

There are seven elements of symmetry which are commonly possessed by molecular systems. These elements of symmetry, their notations and their related symmetry operations are given in Table 4.1.

An element of symmetry is possessed by a molecule if, after the associated symmetry operation is carried out, the atoms of that molecule are not perceived to have moved. The molecule is then in an equivalent configuration. The individual atoms may have moved but only to positions previously occupied by identical atoms.

The Identity, *E*

The symmetry element known as the identity, and symbolized by *E* (or in some texts by *I*), is possessed by all molecules independently of their shape. The related symmetry operation of leaving the molecule alone seems too trivial a matter to have any importance. The importance of *E* is that it is essential, for group theoretical purposes, for a group to contain it. For example, it expresses the result of performing some operations twice, e.g. the double reflexion of a molecule in any particular plane of symmetry. Such action restores every atom of the molecule to its original position so that it is equal to the perfomance of the operation of leaving the molecule alone - expressed by *E*.

Table 4.1 Elements of symmetry and their associated operations

Symmetry element	Symbol	Symmetry operation
Identity	E	Leave the molecule alone
Proper axis	C_n	Rotate the molecule by 360/*n* degrees around the axis
Horizontal plane	σ_h	Reflect the molecule through the plane which is perpendicular to the major axis
Vertical plane	σ_v	Reflect the molecule through a plane which contains the major axis
Dihedral plane	σ_d	Reflect the molecule through a plane which bisects two C_2 axes
Improper axis	S_n	Rotate the molecule by 360/*n* degrees around the improper axis and then reflect the molecule through the plane perpendicular to the improper axis
Inversion centre or centre of symmetry	I	Invert the molecule through the inversion centre

Proper Axes of Symmetry, C_n

A proper axis of symmetry, denoted by C_n, is an axis around which a molecule is rotated by 360°/*n* to produce an equivalent configuration. The trigonally planar molecule, BF_3, may be set up so that the molecular plane is contained by the xy cartesian plane (that containing the *x* and *y* axes) and so that the *z* cartesian axis passes through the centre of the boron nucleus as is shown in Fig. 4.1.

Fig. 4.1 An example of a proper axis of symmetry, in this case of order 3, C_3

If the molecule is rotated around the *z* axis by 120° (360°/3) an equivalent configuration of the molecule is produced. The boron atom does not change its position, and the fluorine atoms exchange places depending upon the direction of the rotation. The rotation described is the symmetry operation associated with the C_3 axis of symmetry, and the demonstration of its production of an equivalent configuration of the BF_3 molecule is what is required to indicate that the C_3 proper axis of symmetry is possessed by that molecule.

There are other proper axes of symmetry possessed by the BF_3 molecule. The three lines joining the boron and fluorine nuclei are all contained by C_2 axes (from hereon the term 'proper' is dropped unless it is absolutely necessary to remove possible confusion) as may be seen from Fig. 4.2. The associated symmetry operation of rotating the molecule around one of the C_2 axes by 360°/2 = 180° produces an equivalent configuration of the molecule. The boron atom and one of the fluorine atoms do not move whilst the other two fluorine atoms exchange places. There are, then, three C_2 axes of symmetry possessed by the BF_3 molecule.

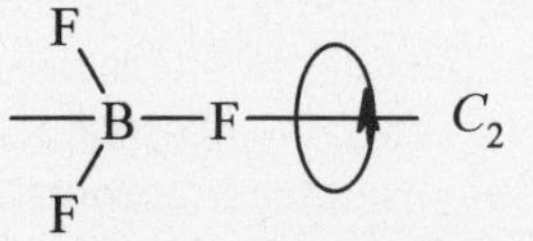

Fig. 4.2 An example of a C_2 axis of symmetry

The value of the subscript, n, in the symbol, C_n, for a proper axis of symmetry, is known as the *order* of that axis. The axis (and there may be more than one) of highest order possessed by a molecule is termed the major axis. The concept of the major axis is important in distinguishing between horizontal and vertical axes of symmetry. It is also important in the diagnosis of whether a molecule belongs to a C group or a D group - terms which are defined later in the chapter. As is the case with the C_n axis of BF_3, the axis of symmetry coincides with one of the cartesian axes (z), but that is because of the manner in which the diagram in Fig. 4.1 is drawn. It is a convention that the major axis of symmetry should be coincident with the z axis. It is not necessary for there to be any coincidences between the axes of symmetry of a molecule and the cartesian axes but it is of considerable convenience if there is at least one coincidence.

Planes of Symmetry, σ

There are three types of planes of symmetry, all denoted by the Greek lower case sigma, σ, and all of them are such that reflexion of the molecule through them produce equivalent configurations of that molecule. Reflexion through the plane is the symmetry operation associated with a plane of symmetry.

(a) Horizontal Planes, σ_h

A horizontal plane of symmetry (denoted by σ_h) is one that is perpendicular to the major axis of symmetry of a molecule. The molecular plane of the BF_3 molecule is an example of a horizontal plane - it is perpendicular to the C_3 axis. Reflexion in the horizontal plane of the BF_3 molecule has no effect upon any of the four atoms. A better example is the PCl_5 molecule (shown in Fig. 4.3) which, since it is trigonally bipyramidal, possesses a C_3 axis (again arranged to coincide with the z axis) and a horizontal plane (the xy plane as it is set up) which contains the phosphorus atom and the three chlorine atoms of the trigonal plane. Reflexion through the horizontal plane causes the apical (out-of-plane) chlorine atoms to exchange places in producing an equivalent configuration of the molecule.

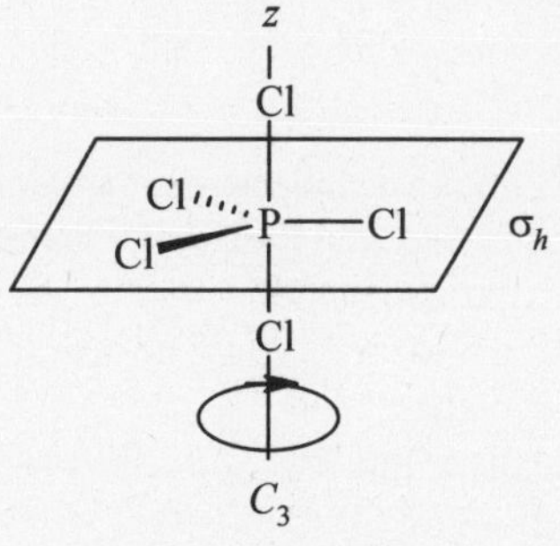

Fig. 4.3 An example of a horizontal plane of symmetry, σ_h

(b) Vertical Planes, σ_v

A vertical plane of symmetry (denoted by σ_v) is one which contains the major axis. The BF_3 and PCl_5 molecules both possess three vertical planes of symmetry. These contain the C_3 axis and, in the BF_3 case, the boron atom and one each of the fluorine atoms respectively. In the PCl_5 case the three vertical planes, σ_v, contain the C_3 axis, the phosphorus atom, the two apical chlorine atoms and one each of the chlorine atoms of the trigonal plane repectively.

(c) Dihedral Planes, σ_d

A dihedral plane of symmetry (denoted by σ_d) is one which bisects two C_2 axes of symmetry. In addition it contains the major axis and so is a special type of vertical plane. An example is shown in Fig. 4.4 which contains a diagram of the square planar ICl_4^- ion.

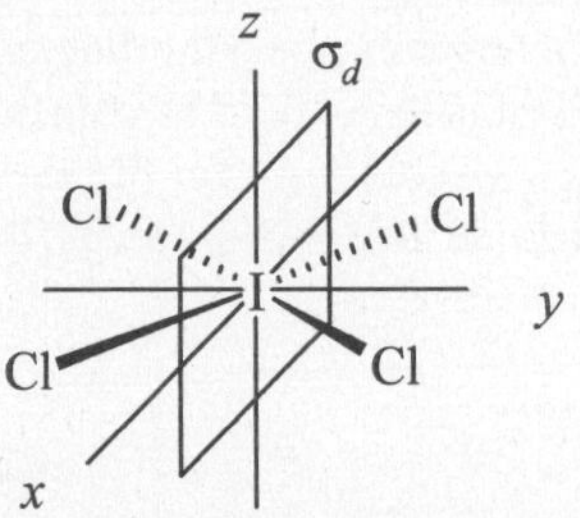

Fig. 4.4 An example of a dihedral plane of symmetry, σ_d

As may be seen from Fig. 4.4 the ICl_4^- ion has a major axis which is a C_4 axis. The C_4 axis is also a C_2 axis in that rotations around it of 90° and 180° both produce equivalent configurations of the molecule. A rotation through 270° also produces an equivalent configuration but that is equivalent to a rotation through 90° in the opposite direction and so does not indicate an extra type of axis of symmetry. The molecular plane contains two C_2' axes (the superscript dash is to distinguish them from the C_2 axis which is coincident with the C_4 axis) both of which contain the iodine atom and two diametrically opposed chlorine atoms. The two C_2' axes are contained respectively by the two vertical planes which also contain the major axis. There are, in addition, two C_2'' axes which are contained by the horizontal plane, are perpendicular to the major axis, and also bisect the respective Cl–I–Cl angles. The double-dash superscript serves to distinguish these axes from the C_2 and C_2' axes. The C_2'' axes are contained by two extra planes of symmetry which are termed dihedral planes since they bisect the two C_2' axes. The dihedral planes contain the major axis.

Improper Axes of Symmetry, S_n

If rotation about an axis by 360°/n followed by reflexion through a plane perpendicular to the axis produces an equivalent configuration of a molecule, then the molecule contains an improper axis of symmetry. Such an axis is denoted by S_n, the associated symmetry operation having been described. The C_3 axis of the PCl_5 molecule is also an S_3 axis. The operation of S_3 on PCl_5 causes the apical chlorine atoms to exchange places.

The operation of reflexion through a horizontal plane may be regarded as a special case of an improper axis of symmetry of order one: S_1. The rotation of a molecule around an axis by 360° produces an identical configuration ($C_1 = E$) and the reflexion in the horizontal plane is the only non-trivial part of the operations associated with the S_1 improper axis. This may be symbolized as:

$$S_1 = C_1 \times \sigma_h = E \times \sigma_h = \sigma_h$$

The Inversion Centre or Centre of Symmetry, *i*

An inversion centre (denoted by *i*) is possessed by a molecule which has pairs of identical atoms which are diametrically opposed to each other about the centre. Any particular atom with the coordinates *x*, *y*, *z* must be partnered by an identical atom with the coordinates -*x*, -*y*, -*z*, if the molecule is to possess an inversion centre. That condition must apply to all atoms in the molecule which are off-centre. The ICl_4^- ion possesses an inversion centre, as does the dihydrogen molecule, H_2. The BF_3 molecule (trigonally planar) does not possess an inversion centre. The symmetry operation associated with an inversion centre is the inversion of the molecule in which the diametrically opposed atoms in each such pair exchange places. The inversion centre is another special case of an improper axis in that the operations associated with the S_2 element are exactly those which produce the inversion of the atoms of a molecule containing *i*. Reference to Fig. 4.5 shows the effects of carrying out the operation, C_2, followed by σ_h, and that the fluorine atoms in each of the three diametrically opposed pairs in the octahedral SF_6 molecule have been exchanged with each other—all of which is what is understood to be the inversion of the molecule. This is symbolized as:

$$S_2 = C_2 \times \sigma_h = i$$

The subscript numbering of the fluorine atoms in Fig. 4.5 is done to make clear the atomic movements which take place as a result of the application of the symmetry operations.

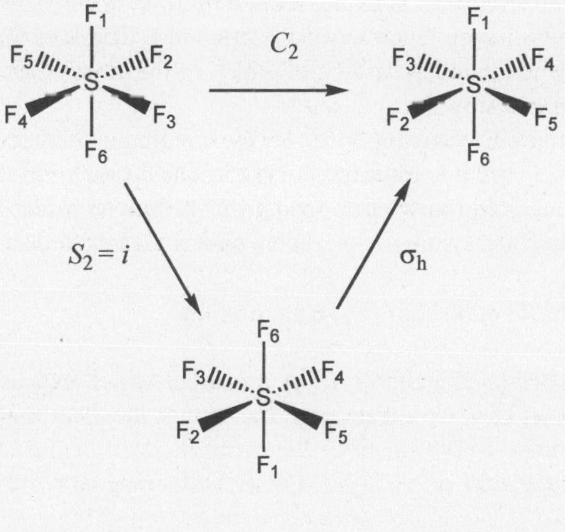

Fig. 4.5 An illustration of the inversion of the SF_6 molecule by two paths

4.2 POINT GROUPS

The symmetry elements which may be possessed by a molecule are defined above and exemplified. The next stage is to decide which, and how many, of these elements are possessed by particular molecules so that the molecules can be assigned to point groups. A point group consists of all the elements of symmetry possessed by a molecule and which intersect at a point. Such elements represent a group according to the rules to be outlined.

A simple example is the water molecule which is a bent triatomic system. Its symmetry elements are easily detected. There is only one proper axis of symmetry which is that which bisects the bond angle and contains the oxygen atom. It is a C_2 axis and the associated operation of rotating the molecule about the axis by 180° results in the hydrogen atoms exchanging places with each other. The demonstration of the effectiveness of the operation is sufficient for the diagnosis of the presence of the element.

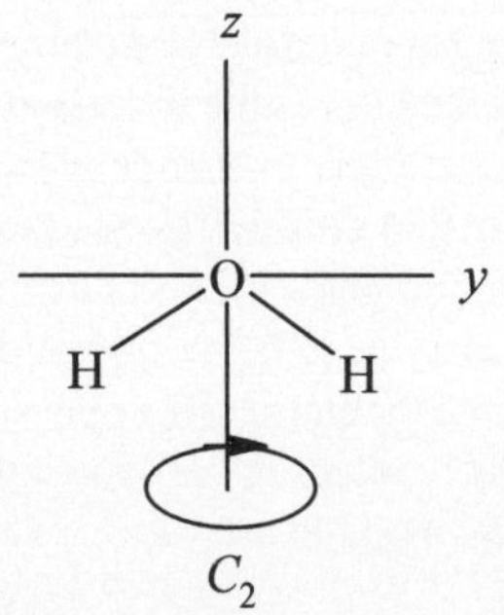

Fig. 4.6 The water molecule set up in the *yz* plane with the *z* axis containing the C_2 axis

Fig. 4.6 shows the molecule of water set up so that the C_2 axis is coincident with the *z* axis as is conventional. The molecular plane is set-up to be in the *yz* plane so that the *x* axis is perpendicular to the paper, as is the *xz* plane. There are two vertical planes of symmetry: the *xz* and the *yz* planes, and these are designated as σ_v and σ_v' respectively, the dash serving to distinguish between the two. The only other symmetry element possessed by the water molecule is the identity, *E*.

The four symmetry elements form a group which may be demonstrated by introducing the appropriate rules. The rules are exemplified by considering the orbitals of the atoms present in the molecule. Such a consideration also develops the relevance of group theory in that it leads to an understanding of which of the atomic orbitals are permitted to combine to form molecular orbitals.

The electrons which are important for the bonding in the water molecule are those in the valence shell: $2s^2 2p^4$. It is essential to explore the character of the 2s and 2p orbitals and this is done by deciding how each orbital transforms with respect to the operations associated with each of the symmetry elements possessed by the water molecule.

The character of the 2s orbital of the oxygen atom

The character of an orbital is symbolized by a number which expresses the result of any particular operation on its wave function. In the case of the 2s orbital of the oxygen atom (whose ψ values are everywhere positive) there is no change of sign of ψ with any of the four operations: E, $C_2(z)$, $\sigma_v(xz)$ and $\sigma_v'(yz)$. These results may be written down in the form:

	E	C_2	$\sigma_v(xz)$	$\sigma_v'(yz)$
2s	1	1	1	1

the '1's indicating that the 2s orbital is symmetric with respect to the individual operations. Such a collection of characters is termed a representation, this particular example being the totally symmetric representation.

The character of the 2p$_x$ orbital of the oxygen atom

The 2p$_x$ orbital of the oxygen atom is symmetric with respect to the identity, but is anti-symmetric with respect to the C_2 operation. This is because of the spatial distribution of ψ values in the 2p orbitals with one positive lobe and one negative lobe. The operation, C_2, causes the positive and negative regions of the 2p$_x$ orbital to exchange places with each other. This sign reversal is indicated as a character of -1 in the representation of the 2p$_x$ orbital as far as the C_2 operation is concerned. The 2p$_x$ orbital is unchanged by reflexion in the *xz* plane but suffers a sign reversal when reflected through the *yz* plane. The collection of the characters of the 2p$_x$ orbital with respect to the four symmetry operations associated with the four elements of symmetry forms another representation of the group currently being constructed:

	E	C_2	$\sigma_v(xz)$	$\sigma_v'(yz)$
2p$_x$	1	−1	1	-1

The character of the 2p$_y$ orbital of the oxygen atom

The 2p$_y$ orbital of the oxygen atom changes sign when the C_2 operation is applied to it, and when it is reflected through the *xz* plane, but is symmetric with respect to the molecular plane, *yz*. The representation expressing the character of the 2p$_y$ orbital is another member of the group:

	E	C_2	$\sigma_v(xz)$	$\sigma_v'(yz)$
2p$_y$	1	-1	-1	1

The character of the 2p$_z$ orbital of the oxygen atom

The 2p$_z$ orbital of the oxygen atom has exactly the same set of characters as the 2s orbital—it is another example of a totally symmetric representation.

The multiplication of representations

At this stage it is important to use one of the rules of group theory which states that the product of any two representations of a group must also be a member of that group. This rule may be used on the examples of the representations for the 2p$_x$ and 2p$_y$ orbitals as deduced above. The product of two representations is obtained by multiplying together the individual characters for each symmetry element of the group. The normal rules of arithmetic apply so that the representation of the product of those of the two 2p orbitals under discussion is given by:

	E	C_2	$\sigma_v(xz)$	$\sigma_v'(yz)$
$2p_x$	1	−1	1	−1
$2p_y$	1	−1	−1	1
$2p_x.2p_y$	1	1	−1	−1

The new representation is, by the rules, a member of the group and is also the only possible addition to the set of representations so far deduced for the molecular shape under consideration. This statement may be checked by trying other double products amongst the four representations. The four representations may be collected together and be given symbols, the collection of characters being termed a *character table*:

	E	C_2	σ_v	σ_v'
A_1	1	1	1	1
A_2	1	1	-1	-1
B_1	1	-1	1	-1
B_2	1	-1	-1	1

The symbols used for the representations are those proposed by Mulliken. The A representations are those which are symmetric with respect to the C_2 operation, and the B's are antisymmetric to that operation. The subscript 1 indicates that a representation is symmetric with respect to the σ_v operation, the subscript 2 indicating antisymmetry to it. No other indications are required since the characters in the σ_v' column are decided by another rule of group theory. This rule is: the product of any two columns of a character table must also be a column in that table. It may be seen that the product of the C_2 characters and those of σ_v give the contents of the σ_v'. The representations deduced above must be described as irreducible representations. They cannot be simpler, but there are other representations (examples follow) which are reducible in that they are sums of irreducible ones. Character tables, in general, are a list of the irreducible representations of the particular group, and as in the ones shown in the Appendix of this book they have an extra column which indicates the representations by which various orbitals transform, and, symbolized by, R_x, R_y, and R_z, the transformation properties of rotations about those respective axes.

In the example under discussion the 2s and $2p_z$ orbitals transform as a_1, the $2p_x$ orbital transforms as b_1 and the $2p_y$ orbital transforms as b_2. In this context 'transform' refers to the behaviour of the orbitals with respect to the symmetry operations associated with the symmetry elements of the particular group. It should be noticed that lower case Mulliken symbols are used to indicate the irreducible representations of orbitals. The upper case Mulliken symbols are reserved for the description of the symmetry properties of electronic states.

Symmetry properties of the hydrogen 1s orbitals in the dihydrogen molecule

Neither of the 1s orbitals of the hydrogen atoms, taken separately, transform within the group of irreducible representations deduced for the water molecule. The two 1s orbitals must be taken together as one or the other of two group orbitals. A more formal treatment of the group orbitals which two 1s orbitals may form is dealt with in Chapter Five.

Whenever there are two or more identical atoms linked to a central atom their wavefunctions must be combined in such a way as to demonstrate their indistinguishability.

This is achieved by making linear combinations of the wavefunctions of the ligand atoms. For the two 1s orbitals of the hydrogen atoms in the water molecule their wave functions may be combined to give:

$$h_1 = 1s_A + 1s_B \quad (4.1)$$

$$h_2 = 1s_A - 1s_B \quad (4.2)$$

where h_1 and h_2 are the wave functions of the two group orbitals and $1s_A$ and $1s_B$ represent the 1s wave functions of the two hydrogen atoms, A and B. The two group orbitals are shown diagrammatically in Fig. 4.7. By inspection they may be shown to transform as a_1 and b_2 representations respectively. The h_1 orbital behaves exactly like the 2s and $2p_z$ orbitals of the oxygen atom, and the h_2 orbital behaves exactly like the $2p_y$ orbital of the oxygen atom, with respect to the four symmetry operations of the point group.

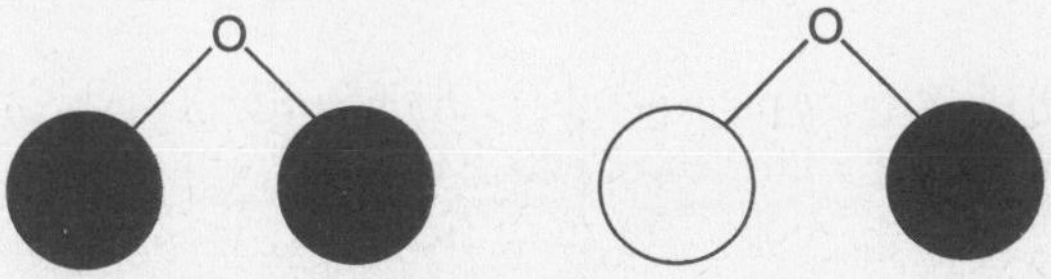

(a) (b)

Fig. 4.7 The (a) a_1 and (b) b_2 group orbitals formed from the 1s atomic orbitals of the two hydrogen atoms of the water molecule

4.3 CHARACTER TABLES

It is clearly necessary to label the various point groups to which molecules may belong. The labelling system used is to use a letter which is related to the major axis, and to use the value of n (the order of the major axis) together with a letter indicating the plane of symmetry (h, v or d) of highest importance for descriptive purposes, as subscripts. The system was suggested by Schönflies and the labels are known as Schönflies symbols. The normal method of deciding the point group of a molecule is that described below and which, after practice with examples, is accurate.

Assignment of point groups to molecules

There are three shapes which are important in chemistry and are easily recognised by the number of their faces, all of which consist of equilateral triangles. They are the tetrahedron (four faces), the octahedron (six faces) and the icosahedron (twenty faces). The first two of these shapes are extremely common in chemistry, whilst the third shape is important in boron chemistry and some other cluster molecules and ions. The three special shapes are associated with point groups and their character tables and are labelled, T_d, O_h and I_h, respectively. The point group to which a molecule belongs may be decided by the answers to four main questions.

1. Does the molecule belong to one of the special point groups, T_d, O_h and I_h? If it does, the point group has been identified.

2. If the molecule does not belong to one of the special groups it becomes necessary to identify the major axis (or axes). The major axis is the one with the highest order - the highest value of n, and is designated, C_n. In a case where there is more than one axis which could be classified as major (of equal values of n), it is conventional to regard the axis placed along the z axis as being the major one. In the trivial cases where there is only a C_1 axis, the point group of the molecule is C_1, unless there is either a plane of symmetry or an inversion centre present, indicating the point groups C_s or C_i respectively.

The main question is: are there n C_2 axes perpendicular to the major axis, C_n? If there are, the molecule belongs to a D_n group, otherwise the molecule belongs to a C_n group.
3. The third question applies only to D_n groups: is there a horizontal plane of symmetry present? If σ_h is present, the molecule belongs to the point group, D_{nh}. If there is no σ_h but there are n σ_v present, the molecule belongs to the point group, D_{nd}. If no σ_v is present then the point group of the molecule is D_n.
4. The fourth question applies only to C_n groups: is there a σ_h present? If there is, the molecule belongs to the point group, C_{nh}. If there is no σ_h, but there are n σ_v present, the molecule belongs to the point group, C_{nv}. If no σ_v is present, but there is an S_{2n} improper axis present, the molecule belongs to the S_{2n} point group. If no S_{2n} is present, the point group of the molecule is C_n.

The procedure described will identify the point groups of the majority of molecules. The procedure may be written in the form of a flow sheet which is shown in Fig. 4.8 and allows the assignment of molecules to point groups. The most frequently used character tables are placed in Appendix II.

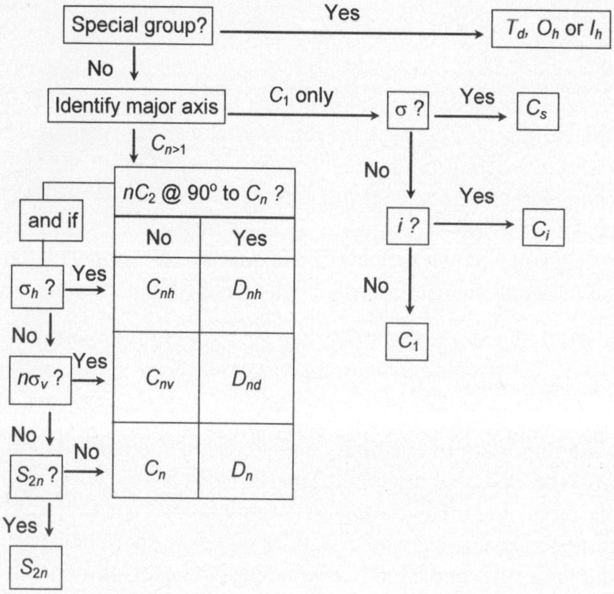

Fig. 4.8 A flow chart for the assignment of molecules, groups of ligands or orbitals to point groups

4.4 PROBLEMS

Allocate the following species to their appropriate point group: (a) N_2O, (b) $HgCl_2$(g), (c) N_3^-, (d) $AuCl_2^-$, (e) O_3, (f) ClO_2, (g) F_2O, (h) $IClBr^-$, (i) S_3^{2-}, (j) H_2Se, (k) HOCl, (l) B_2Cl_4, (m) $[Co(en)_3]^{3+}$, (n) $POCl_3$, (o) NO_2, (p) NMe_2H, (q) *cis*-$[Co(en)_2Cl_2]^+$, (r) SiBrClFI, (s) IF_5, (t) $H_2C{=}C{=}CH_2$, (u) NO_3, (v) $[PtCl_4]^{2-}$, (w) B_8Cl_8.

5

Covalent Bonding in Diatomic Molecules

This chapter consists of the application of the symmetry concepts of Chapter 4 to the construction of molecular orbitals for a range of diatomic molecules. The principles of *molecular orbital theory* are developed in the discussion of the bonding of three sets of compounds. These are (i) dihydrogen molecule-ion, H_2^+, and the dihydrogen molecule, (ii) homonuclear diatomic molecules, A_2, where A is a member of the second period of the periodic classification (Li–Ne) and (iii) some heteronuclear diatomic molecules where differences in electronegativity between the combining atoms are important.

5.1 COVALENT BONDING IN H_2^+ AND H_2

The simplest covalent bond is that which occurs in the molecule-ion, H_2^+, which consists of two protons and one electron. Even so it represents a quantum mechanical *three-body problem* and solutions of the wave equation must be obtained by iterative methods. The molecular orbitals derived from the combination of two 1s atomic orbitals serve to describe the electronic configurations of the four species, H_2^+, H_2, He_2^+ and He_2.

Production of Molecular Orbitals

The basic concept of molecular orbital theory is that molecular orbitals may be constructed from a set of contributing atomic orbitals such that the molecular wavefunctions consist of *linear combinations of atomic orbitals* (LCAO). In the case of the combination of two 1s hydrogen atomic orbitals to give two molecular orbitals the two linear combinations are those already proposed for the two hydrogen atoms of the water molecule and given by equations (4.3) and (4.4). They are re-written below in a modified fashion so that atomic wavefunctions are represented by ψ and molecular wavefunctions by ϕ:

$$\phi_1 = \psi_A + \psi_B \tag{5.1}$$

$$\phi_2 = \psi_A - \psi_B \tag{5.2}$$

where ψ_A and ψ_B are the two hydrogen atomic 1s wavefunctions of atoms A and B respectively. The overlap diagrams of the two 1s orbitals as they are represented in equations 5.1 and 5.2 are shown in Fig. 5.1. Where the two atomic orbitals have the same signs for ψ there is an increase in probability of finding an electron in the internuclear region because the two atomic orbitals both contribute to a build-up of ψ value. If the two 1s orbitals have different signs for ψ the internuclear region has a virtually zero probability for finding the electron because the two wave functions cancel out.

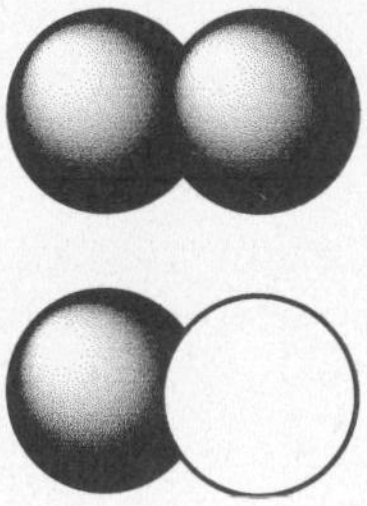

Fig. 5.1 The two ways in which 1s orbitals from two hydrogen atoms are permitted to overlap. In the top diagram both 1s wavefunctions are positive whereas in the lower diagram one 1s wavefunction is positive, the other negative

Excercise. Show by squaring equations (5.1) and (5.2) that there is a build up of probability of finding electrons in the internuclear region in the overlap represented by equation (5.1) and a diminution of probability in the overlap represented by equation (5.2).

The H_2^+ and H_2 molecules belong to the $D_{\infty h}$ point group. The two 1s atomic orbitals, individually, do not transform within the $D_{\infty h}$ point group, but together their character may be elucidated. This is done by considering each of the symmetry elements of the $D_{\infty h}$ group in turn, and writing down under each element the number of orbitals unaffected by the associated symmetry operation:

	E	C_∞^ϕ	σ_v	i	S_∞^ϕ	C_2
$\psi_A + \psi_B$	2	2	2	0	0	0

An explanation of this method is given in Appendix 1. The two 1s orbitals are left alone by the E (identity) operation and are unaffected by it and hence the number 2 is written down in the above table. Rotation by any angle, ϕ, around the C_∞ axis does not affect the orbitals—hence the second 2 appears as the character of the two 1s orbitals. The third 2 appears because the two orbitals are unaffected by reflexion in any of the infinite number of vertical planes which contain the molecular axis. The operation of inversion affects both orbitals in that they exchange places with each other and so a zero is written down in the i column. Likewise an S_∞ operation causes the orbitals to exchange places and a zero is written in that column. There are an infinite number of C_2 axes passing through the inversion centre and are perpendicular to the molecular axis. The associated operation of rotation through 180° around any C_2 axis causes the 1s orbitals to exchange places with each other so that there is a final zero to be placed in the table above.

The sequence of numbers arrived at represent the character of the two 1s orbitals with respect to $D_{\infty h}$ symmetry. Such a combination of numbers is not to be found in the $D_{\infty h}$ character table - it is an example of a *reducible* representation. Its reduction to a sum of irreducible representations is, in this instance, a matter of realizing that the sum of the σ_g^+ and σ_u^+ characters represents the character of the the two 1s orbitals:

	E	C_∞^ϕ	σ_v	i	S_∞^ϕ	C_2
σ_g^+	1	1	1	1	1	1
σ_u^+	1	1	1	-1	-1	-1
$\sigma_g^+ + \sigma_u^+$	2	2	2	0	0	0

Lower case letters are used for the symbols representing the symmetry properties of orbitals. Greek letters are used to symbolize the irreducible representations of the $D_{\infty h}$ point group with g or u subscripts, and with + and − signs as superscripts. The g and u subscripts refer to the character in the i column—g (German *gerade* meaning even) indicating symmetrical behaviour and u (*ungerade* meaning odd) indicating antisymmetrical behaviour with respect to inversion. The + and − signs refer respectively to symmetry and antisymmetry with respect to reflexion in one of the vertical planes. The '2's in the E and some of the other columns are indications of doubly degenerate representations.

By referring to the diagrams in Fig. 5.1 it may be seen that the orbital, ϕ_1, transforms as σ_g^+, and that the orbital, ϕ_2, transforms as σ_u^+. That two molecular orbitals are produced from the two atomic orbitals is an important part of molecular orbital theory - a law of conservation of orbital numbers. The two molecular orbitals differ in energy, both from each other and from the energy of the atomic level. To understand how this arises it is essential to consider the normalization of the orbitals. Normalization is the procedure of arranging for the integral over all space of the square of the orbital wavefunction to be unity. This is expressed by the equation:

$$\int_0^\infty \phi^2 d\tau = 1 \tag{5.3}$$

where $d\tau$ is a volume element (equal to $dx.dy.dz$). The probability of unity expresses the certainty of finding an electron in the orbital. It must be the case that by transforming atomic orbitals into molecular ones that no loss, or gain, in electron probability should occur. For equation (5.3) to be valid, a normalization factor, N, must be introduced into equations (5.1) and (5.2) which then become:

$$\phi_1 = N_1(\psi_A + \psi_B) \tag{5.4}$$

$$\phi_2 = N_2(\psi_A - \psi_B) \tag{5.5}$$

where N_1 and N_2 are normalization factors.

To determine the value of N_1 in equation (5.4) the expression for ϕ_1 must be placed into equation (5.3) giving:

$$N_1^2 \int_0^\infty (\psi_A + \psi_B)^2 d\tau = 1 \tag{5.6}$$

and expanding the square term in the integral gives:

$$N_1^2 \int_0^\infty (\psi_A^2 + 2\psi_A\psi_B + \psi_B^2)d\tau = 1 \tag{5.7}$$

which may be written as three separate integrals:

$$N_1^2 \left(\int_0^\infty \psi_A^2 d\tau + 2\int_0^\infty \psi_A\psi_B d\tau + \int_0^\infty \psi_B^2 d\tau \right) = 1 \tag{5.8}$$

Assuming that the atomic orbital wavefunctions are separately normalized leads to the conclusion that the first and third integrals in equation (5.8) are both equal to unity. The second integral is known as the overlap integral, symbolized by S. This allows equation (5.8) to be simplified to:

$$N_1^2(1 + 2S + 1) = 1 \tag{5.9}$$

which gives a value for N_1 of:

$$N_1 = 1/(2 + 2S)^{\frac{1}{2}} \tag{5.10}$$

A similar treatment of equation (5.5) gives a value for N_2 of:

$$N_2 = 1/(2 - 2S)^{\frac{1}{2}} \tag{5.11}$$

The next stage in the full description of the molecular orbitals is to calculate their energies. This is done by considering the Schrödinger equation for molecular wavefunctions:

$$H\phi = E\phi \tag{5.12}$$

If both sides are pre-multiplied by ϕ (this is essential for equations containing operators such as H) this gives:

$$\phi H\phi = \phi E\phi = E\phi^2 \tag{5.13}$$

there being no difference between $E\phi^2$ and $\phi E\phi$ since E is not an operator. Equation (5.13) may be integrated over all space to give:

$$\int_0^\infty \phi H\phi d\tau = E\int_0^\infty \phi^2 d\tau = E \tag{5.14}$$

since the integral on the right-hand side is equal to unity for normalized orbitals. Equation (5.14) may be used to calculate the energy of the molecular orbital, ϕ_1, by substituting its value from equation (5.4):

$$E(\phi_1) = N_1^2 \int_0^\infty (\psi_A + \psi_B)H(\psi_A + \psi_B)d\tau$$

$$= N_1^2\left(\int_0^\infty \psi_A H \psi_A d\tau + \int_0^\infty \psi_B H \psi_B d\tau + 2\int_0^\infty \psi_A H \psi_B d\tau\right)$$

$$= N_1^2(\alpha + \alpha + 2\beta) = 2N_1^2(\alpha + \beta) \quad (5.15)$$

The first two integrals are entirely concerned with atomic orbitals, ψ_A and ψ_B, respectively and have identical values (since they refer to identical orbitals) which are put equal to α, which is a quantity known as the *Coulomb integral*. In essence it is the energy of an electron in the 1s orbital of the hydrogen atom and equal to the ionization energy of that atom. The third integral is really the sum of two identical integrals (again becauseψ_A and ψ_B are identical) and is put equal to 2β, where β is called the *resonance integral*. β represents the extra energy gained by an electron, over that it possesses in any case by being in the 1s atomic orbital of the hydrogen atom, when it occupies the molecular orbital, ϕ_1. Because the electron is more stable in ϕ_1 than it is in ψ_A or ψ_B, ϕ_1 is called a *bonding* molecular orbital. Occupancy of ϕ_1 by one or two electrons leads to the stabilization of the system.

If the value of N_1 from equation (5.10) is substituted into equation (5.15) the expression for the energy of the bonding orbital becomes:

$$E(\phi_1) = \frac{\alpha + \beta}{1 + S} \quad (5.16)$$

A similar treatment of ϕ_2 produces the equation:

$$E(\phi_2) = \frac{\alpha - \beta}{1 - S} \quad (5.17)$$

which shows it to have a higher energy than ϕ_1, and which is also higher than that of ψ_A (or ψ_B) - it is therefore called an *anti-bonding* molecular orbital. Electrons in anti-bonding orbitals are less stable than in the atomic orbitals from which the molecular orbital was constructed. Such anti-bonding electrons contribute towards a weakening of the bonding of the molecule, or sometimes to complete dissociation of the molecule.

Fig. 5.2 shows the relative energies of the atomic orbitals, ψ_A and ψ_B, and the molecular orbitals, ϕ_1 and ϕ_2, which are involved in the production of H_2^+ and H_2. Both α and β are negative quantities on an energy scale with the ionization limit as the reference zero. The electronic configuration of the H_2^+ molecule-ion may be written as ϕ_1^1, or in symmetry symbols as $(\sigma_g^+)^1$. That of the dihydrogen molecule is ϕ_1^2, or $(\sigma_g^+)^2$, provided that the stabilization of ϕ_1 with respect to the atomic state is sufficiently large to force the electrons to pair up in the bonding orbital. The configuration, $\phi_1^1\ \phi_2^1$, produced by the absorption of a suitable quantum of energy, would lead to dissociation into separate hydrogen atoms. The occupation of the bonding orbital by a pair of electrons is the simplest example of a *single covalent bond.*

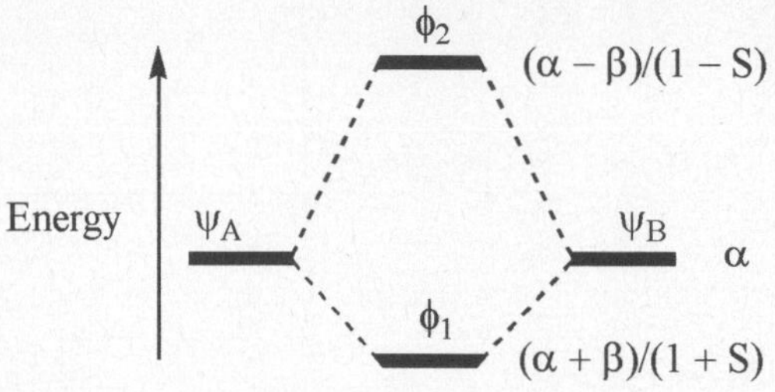

Fig. 5.2 The molecular orbital diagram for the combination of two identical 1s atomic orbitals

The single covalent bond, consisting of two electrons used in the bonding between two atoms is sometimes referred to as a two-electron two-centre (2e2c) bond, the nomenclature being of more general use when considering larger sytems.

The molecular orbitals ϕ_1 and ϕ_2 may be used to describe the electronic configurations of the helium molecule-ion, He_2^+, and the dihelium molecule, He_2. The former is a three-electron case so that two electrons pair up in the bonding orbital, leaving one unpaired electron to occupy the antibonding orbital, $\phi_1^2\phi_2^1$ being the electronic configuration of He_2^+. The single anti-bonding electron offsets some of the bonding effect of the pair of electrons in the bonding orbital to give a bond with a strength about equal to that of the bond in H_2^+.

The electronic configuration of dihelium would be $\phi_1^2\phi_2^2$ which would result in zero bonding. The bond dissociation energies of H_2^+, H_2 and He_2^+ are 264, 436 and 297 kJ mol^{-1}, respectively, and are consistent with the expectations from molecular orbital theory.

Digression about Valence Bond Theory

Valence bond theory, largely developed by Pauling (Nobel Prize for Chemistry, 1954), is not used to any great extent in this book because the alternative molecular orbital theory, largely developed by Mulliken (Nobel Prize for Chemistry, 1966), gives more satisfying explanations of the bonding of molecules and in addition rationalizes their electronically excited states in a way that valence bond theory cannot. Nevertheless, valence bond theory is still used in a general way to describe the bonding in many molecules and some indication here is necessary of why it is used and its relevance to the description of the bonding in any system. Both methods aim to describe the ways in which electrons are distributed in molecules.

The molecular orbital theory of the dihydrogen molecule is dealt with in detail above and describes how the two electrons occupy a bonding molecular orbital so that they are equally shared between the two nuclei. This state of affairs can be written symbolically in the form:

$$\phi = (\psi_a + \psi_b)(1) \times (\psi_a + \psi_b)(2)$$

i.e. both electrons occupy the bonding molecular orbital constructed from the 1s atomic orbitals of the two hydrogen atoms, *a* and *b*. Multiplying out the two terms of the equation together produces the four-term equation:

$$\phi = \psi_a(1)\psi_a(2) + \psi_b(1)\psi_b(2) + \psi_a(1)\psi_b(2) + \psi_a(2)\psi_b(1)$$

The first two terms indicate that the two electrons occupy the 1s orbital of one or the other hydrogen atoms, i.e. they are the *ionic* structures, H^-H^+ and H^+H^-. The third and fourth terms indicate that each hydrogen atom is associated with one of the electrons. All four terms together indicate the equal sharing of the two electrons between the two nuclei, but very much overemphasize the ionic aspect of the bonding. In practice this overemphasis is dealt with by assigning coefficients to the four terms and evaluating them by the process of minimizing the total energy of the system. When minimum energy is achieved the values of the coefficients are such that the ionic terms contribute only about 1% to the overall bonding.

Valence bond theory begins by assigning wavefunctions to the various permutations of electrons and nuclei to give what are known as the *canonical forms* of the molecule, i.e. for the hydrogen molecule these would be $H_a(1)H_b(2)$, $H_a(2)H_b(1)$, $H_a(1)(2)$ and $H_b(1)(2)$, terms similar to those derived from molecular orbital theory. These canonical forms engage in *resonance interaction* (Pauling's terminology), they contribute to the total description of the molecular bonding such that the *resonance energy* stabilizes the system. In practice the ionic terms would again have to be reduced in their participation to arrive at a reasonable description of the bonding.

The conclusion from this short discussion is that both valence bond and molecular orbital theories can describe the bonding of a system and in the limit they both arrive at the same answer. In practice molecular orbital theory is more often used and is much more amenable to giving solutions to more complex systems than does valence bond theory. The latter is discussed further in Chapter 6. Chapter 6 also contains many examples of how molecular orbital theory describes the excited states of molecular systems in a very satisfactory manner, such descriptions being beyond the applicability of valence bond theory. Molecular orbital theory is much more amenable to being handled by computer programs than the alternative theory and is almost exclusively in use at the present time.

5.2 ENERGETICS OF THE BONDING IN H_2^+ AND H_2

It is helpful in the understanding of covalent bond formation to consider the energies due to the operation of attractive and repulsive forces in H_2^+ and H_2, and to estimate the magnitude of the interelectronic repulsion energy in the dihydrogen molecule. Fig. 5.3 shows plots of potential energy against internuclear distance for H_2^+ and H_2. The curves shown are Morse functions which, for diatomic species, have the form:

$$V = D_e\{1 - \exp[-(\mu/2D_e)^{1/2}\omega(r - r_{eq})]\}^2 \qquad (5.18)$$

where V represents potential energy, D_e the electronic dissociation energy corresponding to the minimum of the curve, μ is the reduced mass of the system, ω is the fundamental vibration frequency of the molecule, r is the internuclear distance, and r_{eq} is the equilibrium internuclear distance otherwise known as the bond length. The quantity D_e is related to the dissociation energy, D, of the molecule by the relation:

$$D = D_e - \tfrac{1}{2}h\omega \qquad (5.19)$$

the $\tfrac{1}{2}h\omega$ term representing the *zero-point vibrational energy* of the molecule.

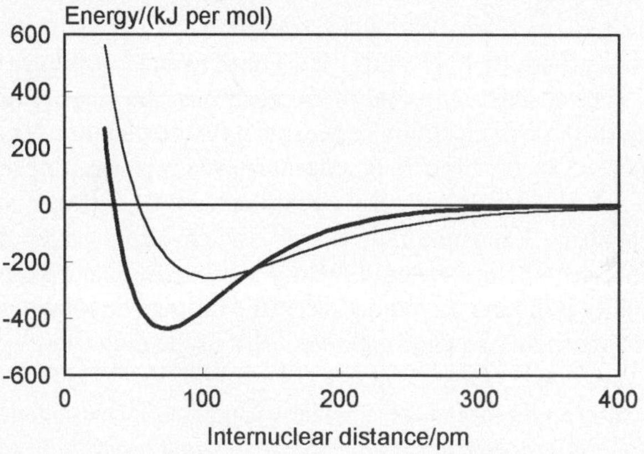

Fig. 5.3 Morse curves for H_2^+(thin line) and H_2 (thick line)

Under standard conditions both molecules exist in their lowest vibrational energy levels. These are known as their zero-point vibrational states, in which the value of the vibrational quantum number is zero. The fact that molecules in their zero-point vibrational states possess vibrational energy is a consequence of the uncertainty principle; this would be violated if the internuclear distance was unchanging. The dissociation limits for both species are identical—the complete separation of the two atoms, which is taken as an arbitrary zero of energy. The difference between the zero of energy and the zero-point vibrational energy in both cases represents the bond dissociation energies, respectively, of H_2^+ and H_2.

To obtain an accurate assessment of the interelectronic repulsion energy of the H_2 molecule it is essential to carry out calculations in which the hydrogen nuclei are a constant distance apart. The following calculations are for an internuclear distance of 74 pm for both molecules which is the equilibrium internuclear distance in the dihydrogen molecule.

The H_2^+ molecule-ion

There are only two forces operating in H_2^+: the attractive force between the nuclei and the single electron, and the repulsive force between the two nuclei. The interproton repulsion energy may be calculated from Coulomb's law:

$$E(p\text{-}p) = \frac{N_A e^2}{4\pi\epsilon_0 r} \qquad (5.20)$$

where e is the charge on the proton, r is the interproton distance, ϵ_0 is the permittivity of a vacuum and N_A the Avogadro number. For two protons separated by 74 pm the force of repulsion between them causes an increase in energy of 1877 kJ mol^{-1} compared to the infinite separation of H^+ and H (the arbitrary zero of energy). From the Morse curve for H_2^+ in Fig 5.3 it may be estimated that if H^+ and H are brought from infinite separation to an interproton distance of 74 pm there is a stabilization of 180 kJ mol^{-1}. This represents the resultant energy of the system with both forces operating. It means that the attractive force operating between the electron and the two protons produces a stabilization which is in excess of 1877 kJ mol^{-1} by 180 kJ mol^{-1}, so that the quantity known as the *electronic binding energy* is calculated to be 1877 + 180 = 2057 kJ mol^{-1}. The interrelationship of these energies is shown in the diagram of Fig. 5.4. Notice that the actual dissociation energy is a relatively small quantity compared to the energies representing the effects of the attractive and repulsive forces operating in the H_2^+ system.

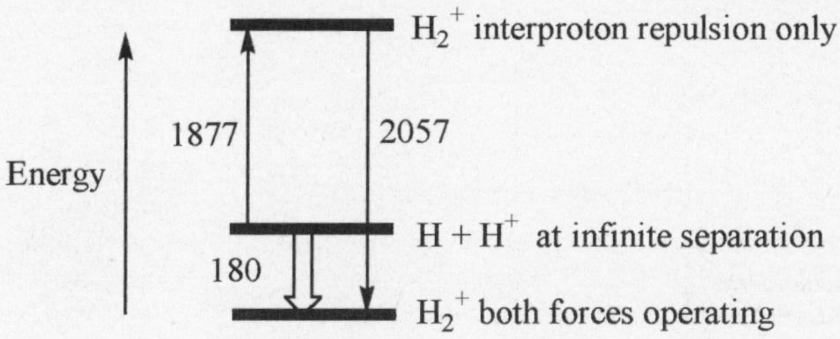

Fig. 5.4 Energetics of formation of the H_2^+ molecule-ion

The dihydrogen molecule, H_2

In the dihydrogen molecule there are three forces operating:
(i) interproton repulsion,
(ii) proton–electron attraction and
(iii) interelectronic repulsion.

The force of interproton repulsion produces a destabilization of the dihydrogen system equal to that of the H_2^+ molecule-ion since the interproton distance is taken to be 74 pm. The resultant stabilization of all three forces is equal to the bond dissociation energy of H_2 which is 436 kJ mol^{-1}. For a comparison with H_2^+ the electronic binding energy may be calculated as: 1877 + 436 = 2313 kJ mol^{-1}. It is 12% greater than that in H_2^+ indicating that two bonding electrons are only marginally better than one at binding the two nuclei together. The reason for this is that, with two electrons present, there is a substantial destabilization of the system as a result of the interelectronic repulsion.

The magnitude of this may be calculated by assuming that the electronic binding energy per electron is as calculated for the H_2^+ system (2057 kJ mol^{-1}). For two electrons the stabilization from the electronic binding energy is 2 × 2057 = 4114 kJ mol^{-1} and that amount is offset by the interelectronic repulsion energy so that the resultant is 2313 kJ mol^{-1} to give 4114 - 2313 = 1801 kJ mol^{-1} as representing the interelectronic repulsion energy in H_2. Notice that all three energy quantities in H_2 are large by comparison with the resultant bond dissociation energy. The above calculations are represented diagrammatically in Fig. 5.5.

It is of interest to calculate the magnitude of the interelectronic repulsion energy in the hydride ion, H^-, which possesses an ionic radius of 208 pm. The *electron attachment energy* of the proton is −1312 kJ mol^{-1} (numerically equal to the ionization energy of the hydrogen atom, but of opposite sign) and represents the energy released when an electron enters the 1s orbital of the hydrogen atom. The electron attachment energy of the hydrogen atom is considerably smaller than that of the proton and is −71 kJ mol^{-1}. Since the electron enters the same 1s orbital the difference between the two electron attachment energies gives an estimate of the interelectronic repulsion energy of −71 + 1312 = 1241 kJ mol^{-1}. This is appreciably smaller than the value calculated for the two electrons occupying the bonding orbital of the dihydrogen molecule. The increased size of the hydride ion is one reason for this, the other being that in H_2 there are two attracting protons which draw the electrons closer to each other in spite of their like charges. The other case for which interelectronic repulsion energy has been estimated is for the helium atom (Chapter 3) where the value was calculated to be 2878 kJ mol^{-1}. Such a high value arises from the double charge on the single nucleus which has a greater attractive effect upon the two electrons than do the two singly charged and spaced out nuclei in the case of H_2.

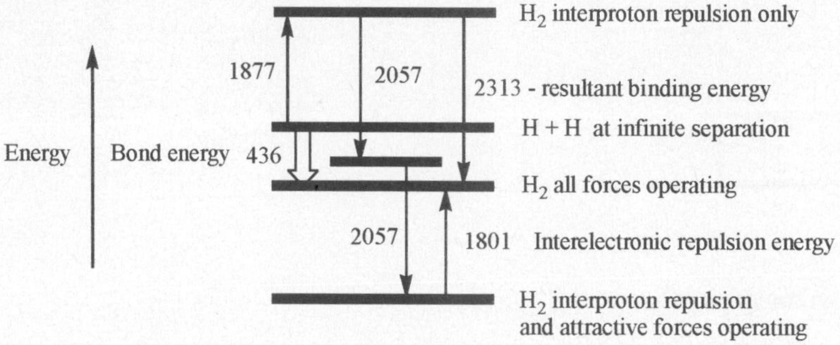

Fig. 5.5 Energetics of formation of the H_2 molecule

5.3 SOME EXPERIMENTAL OBSERVATIONS

In the proper development of a scientific theory it is to bc expected that theory and experimental observations should be consistent with one another, preferably the theory being refined by further observations. It is essential to obtain experimental observations by which ideas such as molecular orbital theory may be tested and possibly refined.

A direct method of obtaining experimental measurement of the energies of electrons in molecules is known as photoelectron spectroscopy (p.e.s.). The basis of the method is to bombard an atomic or molecular species with radiation of sufficient energy to cause its ionization. If the quantum energy of the radiation is high enough ionizations may be caused from one or other of the permitted levels within the bombarded atom or molecule. In addition to causing ionization of the target species the radiation (in the case of molecules) may cause changes in the vibrational, ΔE_{vib}, and rotational, ΔE_{rot}, energies of the resulting positive ion. This may be written as:

$$M(g) + h\nu \rightarrow M^+ + e^- \text{ (photoelectron)} \quad (5.21)$$

the energy balance being:

$$h\nu = I_M + \Delta E_{vib} + \Delta E_{rot} + \text{Kinetic energy of the electron} \quad (5.22)$$

The kinetic energies (K.E.) of the photoelectrons are measured by the use of a modification of a conventional β-ray spectrometer as used in the study of β-particle (electron) emissions from radioactive nuclei. The photoelectron spectrum of a molecule is presented as a plot of the count rate (the intensity of the photoelectrons detected) against the electron energy. The p.e.s. for the dihydrogen molecule is shown in Fig. 5.6.

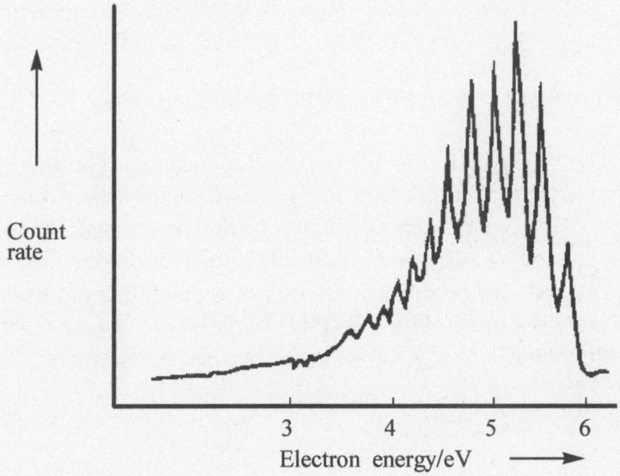

Fig. 5.6 The photoelectron spectrum of the H_2 molecule. Redrawn from D. W. Turner and D. P. May, *J. Chem. Phys.*, **45**, 471, 1966

The energy of the quantum used to cause the ionization is 21.22 eV (1 eV = 96.485 kJ mol^{-1}) so that the ionization energy of the dihydrogen molecule is 21.22 eV minus the energy of the photoelectron with the energy of 5.8 eV, corresponding to the peak on the far right-hand side of the spectrum, which gives 15.42 eV as the first ionization energy of H_2. The additional peaks in the photoelectron spectrum represent ionizations of the molecule with additional energy being used to vibrationally and rotationally excite the product H_2^+ molecule-ion. The rotational 'fine structure', which arises from variations in the rotational energy of the ion and its neutral molecule (ΔE_{rot}), is not observed at the resolution at which the spectrum was measured, but contributes to the width of the vibrational bands which are observed. Vibrational excitations up to the eleventh level can be discerned from the spectrum shown.

The second peak from the right in the dihydrogen p.e.s. corresponds to the energy required for its ionization plus that required to give H_2^+ one quantum of vibrational excitation. The difference in energy between the first two peaks is a measure of the magnitude of one quantum of vibrational excitation energy of H_2^+. The difference amounts to 0.3 eV or 29 kJ mol^{-1}. It is useful to compare such a value with the energy of one quantum of vibrational excitation of the dihydrogen molecule which is 49.8 kJ mol^{-1}. The frequency (and, in consequence, the energy) of the vibration of two atoms bonded together is related to the bond strength so it may be concluded that the removal of an electron from a dihydrogen molecule causes the remaining bond (in H_2^+) to be considerably weaker than that in the parent molecule. This is a confirmation of the bonding nature of the electron removed in the ionization process. Diagnosis of the bonding, non-bonding or anti-bonding nature of electrons in molecules may be made from a study of the effects of their ionization upon the vibrational frequency of the resulting positive ion. The removal of a non-bonding electron would have very little effect upon the bond strength and the vibrational frequencies of the neutral molecule and its unipositive ion would be expected to be similar. The vibrational frequency of the ion would be greater than that of the neutral molecule if the electron removed originally occupied an anti-bonding orbital.

Experimental confirmation of the energy of the anti-bonding level in the dihydrogen molecule comes from the observation of its absorption spectrum in the far ultra-violet region.

Dihydrogen absorbs radiation of a wavelength of 109 nm which is equivalent to a quantum energy of 1052 kJ mol^{-1}. In terms of an electronic transition the process of absorbing the quantum is:

$$\phi_1^2 \text{ (ground state configuration)} \xrightarrow{h\nu} \phi_1^1\phi_2^1 \text{ (excited state configuration)}$$

which would lead to the dissociation of the molecule into two hydrogen atoms. The energy of the transition should not be equated to the difference in energy between the two molecular orbitals, ϕ_1 and ϕ_2, but rather to a difference in energy between the ground and excited electronic states of H_2. In the ground state there is considerable interelectronic repulsion since the ϕ_1 orbital is doubly occupied. In the excited state there is much less repulsion between the electrons, which occupy separate orbitals, so that the difference in energy, $\phi_2 - \phi_1$, is greater than the actual difference in energy between the two electronic states:

$$E_{\text{excited}} - E_{\text{ground}} < \phi_2 - \phi_1 \tag{5.23}$$

5.4 HOMONUCLEAR DIATOMIC MOLECULES OF THE SECOND ROW ELEMENTS

The extension of molecular orbital theory to the homonuclear diatomic molecules of the first short period elements, A_2, involves the arrangement of the 2s and 2p orbitals as group orbitals, and their classification within the $D_{\infty h}$ point group to which the molecules belong. The molecular axis of an A_2 molecule is arranged to be coincident with the *z* axis by convention. It is necessary to look at the 2s and 2p orbitals separately.

Classification of the 2s orbitals of A_2 molecules

The classification of the two 2s orbitals of an A_2 molecule is very similar to that of the two 1s orbitals of dihydrogen and is not be repeated here. The two combinations of the 2s orbitals are:

$$\phi(2\sigma_g^+) = \psi(2s)_A + \psi(2s)_B \tag{5.24}$$

and

$$\phi(2\sigma_u^+) = \psi(2s)_A - \psi(2s)_B \tag{5.25}$$

the A and B subscripts referring to the two atoms contributing to the molecule. It is assumed that all the wavefunctions are normalized, although the normalization factors are omitted.

There is a bonding combination (equation (5.24)) which transforms within the $D_{\infty h}$ point group as a σ_g^+ irreducible representation, the prefix 2 being assigned because of the bonding combination of the 1s orbitals having the same symmetry (and termed $1\sigma_g^+$). Likewise the anti-bonding combination (equation (5.25)) is termed $2\sigma_u^+$.

Classification of the 2p orbitals of A_2 molecules

The character of the reducible representation of the 2p orbitals of A_2 molecules may be obtained by writing down, under each symmetry element of the $D_{\infty h}$ group, the number of such orbitals which are unchanged by each symmetry operation. This produces:

	E	C_∞^ϕ	σ_v	i	S_∞^ϕ	C_2
6 × 2p	6	2 + 4cosϕ	2	0	0	0

This result requires some explanation, particularly with regard to the character of the 2p orbitals with respect to the C_∞ operation. The two $2p_z$ orbitals, lying along the C_∞ axis with their positive lobes overlapping, are unaffected by the associated operation and account for the 2 in the character column. The term, 4cosϕ, arises because of the two $2p_x$ and two $2p_y$ orbitals which are perpendicular to the C_∞ axis. Although a rotation through ϕ degrees around that axis does not move any of the orbitals to another centre it does alter their disposition with regard to the *xz* and *yz* planes.

If the angle, ϕ, was chosen to be 180°, for instance, it would have the effect of inverting the $2p_x$ and $2p_y$ orbitals, and it would be necessary to place −1 in the above table for each orbital as their characters (note that cos 180° = −1). To take into account all possible values of ϕ it is essential to express the character of each orbital as the cosine of the angle of rotation, ϕ. Effectively this implies that the character of each orbital is represented by the resolution of the orbital on to the plane it occupied before the symmetry operation was carried out. This ensures that for a rotation through 180° the character of a $2p_x$ or $2p_y$ orbital will be −1; in effect such an orbital, whilst not moving from its original position, changes the signs of its ψ values.

The character with respect to reflexion in one of the infinite number of vertical planes requires some explanation. It is best to choose a particular vertical plane such as that represented by the *xz* plane. Reflexion in any of the vertical planes has no effect upon the two $2p_z$ orbitals, which gives 2 as their character. Reflexion of the two $2p_x$ orbitals in the *xz* plane does not change them in any way—their character is 2. The reflexion of the two $2p_y$ orbitals in the *xz* plane causes their ψ values to change sign, and because they are otherwise unaffected, their character is −2. The resultant character of the six 2p orbitals, with respect to the operation, σ_v, is given by 2 + 2 − 2 = 2.

The reducible representation of the six 2p orbitals may be seen, by inspection of the $D_{\infty h}$ character table and carrying out the following exercise, to be equivalent to the sum of the irreducible representations:

$$6 \times 2p = \sigma_g^+ + \sigma_u^+ + \pi_g + \pi_u \tag{5.26}$$

Excercise. Demonstrate the truth of equation (5.26) by summing the appropriate characters of the $D_{\infty h}$ character table (Appendix II).

It is important to realize which orbital combinations are represented by the above irreducible representations. This is best achieved by looking at the diagrams in Fig. 5.7 for the overlaps represented by the equations:

$$\phi(3\sigma_g^+) = \psi(2p_z)_A + \psi(2p_z)_B \tag{5.27}$$

$$\phi(3\sigma_u^+) = \psi(2p_z)_A - \psi(2p_z)_B \tag{5.28}$$

$$\phi(1\pi_u) = \psi(2p_x)_A + \psi(2p_x)_B;\ \psi(2p_y)_A + \psi(2y_z)_B \tag{5.29}$$

$$\phi(1\pi_g) = \psi(2p_x)_A - \psi(2p_x)_B;\ \psi(2p_y)_A - \psi(2y_z)_B \tag{5.30}$$

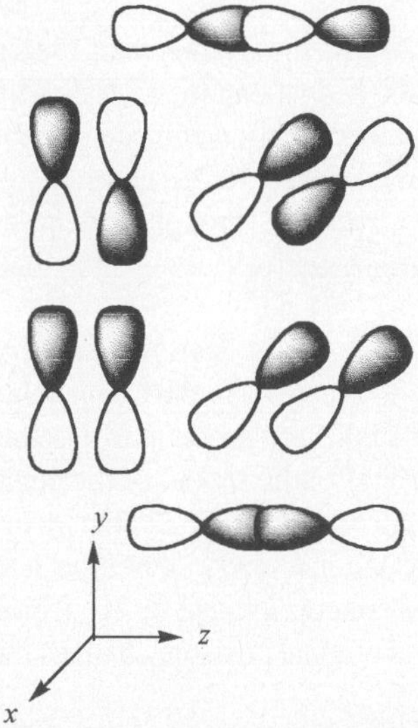

Fig. 5.7 The orbital overlaps between 2p atomic orbitals in the formation of some diatomic molecules

The convention used throughout this text to express the form of a molecular orbital is to use a **plus sign** in the equations to indicate a **bonding** combination and a **minus sign** to indicate an **anti-bonding** combination.

The molecular orbital, $\phi(3\sigma_g^+)$, is bonding and is the *third* highest energys σ_g^+ orbital —hence the prefix 3. The $\phi(3\sigma_u^+)$ orbital is the antibonding combination of the two $2p_z$ orbitals, and the third highest energys σ_u^+ orbital. The π_u and π_g orbitals are both *doubly degenerate*, the π_u combination being bonding, the π_g combination being anti-bonding. They both are prefixed by the figure '1' since they are the lowest energy orbitals of their type.

In the forthcoming discussion of the bonding of the A_2 molecules the orbitals will be referred to by their symmetry symbols with the appropriate numerical prefixes.

Fig. 5.8 is a diagram of the relative energies of the molecular orbitals of the A_2 molecules, together with those of the atomic orbitals from which they were constructed. The 'sideways' overlap involved in the production of π orbitals is by no means as effective as the 'end-on' overlap which characterizes the production of σ orbitals. For a given interatomic distance the overlap integral for σ-type overlap is generally higher than that for π-type overlap between two orbitals. The consequence of this is that the bonding stabilization and the anti-bonding destabilization associated with π orbitals are significantly less than those associated with σ orbitals. This accounts for the differences in energy, shown in Fig. 5.8(a), between the σ and π orbitals which originate from the 2p atomic orbitals of the A atoms. The order of energies of the orbitals of A_2 molecules is dependent upon the assumption that the energy difference between the 2p and 2s atomic orbitals is sufficient to prevent significant interaction between the molecular orbitals.

If two molecular orbitals have identical symmetry, such as the $2\sigma_g^+$ and $3\sigma_g^+$ orbitals, they may interact by the formation of linear combinations. The resulting combinations still have the same symmetry (and retain the nomenclature) but the lower orbital is stabilized at the expense of the upper one. Such interaction is also possible for the $2\sigma_u^+$ and $3\sigma_u^+$ orbitals. The extent of such interaction is determined by the energy gap between the two contributors. If the energy gap between the 2p and 2s atomic orbitals is small enough the interaction between the $2\sigma_g^+$ and $3\sigma_g^+$ molecular orbitals may be so extensive as to cause the upper orbital $3\sigma_g^+$ to have an energy which is greater than that of the $1\pi_u$ set. Such an effect is shown in Fig. 5.8(b). The magnitude of the 2p−2s energy gap varies along the elements of the second period (Li – Ne) as is shown in Fig. 5.9. The energy gaps in the elements lithium to nitrogen are sufficiently small to make significant $2\sigma_g^+$ and $3\sigma_g^+$ interaction possible such that Fig. 5.8(b) is relevant in determining the electronic configurations of the molecules, Li_2, Be_2, B_2, C_2 and N_2. Fig. 5.8(a) is to be used to determine the electronic configurations of the molecules, O_2, F_2 and Ne_2, since the 2p–2s energy gaps in O, F and Ne, are sufficiently large to preclude significant molecular orbital interaction.

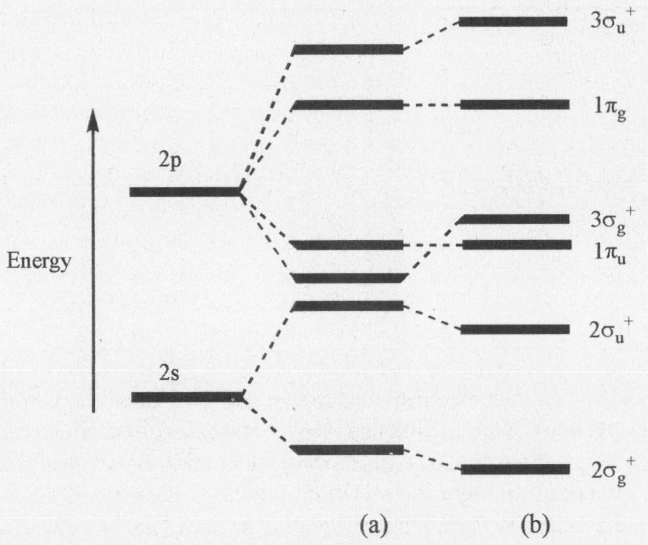

Fig. 5.8 The molecular orbital diagrams for homonuclear diatomic molecules of the second short period, Li_2 to Ne_2. Diagram (a) is appropriate for O_2, F_2 and Ne_2, diagram (b) for the molecules Li_2 to N_2

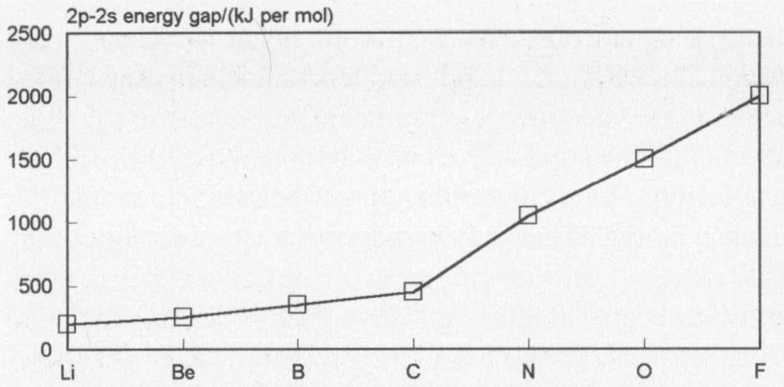

Fig. 5.9 Variation of the 2p–2s energy gap along the second short period

The interaction of molecular orbitals of the same symmetry has important consequences for all systems where it occurs and more examples will be referred to in later chapters. It is possible to carry out the mixing of the original atomic orbital (known as hybridization) before the m.o.'s are formed. Some examples of this approach are included in later chapters. Both approaches give the same eventual result for the contributions of atomic orbitals to the molecular ones.

Electronic configurations of homonuclear diatomic molecules

The electronic configurations of the homonuclear diatomic molecules of the elements of the second period, and some of their ions, are given in Table 5.1.

Table 5.1 The electronic configurations of some homonuclear diatomic molecules and ions

Molecule/ion	$2\sigma_g^+$	$2\sigma_u^+$	$3\sigma_g^+$	$1\pi_u$	$1\pi_g$	$3\sigma_u^+$
Li_2	2					
Be_2	2	2				
B_2	2	2		2		
C_2	2	2		4		
N_2	2	2	2	4		
N_2^+	2	2	1	4		
N_2^-	2	2	2	4	1	
O_2	2	2	2	4	2	
O_2^+	2	2	2	4	1	
O_2^-	2	2	2	4	3	
O_2^{2-}	2	2	2	4	4	
F_2	2	2	2	4	4	
Ne_2	2	2	2	4	4	2

In all cases the four electrons occupying the $1\sigma_g^+$ and $1\sigma_u^+$ orbitals are not indicated. The interactions of the 1s orbitals of the atoms under discussion are minimal and, although they may be regarded as occupying the molecular orbitals previously indicated, they are virtually non-bonding. In any case the slight bonding character of the two electrons occupying $1\sigma_g^+$ is cancelled by the slight anti-bonding nature of the two electrons in $1\sigma_u^+$. In some texts the four electrons are indicated symbolically by KK—a reference to the 'K shell'—now recognized as the 1s atomic orbital.

Dilithium, Li_2

This molecule exists in the gas phase and has a bond dissociation energy of 107 kJ mol^{-1} and a bond length of 267 pm. The weak, and very long, bond is understandable in terms of the two electrons in $2\sigma_g^+$ being the only ones having bonding character. The bond may be described as having a bond order of one. The bond order is defined as a half of the resultant excess of the number of bonding electrons above the number of anti-bonding electrons. The four 1s electrons have no resultant bonding effect and yet contribute considerably to the interelectronic repulsion energy. The electron attachment energy of the lithium atom is −59.8 kJ mol^{-1} which indicates that the nuclear charge of $+3e$ is not very effective in attracting more electrons. The atom also has a very low ionization energy (513 kJ mol^{-1}) which is another indication of the low effectiveness of the nuclear charge. That is discussed in Chapter 3 in terms of there being a considerable amount of interelectronic repulsion between the three electrons possessed by the lithium atom. Another approach is to use the concept of the nuclear charge being shielded by the various occupied atomic orbitals so as to reduce its effectiveness in attracting extra electrons. For bonding to be achieved, the attraction between the two shielded nuclei and the two bonding electrons must outweigh the two repulsive interactions—internuclear and interelectronic. The efficient shielding of the lithium nuclei by their $1s^2$ 'core' configurations contributes to the weakness of the bond in the Li_2 molecule. Solid lithium does not contain discrete molecules and has a metallic lattice. The bonding in metals is discussed in Chapter 6.

Diberyllium, Be_2

Since the electronic configuration of the diberyllium molecule, Be_2, would be $(2\sigma_g^+)^2\,(2\sigma_u^+)^2$ with two bonding electrons being counterbalanced by two anti-bonding electrons, it is not surprising that the molecule does not exist with that configuration. It would possess a bond order of zero.

Diboron, B_2

The diboron molecule, B_2, has a transient existence in the vapour of the element, and it is known that its bond dissociation energy is 291 kJ mol^{-1}, the bond length being 159 pm. The bond is stronger and shorter than that in Li_2. The first ionization energy of the boron atom (800 kJ mol^{-1} indicates that the nuclear charge is considerably more effective than that of the lithium atom and a somewhat stronger (and shorter) bond is to be expected for B_2 as compared to that in Li_2. The two pairs of sigma electrons have a zero resultant bonding effect, leaving the stability of the bond to the two electrons which occupy the $1\pi_u$ orbitals. It is of interest that, since the $1\pi_u$ level is doubly degenerate (so that Hund's rules apply to their filling), the two orbitals are singly occupied and the two π_u electrons have parallel spins. The bonding in B_2 consists of two 'half-π' bonds if the term 'bond' is understood to indicate a pair of bonding electrons. The bonding may be described in terms of the bond order being unity.

Evidence which is consistent with the above description of the bonding in B_2 is the observation that the molecule is *paramagnetic*—the property associated with an unpaired electron (or with more than one unpaired electrons with parallel spins). In the B_2 case, the evidence for the presence of unpaired electrons comes from the observation of its *electron spin resonance* (esr) spectrum. The detailed theory of electron spin resonance spectra is not dealt with in this book. The essentials of the method depend upon there being a difference in the energy of an unpaired electron when subjected to a magnetic field. The electron spin is aligned either in the same direction as the applied field or against it. The difference in energy between the two quantized alignments corresponds to the energies of radio-frequency photons. No esr signal is obtained from paired-up electrons since neither of the electrons may change its spin without violating the Pauli exclusion principle.

Dicarbon, C_2

This molecule exists transiently in flames and has a bond dissociation energy of 590 kJ mol^{-1} and a bond length of 124 pm. The bond order is 2 since both the orbitals of the bonding $1\pi_u$ set are filled. The atoms are held together by two π bonds with no overall σ-bonding: a very unusual example. The carbon nucleus is more effective than that of the boron atom and, combined with there being twice as many resultant bonding electrons, serves to produce a much more stable molecule (with respect to the constituent atoms) than in the case of B_2. The average bond energy for C–C σ bonds is generally accepted to be 348 kJ mol^{-1} and that for

C=C double ($\sigma + \pi$) bonds is 612 kJ mol^{-1}. These energies are consistent with the view expressed above that π bonding is weaker than σ. The extra bond energy of the C=C double bond is given by 612 − 348 = 264 kJ mol^{-1} and is considerably smaller than that representing the strength of the C−C single σ bond.

Dinitrogen, N_2, and the ions, N_2^+ and N_2^-

In the dinitrogen molecule, N_2, two electrons occupy the $3\sigma_g^+$ orbital, and the bond order is three—one sigma pair plus the two pi pairs. The electronic configuration is consistent with the very high bond dissociation energy of 942 kJ mol^{-1} and the short bond length of 109 pm. The molecule is chemically inert to oxidation and reduction, although it does easily form some complexes when it acts as a ligand as, for example, in $[Ru(NH_3)_5N_2]^{2+}$. It undergoes reaction with dihydrogen only under conditions of high temperature and pressure (e.g. 380−450°C and 200 atm. In the Haber-Bosch process) in the presence of a catalyst. Some bacteria possess the capability of reducing dinitrogen at ambient temperature. The great strength of the bond, in dinitrogen, is associated with the presence of an excess of six bonding electrons together with the greater effectiveness of the nuclear charge compared to that of carbon.

The ionization of the molecule to give the N_2^+ ion causes the bond order to be reduced to 2.5, with consequent weakening (bond dissociation energy = 841 kJ mol^{-1}) and lengthening (bond length = 112 pm) of the bond as compared to that in N_2. The electron removed in the ionization comes from the $3\sigma_g^+$ orbital, which, because of the interaction with the $2\sigma_g^+$ orbital, is only moderately bonding. The effects upon the bond strength and length are, therefore, relatively slight.

The N_2^- ion is produced by adding an electron to the anti-bonding $1\pi_g$ level and, as does the N_2^+ ion, has a bond order of 2.5. Since the $1\pi_g$ level is anti-bonding, with no off-setting effects, the addition of an electron causes the bond in N_2^- to be significantly weaker (bond dissociation energy = 765 kJ mol^{-1}) and longer (bond length = 119 pm) than the one in N_2.

Dioxygen, O_2, and the ions, O_2^+, O_2^- and O_2^{2-}

The dioxygen molecule, O_2, has the dinitrogen configuration (except that the $1\pi_u$ level is higher in energy than the $3\sigma_g^+$ orbital) with an extra two electrons which occupy the $1\pi_g$ level. Since $1\pi_g$ is doubly degenerate the orbitals are singly occupied by electrons with parallel spins (Hund's rules). The molecule is paramagnetic as would be expected, and since the additional electrons occupy anti-bonding orbitals, the bond order decreases to two as compared with the N_2 molecule. In consequence the bond dissociation energy (494 kJ mol^{-1}) is considerably lower, and the bond length (121 pm) is considerably larger, than in the N_2 case.

The ionization of an electron from the highest energy level ($1\pi_g$) of the O_2 molecule produces the positive ion, O_2^+, which has a bond order of 2.5 (as does N_2^-). The bond dissociation energy of O_2^+ is 644 kJ mol^{-1} and its bond length is 112 pm. The nuclear charge of the oxygen is not as effective as that of nitrogen and the anti-bonding electron in O_2^+ has a very significant weakening effect.

In the superoxide ion, O_2^-, there are three electrons occupying the $1\pi_g$ orbitals. The bond order is 1.5, which is consistent with its observed bond dissociation energy of 360 kJ mol^{-1} and bond length of 132 pm.

The peroxide ion, O_2^{2-}, has a filled set of $1\pi_g$ orbitals and the bond is weaker (bond dissociation energy = 149 kJ mol^{-1}) and longer (bond length = 149 pm) than that of O_2^-. The bond order of the O_2^{2-} ion is 1.0.

Difluorine, F_2

The difluorine molecule, F_2, has an identical electronic configuration to that of the peroxide ion. The bond order is 1.0 and the bond dissociation energy of 155 kJ mol^{-1} and bond length of 144 pm being very similar to the values for O_2^{2-}.

The species with any occupancy of the $1\pi_g$(anti-bonding) level (O_2^+, O_2, O_2^-, O_2^{2-} and F_2) exhibit considerable chemical reactivity. The bonds are relatively weak and, therefore, easily cleaved and the species with unpaired electrons can easily form linkages to other atoms.

Dineon, Ne_2

The dineon molecule, Ne_2, with all its molecular orbitals filled would not be expected to exist with that particular electronic configuration.

5.5 SOME HETERONUCLEAR DIATOMIC MOLECULES

The heteronuclear diatomic molecules, nitrogen monoxide, NO, carbon monoxide, CO, and hydrogen fluoride, HF, are dealt with in this section. They belong to the point group, $C_{\infty v}$, and possess a C_∞ axis of symmetry and an infinite number of vertical planes all containing that axis. The orbitals of the molecules, NO and CO, are similar to those of the A_2 molecules of the previous section but have different terminology. In addition to the different symmetry there are effects due to the participating atoms having different electronegativities, the energies of the combining atomic levels not being identical.

Nitrogen monoxide, (nitric oxide) NO

The molecular orbital diagram for the nitrogen monoxide molecule is shown in Fig. 5.10. The orbitals are produced from the same pairs of atomic orbitals as in the cases of the homonuclear diatomic molecules of Section 5.7. The first ionization energies of nitrogen (1400 kJ mol^{-1}) and oxygen (1314 kJ mol^{-1}) are quite similar so the atomic orbitals match up reasonably well. The terminology is different and because of the absence of an *inversion centre* there are no *g* or *u* subscripts and this alters the numbering of the σ and π orbitals. The two pairs of 1s electrons (or KK) form the $1\sigma^+$ and $2\sigma^+$ orbitals, the other $2\sigma^+$ orbitals following on in order of increasing energy. The two sets of π orbitals become 1π (bonding) and 2π (anti-bonding) respectively. The electronic configuration of the nitrogen monoxide molecule is thus: KK $(3\sigma^+)^2(4\sigma^+)^2(1\pi)^4(5\sigma^+)^2(2\pi)^1$. The bond order is 2.5, consistent with the bond dissociation energy (626 kJ mol^{-1}) and bond length (114 pm). The molecule is chemically very reactive and is paramagnetic because of the single unpaired anti-bonding electron. The bond order in the NO^+ ion is 3.0 in the absence of any electrons in the 2p level.

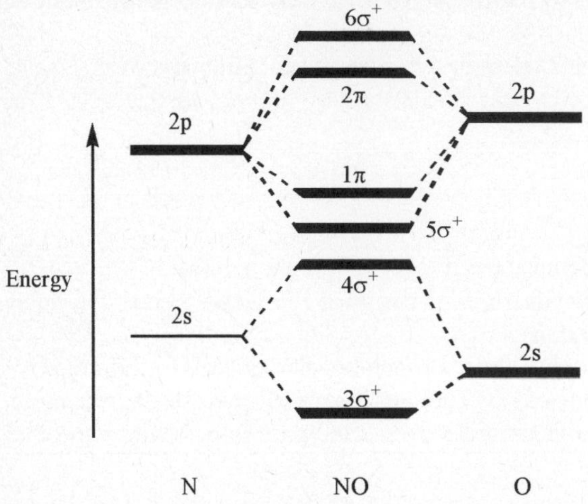

Fig. 5.10 Molecular orbital diagram for nitrogen monoxide, NO

In the NO case (ignoring the $1s^2$ pairs) there are four orbitals possessing the same symmetry, σ^+, and which can therefore mix to an extent which depends upon energy differences. It would be expected that all four orbitals would possess 2s and 2p contributions although it is not possible to quantify those in qualitative molecular orbital theory. The 2p–2s gaps in nitrogen and oxygen are relatively large so that the 2p-2s admixtures in the σ^+ m.o.'s would not be expected to approach equivalence. The NO molecule, because of its single unpaired electron, is sometimes written as NO$^\bullet$, the 'dot' signifying the 'odd' electron. This has given rise in the (bio)chemical literature of some serious misunderstandings with the formulae NO and NO$^\bullet$ being ascribed to species having different chemical properties!

Carbon monoxide

As the first ionization energies imply, the energies of the atomic orbitals of carbon (first I.E. = 1086 kJ mol^{-1}) and oxygen (first I.E. = 1314 kJ mol^{-1}) do not match up very well. When the 2p–2s energy gaps are compared (C, 386 and O, 1544 kJ mol^{-1}) it is obvious that there is a great mismatch between the respective 2s levels. That of the oxygen atom is too low to interact with the 2s atomic orbital of the carbon atom in any significant manner. The 2s atomic orbital of the oxygen atom is to be regarded as a virtually non-bonding orbital ($3\sigma^+$). The small 2p–2s energy gap in the carbon atom facilitates the mixing, or *hybridization*, of its 2s and $2p_z$ orbitals (assuming, as is conventional, that the molecular axis is coincident with the z axis). The two orbitals participate in σ^+ orbitals of the molecule and can mix. Because of the relatively small 2p–2s energy gap they do mix, the new orbitals being written as:

$$h_1(\mathrm{C}) = \psi(2s)_\mathrm{C} + \psi(2p_z)_\mathrm{C} \tag{5.31}$$

and

$$h_2(\mathrm{C}) = \psi(2s)_\mathrm{C} - \psi(2p_z)_\mathrm{C} \tag{5.32}$$

The hybrid orbitals, h_1 and h_2, are shown diagrammatically in Fig. 5.11. They have two unequal lobes of oppositely signed ψ values, the large positive lobe of h_1 being directed at the oxygen atom, with that of h_2 pointing in the opposite direction. It would be difficult for h_2 to play any significant part in the bonding to the oxygen atom and it is to be regarded as being non-bonding ($5\sigma^+$).

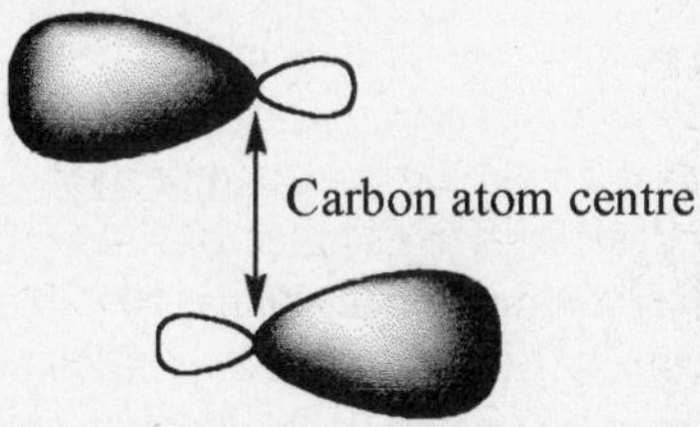

Fig. 5.11 The two sp hybrid orbitals of the carbon atom used in the production of molecular orbitals in the CO molecule. h_1 is the lower orbital which is arranged so that it is directed towards the oxygen atom, h_2 is the higher orbital which is directed away from the CO bonding region

The interaction of h_1(C) and the $2p_z$ orbital of the oxygen atom gives a bonding combination:

$$\phi(4\sigma^+) = h_1(\mathrm{C}) + 2p_z(\mathrm{O}) \tag{5.33}$$

and an anti-bonding combination:

$$\phi(6\sigma^+) = h_1(\mathrm{C}) - 2p_z(\mathrm{O}) \tag{5.34}$$

The molecular orbitals are completed by the interaction of the two sets of $2p_x$ and $2p_y$ orbitals to give doubly degenerate 1π (bonding) and 2π (anti-bonding) levels:

$$\phi(1\pi) = \psi(2p_{x,y})_{\mathrm{C}} + \psi(2p_{x,y})_{\mathrm{O}} \tag{5.35}$$

$$\phi(2\pi) = \psi(2p_{x,y})_{\mathrm{C}} - \psi(2p_{x,y})_{\mathrm{O}} \tag{5.36}$$

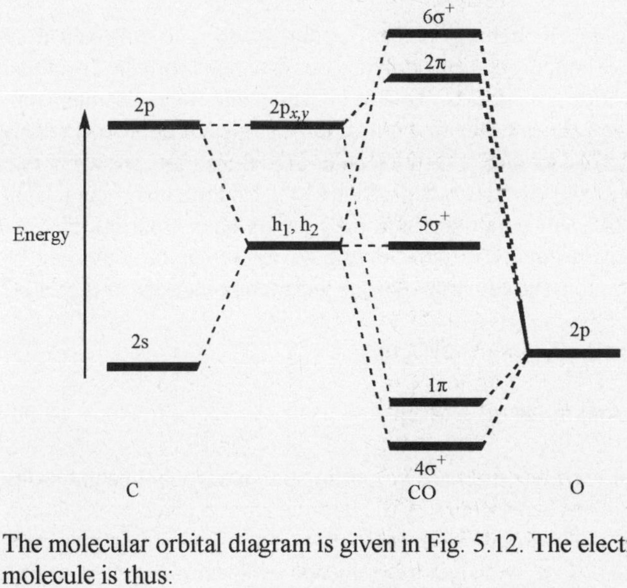

Fig. 5.12 The molecular orbital diagram for the CO molecule

The molecular orbital diagram is given in Fig. 5.12. The electronic configuration of the CO molecule is thus:

$$\mathrm{KK}(3\sigma^+)^2\,(4\sigma^+)^2\,(3\pi)^4\,(5\sigma^+)^2$$

and the bond order is 3 being consistent with the high bond dissociation energy of 1090 kJ mol^{-1} and the short bond length of 113 pm. Carbon monoxide is a relatively inert chemical substance but it does have an extensive involvement with the lower oxidation states of the transition elements with which it forms a great many carbonyl complexes, in which it acts as a ligand. The bonding of CO to a transition metal involves the use of the otherwise non-bonding electron pair in the $5\sigma^+$ orbital. The vacant 2π orbital is also important in the bonding of CO to transition metals, the details being dealt with in Chapter 10.

The cyanide ion, CN^-, is isoelectronic with carbon monoxide and has an extensive chemistry of involvement with transition metals but, unlike CO, it exhibits a preference for the positive oxidation states of the elements. This is mainly because of its negative charge.

Hydrogen fluoride, HF

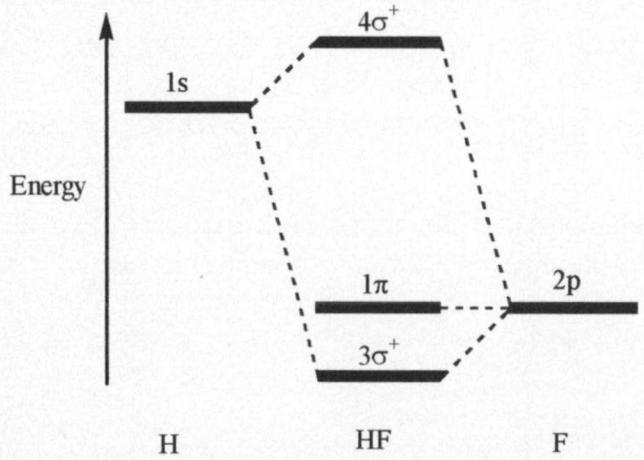

Fig. 5.13 The molecular orbital diagram for the HF molecule

The molecule of hydrogen fluoride, HF, belongs to the $C_{\infty v}$ point group. The hydrogen atom uses its 1s atomic orbital to make bonding and anti-bonding combinations with the $2p_z$ orbital of the fluorine atom (by convention the z axis is made to coincide with the molecular axis).

Because of the different values of the first ionization energies of the two elements (hydrogen, 1312, and fluorine, 1681 kJ mol^{-1}) the molecular orbital diagram, shown in Fig. 5.13, is considerably skewed with the lower, bonding orbital ($3\sigma^+$), having a major input from the fluorine $2p_z$ orbital. The higher, anti-bonding orbital ($4\sigma^+$), has its major contribution from the hydrogen 1s orbital. Such inequality in contributing to molecular orbitals may be expressed by including a factor, λ, in the equations for the linear combinations of the 1s(H) and $2p_z$(F) wavefunctions:

$$\phi(3\sigma^+) = \psi(1s)_H + \lambda\psi(2p_z)_F \tag{5.37}$$

and

$$\phi(3\sigma^+) = \lambda\psi(1s)_H - \psi(2p_z)_F \tag{5.38}$$

Because the square of wavefunctions are proportional to electron probabilities the 'share' of the two bonding electrons experienced by the fluorine atom is given by the factor, $\lambda^2/(1+\lambda^2)$. That experienced by the hydrogen atom is $1/(1+\lambda^2)$. The HF molecule possesses a dipole moment of 6.33×10^{-30} C m, which is experimental evidence for such a charge separation. The dipole moment (μ) in units of C m is given by the product of the charge (Q in coulombs) at the positive end of the molecule and the distance (in metres) between the two ends of the molecule (the bond length in the case of a diatomic molecule), $\mu = Qr$. The charge separation in HF is thus given by:

$$Q = \mu/r = (6.33 \times 10^{-30}) \div (0.092 \times 10^{-9}) = 6.88 \times 10^{-20}\ \mathrm{C}$$

which is equal to an electronic charge of $(6.88 \times 10^{-20}) \div (1.6 \times 10^{-19}) = 0.43\ e$. The hydrogen end of the molecule is, in effect, positive to the extent of 0.43 *e*, whilst the fluorine end is negative to the extent of $-0.43\ e$. Another viewpoint is that the 'covalent' bond is really 43% ionic.

The other orbitals in Fig. 5.13 are the non-bonding $2\sigma^+$(the 2s atomic orbital of the fluorine atom), and the non-bonding 1π (the $2p_x$ and $2p_y$ atomic orbitals of the fluorine atom). The $1s^2$ pair of fluorine electrons ($1\sigma^+$) are omitted from the diagram.

The discrepancy in matching up of the 1s(H) and $2p_z$(F) orbitals is 1681 - 1312 = 369 kJ mol^{-1} and leads to a bond with considerable ionic character. The difference in energy between the orbitals of the sodium atom (3s ionization energy = 496 kJ mol^{-1}) and the fluorine atom ($2p_z$ ionization energy = 1681 kJ mol^{-1}) amounts to 1185 kJ mol^{-1} and leads to the conclusion that covalent bond formation is impossible between Na and F. If the elements are to combine at all it has to be by an alternative method (ionic bonding) which is dealt with in Chapter 7.

5.6 PROBLEMS

1. The dissociation energies, *D*, of the M_2 Group 1 molecules in the gas phase are:

	Li_2	**Na_2**	**K_2**
D/kJ mol^{-1}	107.8	73.3	49.9
r/pm	134	154	196

Correlate the variation of *D* down the Group with the listed covalent radii, *r*, of the three metals.

2. The photoelectron spectra of the isoelectronic molecules, N_2 and CO, exhibit peaks corresponding to the following ionization energies, *I*. These and the vibrational wavenumbers associated with the appropriate ions formed are as follows: υ/cm^{-1}

N_2	**I/kJ mol^{-1}**	**1505**	**1639**	**1812**
	υ/cm^{-1}	2150	1810	2390
CO	I/kJ mol^{-1}	1352	1632	1903
	υ/cm^{-1}	2184	1535	1678

The fundamental vibration frequencies of the neutral molecules, N_2 and CO, are 2345 and 2143 cm^{-1} respectively. Derive the electronic configurations of the three ions produced in the two cases. What conclusions can be drawn from the values of the vibrational frequencies of the ions? Correlate the data with the abilities of the molecules to form complexes with transition metal species.

3. What conclusions can be drawn from the S–N distances in the following species?

Species	***d*/pm**
NSF	145
$N_3S_3F_3$	160
$S_4N_4^{2+}$	161 and 152

4. Use the data given in the text to plot the O–O bond dissociation energy of the species, O_2, O_2^+, O_2^-, O_2^{2-}, against their bond order. On the same graph plot the O–O bond length against bond order for the same species. Specify the magnetic behaviour of each species, using a suitable molecular orbital diagram.

6

Polyatomic Molecules and Metals

This chapter consists of the application of molecular orbital theory to the bonding in polyatomic molecules with its extension to the factors responsible for the determination of bond angles and molecular shapes. Additionally, a straightforward method of predicting molecular shape, the Valence Shell Electron Pair Repulsion theory, VSEPR, is described. The discussions are restricted to (i) three triatomic molecules; H_2O, CO_2 and XeF_2, (ii) a small selection of higher polyatomic molecules; NH_3, CH_4, BF_3, NF_3, ClF_3 and B_2H_6 with other two-centre molecules and (iii) the HF_2^- ion as an example of strong hydrogen bonding.

In the final section the factors which are responsible for determining the forms of elements (metallic or non-metallic) are discussed, together with an introductory treatment of metallic bonding and electrical conduction.

6.1 TRIATOMIC MOLECULES

Triatomic molecules may be linear or bent (i.e. V-shaped). The shape adopted by any particular molecule is that which is consistent with the minimization of its total energy.

The VSEPR theory is based upon the original ideas of Sigwick and Powell and was extended by Gillespie and Nyholm. The basis of the method is that the shape of a molecule results from the minimization of the repulsions between the pairs of electrons in the valence shell. The term 'valence shell' refers to the orbitals of the central atom of the molecule which could possibly be involved in bonding. The theory is operated by counting the number of electrons in the valence shell of the central atom together with suitable contributions from the ligand atoms - one electron per bond formed. This sum is then divided by two to give the number of pairs of electrons. These electron pairs are then considered to repel each other (ignoring the interelectronic repulsion between the electrons in each pair) to give a spatial distribution which is that corresponding to the minimization of the repulsive forces. This method of predicting molecular shape affords a very rapid and easy-to-use set of arguments which almost always produces an answer which is consistent with observation.

The basis of molecular orbital theory is described in Chapter 5 and its extension to the treatment of triatomic molecules is a major part of this chapter. It gives a very satisfactory description of the shapes and the bonding of molecules in general, and is consistent with observations of photoelectron and electronic absorption spectra. It is not possible for the VSEPR theory to explain these latter observations.

Valence shell electron pair repulsion theory

The shape of a molecule (or the geometry around any particular atom connected to at least two other atoms) is assumed to be dependent upon the minimization of the repulsive forces operating between the pairs of valence electrons. There is an important restriction upon the type of electron pairs in that they must be 'sigma' (σ) pairs. Any 'pi' (π) or 'delta' (δ) pairs must be discounted. The terms 'sigma', 'pi' and 'delta', refer to the type of overlap undertaken by the contributory atomic orbitals in producing the molecular orbitals and are referred to by their Greek letter symbols in the remainder of the book.

The overlap of two s orbitals on separate atoms and that of two p orbitals along the bond axis give rise to σ orbitals. These are cylindrically symmetrical with respect to the bond axis. The sigma overlap of two p orbitals is shown in Fig. 5.7. The sideways overlap of two p orbitals gives rise to π orbitals as is also shown in Fig. 5.7. Two d_{z^2} orbitals overlapping along the z axis give rise to σ orbitals.

Fig. 6.1 Two d_{xz} orbitals overlapping in the xz plane in a π manner and two d_{xy} orbitals overlapping along the z axis in a δ manner

Fig. 6.1 demonstrates the different types of overlap for $d_{xz}-d_{xz}$ (in the xz plane) and $d_{xy}-d_{xy}$ (along the z axis). The 'sideways' (two lobes) overlap by the d_{xz} orbitals is π-type, the four-lobe overlap by the two d_{xy} orbitals being termed δ-type. The Greek letters are used loosely for such orbitals—strictly they should be used for molecules possessing either $D_{\infty h}$ or $C_{\infty v}$ symmetry, but they are used generally to describe the type of overlap. Only σ pairs are counted up in the VSEPR approach. This often means that some assumptions about the bonding have to be made before the theory may be applied.

Table 6.1 contains the basic shapes adopted by the indicated numbers of σ electron pairs and Fig. 6.2 shows representations of the basic shapes assumed by the various electron pair distributions.

Table 6.1 The basic shapes adopted by various numbers of σ pairs of electrons

Number of σ electron pairs	Shape of electron pair distribution
2	Linear
3	Trigonally planar
4	Tetrahedral
5	Trigonally bipyramidal
6	Octahedral
7	Pentagonally bipyramidal

The application of VSEPR theory to triatomic molecules is exemplified by considering water, carbon dioxide and xenon difluoride.

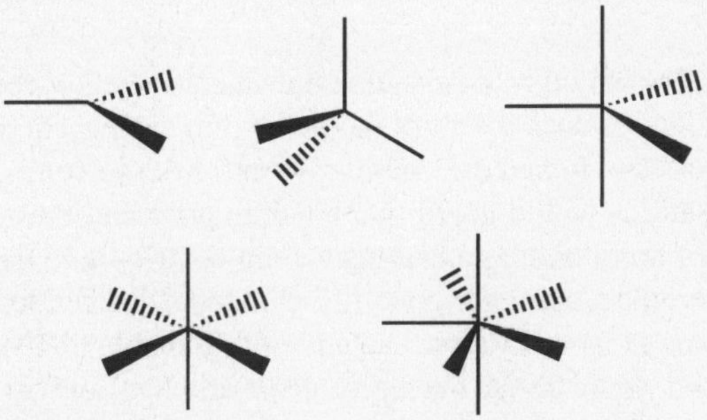

Fig. 6.2 The basic shapes for the spatial distribution of 3 (trigonally planar), 4 (tetrahedral), 5 (trigonally bipyramidal), 6 (octahedral) and 7 (pentagonally bipyramidal) pairs of electrons around a central atom

Water, H_2O

The isolated water molecule has a bond angle of 104.5°. The central oxygen atom has the valence shell electron configuration: $2s^22p^4$, with two of the 2p orbitals singly occupied. The hydrogen atoms supply one electron each to the valence shell of the central atom which makes a total of eight σ electrons (since the 1s–2p overlaps are σ-type). The four electron pairs are most stable when their distribution is tetrahedral. Two of the pairs are bonding pairs, the other two being non-bonding or lone pairs. This VSEPR picture of the water molecule would imply that the bond angle should be that of the regular tetrahedron; 109°28'. The theory predicts that the water molecule should be bent and so it is. To refine the prediction of the bond angle it is possible to argue that the non-bonding pairs of electrons, which are more localized on the central oxygen atom than the bonding pairs, have a greater repulsive effect upon the bonding pairs than that exerted by the bonding pairs upon each other. The result would be a reduction of the angle between the two bonding pairs making the bond angle somewhat less than the regular tetrahedral value. It is not possible to quantify this aspect of the theory.

Carbon Dioxide, CO_2

The carbon dioxide molecule is linear and belongs to the $D_{\infty h}$ point group. The carbon atom has the valence shell configuration of $2s^22p^2$. Ostensibly the carbon atom should be divalent—there are two unpaired electrons which occupy two of the three 2p atomic orbitals. In order for carbon to be tetravalent and to arrange for there to be four unpaired electrons in its valence shell it is possible (as a thought experiment) to cause the excitation of one of the 2s electrons to the previously unoccupied 2p orbital. It is then possible to feed two σ electrons into the 2s and one of the 2p orbitals, leaving the other two 2p orbitals to accept the two π electrons. Prior knowledge is used of the normal divalency of the oxygen atom and the impossibility of there being more than one σ bond between any pair of atoms—the spatial distribution of atomic orbitals decides that. The count of σ electron pairs comes to two and quite correctly the theory predicts the shape of the CO_2 molecule to be linear, the two σ pairs repelling each other to a position of minimum repulsion. The two π pairs are used to complete the bonding picture.

Xenon Difluoride, XeF_2

The XeF_2 molecule is linear. VSEPR theory would recognise that the valence shell of the xenon atom has the configuration: $5s^25p^6$, and as such the atom should be zero valent. There is the possibility of causing a 5p to 5d excitation to make the atom divalent. Addition of the two valence electrons from the fluorine atoms would make two bonding pairs and these, together with the remaining three pairs of non-bonding electrons, would contribute to the total of five electron pairs. The five pairs would assume a trigonally bipyramidal distribution and it is logical for the fluorine atoms to be as far away from each other as possible to give a linear molecule. The alternative approach would divide the molecule up into Xe^{2+} and two ligand fluoride ions. The Xe^{2+}, $5s^25p^4$, would then accept two electron pairs from the fluorine atoms, one which would fill the third 5p orbital and the second would enter the next available level, 5d. The same (correct) conclusion would follow.

Molecular orbital theory

The application of molecular orbital theory, even on a qualitative basis, is a more lengthy procedure than using VSEPR theory. It begins with the atomic orbitals which are available for bonding and the number of electrons which have to be accommodated. The understanding of the bonding of a molecule arises from the proper application of symmetry theory as is demonstrated for the H_2O, CO_2 and XeF_2 molecules. Quantitative results may be obtained for quite complicated molecules using modern computer methods. The results of solutions of the wave equation for molecules are referred to when necessary.

Water

The groundwork for the application of molecular orbital theory to the water molecule has been carried out to a large extent in Chapter 4. The procedure is to identify the point group to which the molecule belongs. If the normal state of the molecule is to be treated then that is a simple matter providing that the shape is known. To demonstrate the power of molecular orbital theory both of the extreme geometries of the molecule, the bent (bond angle, 90°) and linear (bond angle, 180°) forms, are treated.

90° Water Molecule

The general protocol for the derivation of the molecular orbital energy diagram for any molecule is outlined in this section. It is applied to the 90° water molecule as follows.

1. *Identify the point group to which the molecule belongs.*

The 90° form of the water molecule belongs to the C_{2v} point group. There is only one axis of symmetry, C_2, and by convention this is arranged to coincide with the z axis. The position of the molecule with respect to the coordinate axes is as shown in Fig. 4.6.

2. *Classify the atomic orbitals in the valency shell of the central atom with respect to the point group of the molecule.*

The classification of the orbitals of the oxygen atom is a matter of looking them up in the C_{2v} character table, a full version of which is included in the Appendix. The results are:

the 2s(O) orbital transforms as an a_1 representation,
the $2p_x$(O) orbital transforms as a b_1 representation,
the $2p_y$(O) orbital transforms as a b_2 representation, and
the $2p_z$(O) orbital transforms as another a_1 representation.

3. *Classify the valency orbitals of the ligand atoms with respect to the point group of the molecule and identify their group orbitals.*

The two hydrogen 1s orbitals have the character shown in the following table - the individual characters are the number of orbitals *unaffected* by the particular symmetry operation carried out upon the two orbitals.

C_{2v}	E	C_2	$\sigma_v(xz)$	$\sigma_v'(yz)$
1s + 1s	2	0	0	2

This representation may be seen, by inspection of the C_{2v} character table, to be equivalent to the sum of the characters for the a_1 and b_2 irreducible representations. Those particular hydrogen group orbitals are very similar to the ones used to describe the bonding of the H_2 molecule (equations 5.1 and 5.2 respectively and shown in Fig. 5.1)—the nuclei are further apart in the water case. The a_1 hydrogen group orbital (labelled as h_1) is H−H bonding whilst the b_2 group orbital (labelled as h_2) is H−H anti-bonding. Because of the greater distance between the hydrogen atoms in the water molecule such bonding and anti-bonding interactions are not as great as those in the H_2 molecule.

4. *Draw the molecular orbital diagram for the molecule.*

Two considerations are of importance in drawing the molecular orbital diagram for a molecule. First, the relative energies of the orbitals must be borne in mind. The first ionization energies of the atoms involved in molecule formation give a good indication of how to position the atomic levels, and information about the magnitude of p−s energy gaps is useful. Second, and highly important, is the guiding principle that only atomic orbitals (single or group) belonging to the same irreducible representation may combine to give bonding and anti-bonding molecular orbitals. It is also helpful to have knowledge of the photoelectron and absorption spectra to assist with the exact placement of the molecular levels. This information is not necessary for the production of a qualitative m.o. diagram. When such a diagram is seen to be consistent with the spectroscopic information it may be transformed into a reasonably scaled presentation. In addition it is very instructive to obtain the details of any proper quantum mechanical calculations which have been carried out on the particular molecule. Such calculations are possible for quite complicated molecules and software packages exist which would allow the reader to carry out calculations of the molecular orbitals of systems of interest.

With regard to the water molecule the first ionization energies of the hydrogen and oxygen atoms are almost identical (H, 1312; O, 1314 kJ mol^{-1}), and there is a large energy difference between the 2p and 2s levels of the oxygen atom (1544 kJ mol^{-1}). The latter piece of information indicates that the 2s(O) orbital does not participate in the bonding to a major extent. The 2s(O) and $2p_z$ orbitals do have identical symmetry properties and so have the possibility of mixing to give two hybrid orbitals but the large energy gap between them precludes very much interaction. That means that the main a_1 combination is between the $2p_z$ and h_1 orbitals. Combination occurs between the b_2 ($2p_y$(O)) and h_2 orbitals, with the b_1 ($2p_x$(O)) orbital remaining as a non-bonding m.o. The a_1 interaction produces a bonding orbital which is O-H bonding and which is also H–H bonding. The bonding orbital from the b_2 combination is O-H bonding but is H–H anti-bonding.

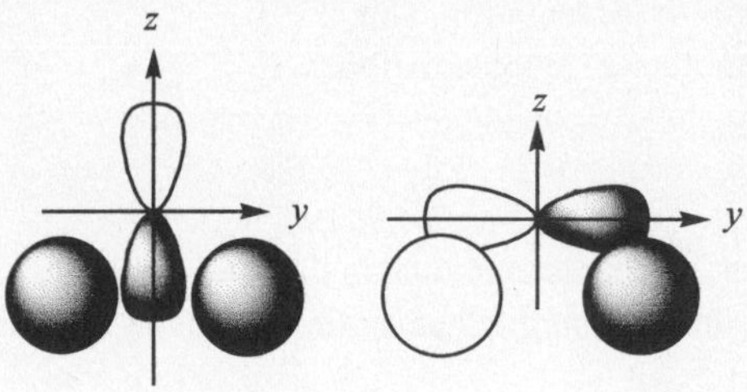

Fig. 6.3 Overlap diagrams for the a_1 (left) and b_2 (right) bonding orbitals of the C_{2v} water molecule

The above interactions are made obvious in the pictorial representation of the a_1 and b_2 orbital combination diagrams shown in Fig. 6.3. The molecular orbital energy diagram for the 90° water molecule is shown in Fig. 6.4.

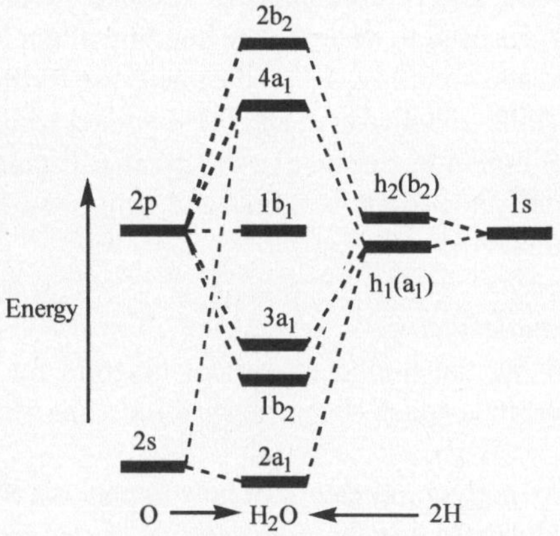

Fig. 6.4 Molecular orbital diagram for the 90°water molecule

The $2a_1$ orbital gains some stability from a small interaction with the bonding $3a_1$ orbital which becomes destabilized to a similar extent. This causes the $3a_1$ orbital to have a higher energy than that of the $1b_2$ orbital. Such matters cannot be predicted by the qualitative approach used here, quantitative calculations being necessary. The ‘tie lines’ between the atomic and molecular orbitals in Fig. 6.4 (and all other m.o. diagrams in this text) do not take the mixing of molecular orbitals into account. The m.o. diagrams in this book are simplified as far as possible and, wherever there is orbital mixing, the contributions from the various atomic orbitals are not always fully indicated by the tie lines.

One very important difference between VSEPR theory and molecular orbital theory should be noted. The molecular orbitals of the water molecule which participate in the bonding are three centre orbitals. They are associated with all three atoms of the molecule. There are no localized electron pair bonds between pairs of atoms as used in the application of VSEPR theory. The existence of three-centre orbitals (and multi-centre orbitals in more complicated molecules) is not only more consistent with symmetry theory - it also allows for the reduction of interelectronic repulsion effects when more extensive, non-localized, orbitals are doubly occupied.

180° Water Molecule

1. The linear water molecule belongs to the $D_{\infty h}$ point group. The classifications of atomic and group orbitals must be carried out using the $D_{\infty h}$ character table. The molecular axis, C_∞, is arranged to coincide with the z axis.
2. The classification of the 2s and 2p atomic orbitals of the central oxygen atom.

The 2s(O) orbital transforms as σ_g^+,
the $2p_x$ and $2p_y$ orbitals transform as the doubly degenerate π_u representation, and
the $2p_z$ orbital transforms as σ_u^+.

In some texts the '+' and '−' superscripts are omitted, but throughout this one there is strict adherence to the use of the full symbols for all orbital symmetry representations.

Unlike the 90° case the 2s and $2p_z$ orbitals have different symmetry properties so there is no question of their mixing.

3. Classification of the hydrogen group orbitals.

The H−H bonding group orbital, h_1, transforms as σ_g^+, and the H−H anti-bonding group orbital, h_2, transforms as σ_u^+. The detailed conclusions are dealt with in section 3.2.

4. The molecular orbitals may now be constructed by allowing the orbitals of the oxygen and hydrogen atoms, belonging to the same representations, to combine. Thus the interaction between σ_g^+ and h_1 is possible, by symmetry, but is very restricted in extent by the large difference in energy between the two orbitals. The major interaction is between σ_u^+(O) and h_2 which possess very similar energies. The $2p_x$ and $2p_y$ orbitals of the oxygen atom remain as a doubly degenerate pair of π_u orbitals. The m.o. diagram for the linear water molecule is shown in Fig. 6.5.

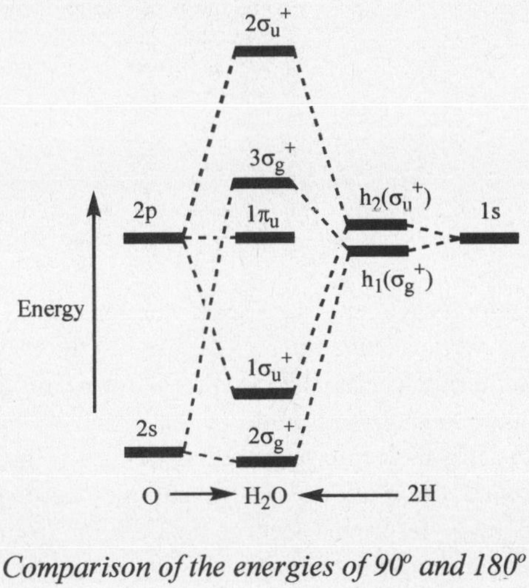

Fig. 6.5 Molecular orbital diagram for the linear water molecule

Comparison of the energies of 90° and 180° water molecules

In the water molecule there are eight valency electrons to be distributed in the lowest four molecular orbitals. In the 90° case this produces an electronic configuration which may be written as: $2a_1^2 3a_1^2 1b_2^2 1b_1^2$ (ignoring the $1s^2$ pair of the oxygen atom which could be written as $1a_1^2$ or K. The $2a_1$ orbital is slightly bonding because of the interaction with the $3a_1$ orbital. The latter orbital has considerable bonding character, as has the $1b_2$, so that there are two bonding pairs of electrons responsible for the cohesion of the three atoms. The $1b_1$ orbital is non-bonding, there being no hydrogen orbitals of that symmetry. The 90° molecule would have two similar, strongly bonding, electron pairs (in the $3a_1$ and $1b_2$ three-centre m.o.'s) and two virtually non-bonding electron pairs (in the $1b_1$ and $2a_1$ orbitals) of very different energies.

The electronic configuration of the linear water molecule is $(2\sigma_g^+)^2(1\sigma_u^+)^2(1\pi_u)^4$ (the $1s^2$(O) pair of electrons occupy the $1\sigma_g^+$ molecular orbital). The $2\sigma_g^+$ pair of electrons are only weakly bonding with the $1\sigma_u^+$ pair supplying practically the only cohesion for the three atoms, the other four electrons being non-bonding.

A comparison of the m.o. diagrams for the two forms of the water molecule is given in the correlation diagram of Fig. 6.6 and shows that the 90° angle confers the extra stability of two bonding pairs of electrons as opposed to the single pair of bonding electrons in the linear molecule.

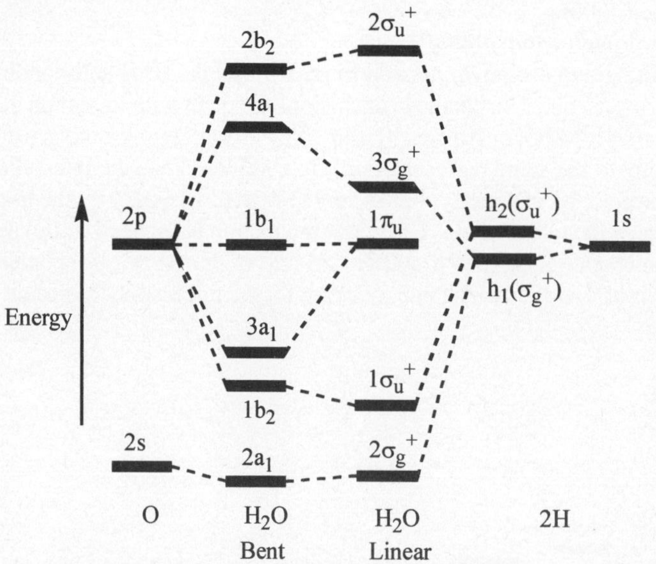

Fig. 6.6 A molecular orbital correlation diagram for the bent and linear forms of the water molecule

Neither interelectronic repulsions nor internuclear repulsions have been considered. The ignoring of interelectronic repulsions is not serious since the orbitals used in the two forms of the molecule are extremely similar. The internuclear repulsion in the 90° form would be larger than in the linear case and contributes to the bond angle in the actual water molecule being greater than 90°. The actual state of the molecule, as it normally exists, is that with the lowest total energy and only detailed calculations can reveal the various contributions. At a qualitative level, as carried out so far in this section, the decision from molecular orbital theory is that the water molecule should be bent.

The $1\pi_u$ orbitals in the linear case (these are the non-bonding $2p_x$ and $2p_y$ orbitals of the oxygen atom) lose their degeneracy when the molecule bends. The $2p_x$ orbital retains its non-bonding character but the $2p_y$ orbital makes a very important contribution to the $1b_2$ bonding m.o. of the bent molecule. It is this factor which is critical in determining the shape of the water molecule. Fig. 6.7 shows how the energy of the filled, $1\pi_u$—$3a_1$, $1b_1$, $1\sigma_u^+$—$1b_2$ and $2\sigma_g^+$—$2a_1$, orbitals vary with bond angle. It also shows the variation of the internuclear repulsion energy due to the varying interproton distance. The information is derived from quantum mechanical calculations and it is the balancing between these energies which is largely responsible for the observed bond angle of the ground state water molecule.

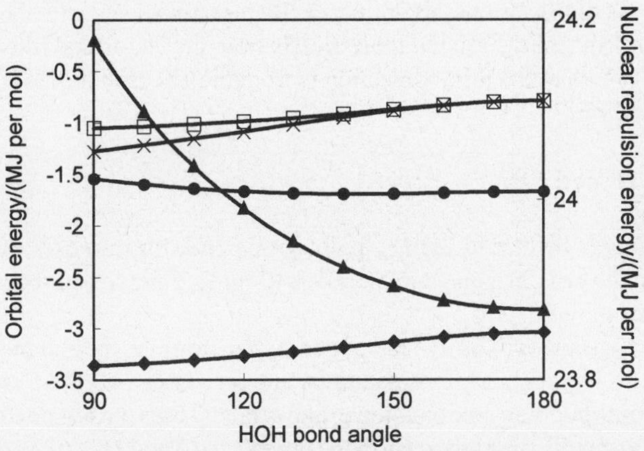

Fig. 6.7 The variation with bond angle of the energies of the $1b_1$ (open squares), $3a_1$ (crosses), $1b_2$ (filled circles) and $2a_1$ (filled diamonds) orbitals [left-hand axis] and the nuclear repulsion energy (filled triangles) [right-hand axis] for the water molecule

Qualitative orbital correlations similar to those of Fig. 6.7 were used by Walsh in 1953 to make predictions of whether AH_2 molecules (A representing any main group element) would be linear or bent. The qualitative Walsh diagrams for various systems are still of use in the predictions of molecular shape and electronic spectra but the underlying ideas are not fully justifiable. They seem to give generally good results because of the cancelling out of various factors. Most small molecules can now be treated on an *ab initio* (from the beginning) basis in which the appropriate wave equation is solved to give reliable m.o. information. This includes the nuclear repulsion energy as well as the electronic and total energies.

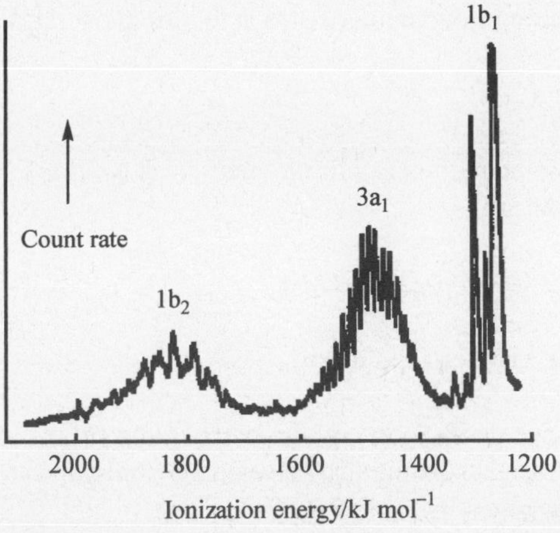

Fig. 6.8 The photoelectron spectrum of water vapour; ionizations from the $1b_1$, $3a_1$ and $1b_2$ orbitals are indicated. Ionizations from the more stable $2a_1$ orbital are not produced by the helium radiation used. Modified from the diagram given in C. R. Brundle, M. B. Brown, N. A. Kuebler and H. Basch, *J. Am. Chem. Soc.*, **94**, 1455, 1972.

Experimental confirmation of the order of molecular orbital energies for the water molecule is given by its photoelectron spectrum. Fig. 6.8 shows the helium line photoelectron spectrum of the water molecule. There are three ionizations at 1216, 1322 and 1660 kJ mol^{-1}. A fourth ionization at 3474 kJ mol^{-1} has been measured by using suitable X-ray quanta instead of the helium emission. That there are the four ionization energies is consistent with expectations from the m.o. levels for a bent C_{2v} molecule (see Fig. 6.4).

A study of the magnitudes of the various vibrational excitations associated with the second and third ionizations confirm that the second ionization is from the $3a_1$ orbital (this is both O–H and H–H bonding) and that the third ionization is from the $1b_2$ orbital (this is O–H bonding but H–H anti-bonding).

Molecular orbital theory of CO_2 and XeF_2

The linear CO_2 and XeF_2 molecules belong to the $D_{\infty h}$ point group and, for purposes of classifying their atomic orbitals, the principal quantum numbers (2 for C, 5 for Xe) of their valence electrons may be dispensed with.

The s orbital of C and Xe transform as σ_g^+, the p_x and p_y transform as π_u, and the p_z as σ_u^+. The p–s energy gaps in oxygen and fluorine atoms are very large and their 2s orbitals (as formal bonding and anti-bonding combination group orbitals) must be regarded as being virtually non-bonding. Formally, they give rise to the orbitals, $1\sigma_g^+$ and $1\sigma_u^+$ (the 1s orbitals of the atoms being ignored). The 2p orbitals of the ligand atoms may be dealt with in two sets of group orbitals - the σ orbitals being the $2p_z$ lying along the molecular axis, and the π orbitals being made up from the $2p_x$ and $2p_y$ orbitals perpendicular to the molecular axis. The s orbital group combinations may be written as:

$$\phi(\sigma_g^+) = \psi(2p_z)(O,F)_A + \psi(2p_z)(O,F)_B \quad (6.1)$$

the plus sign being used to indicate a bonding interaction (a plus ψ to plus ψ overlap), and

$$\phi(\sigma_u^+) = \psi(2p_z)(O,F)_A - \psi(2p_z)(O,F)_B \quad (6.2)$$

the minus sign indicating an anti-bonding interaction (a plus ψ to minus ψ overlap).

The π group orbital combinations are:

$$\phi(\pi_u) = \psi(2p_{x,y})(O,F)_A + \psi(2p_{x,y})(O,F)_B \quad (6.3)$$

and

$$\phi(\pi_g) = \psi(2p_{x,y})(O,F)_A - \psi(2p_{x,y})(O,F)_B \quad (6.4)$$

with the signs having the same significance as those in equations (6.1) and (6.2). In the above equations the subscripts A and B are used to distinguish between the two ligand atoms. A summary of the above classifications is given in Table 6.2.

Taking into account the first ionization energies of the atoms participating in the CO_2 and XeF_2 molecules it is possible to use one qualitative m.o. energy diagram to discuss the disposition of the electrons which they contain. The diagram is shown in Fig. 6.9 and ignores the two orbitals which are constructed from the two 2s orbitals of the ligand atoms.

Table 6.2 The orbitals of C and Xe and the group orbitals of O and F which can combine to give the m.o.'s of CO_2 and XeF_2

C or Xe atomic orbital	**O_2 or F_2 group orbitals**	
σ_g^+(s)	σ_g^+(2s ~ non-bonding)	σ_g^+(2p_z bonding)
σ_u^+(p_z)	σ_u^+(2s ~non-bonding)	σ_u^+(2p_z anti-bonding)
π_u($p_{x,y}$)	π_u(2$p_{x,y}$ bonding)	
	π_g(2$p_{x,y}$ anti-bonding)	

The sixteen valence electrons give the electronic configuration of the CO_2 molecule as $(1\sigma_g^+)^2(1\sigma_u^+)^2\,(2\sigma_g^+)^2(2\sigma_u^+)^2\,(1\pi_u)^4\,(1\pi_g)^4$ (the 1s pairs of the three atoms being ignored in the numbering). The first two pairs of electrons are non-bonding (they are the $2s^2$ pairs of the oxygen atoms) but the next four pairs (two σ and two π) occupy three-centre (all three atoms) bonding molecular orbitals and are responsible for the high bond strength of the so-called 'double bonds' of the molecule (743 kJ mol^{-1}). The remainder of the electrons occupy the three-centre non-bonding π_g orbitals. The photoelectron spectrum of the CO_2 molecule indicates ionization energies of 1330, 1671, 1744 and 1872 kJ mol^{-1}, for the removal of electrons from the $1\pi_g$, $1\pi_u$, $2\sigma_u^+$ and $2\sigma_g^+$ molecular orbitals respectively.

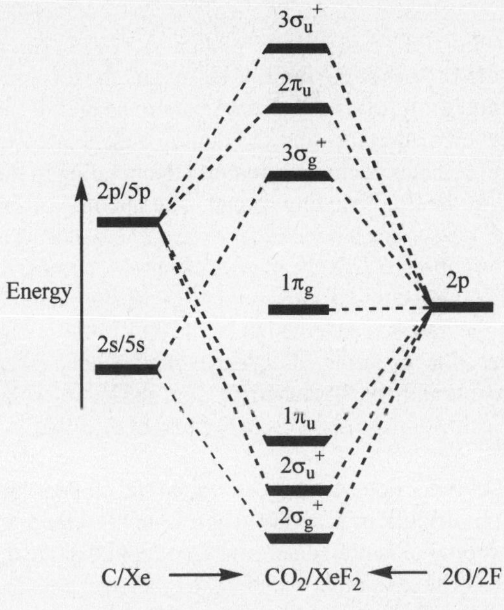

Fig. 6.9 A molecular orbital diagram for the CO_2 and XeF_2 molecules

Some comment is necessary here to stress the three-centre nature of the bonding in CO_2, particularly with regard to the π orbitals. The three p orbitals in the *xz* plane and the three in the *yz* plane (remember that the molecular axis is conventionally taken to coincide with the *z* axis) form the three-centre set of three π molecular orbitals, $1\pi_u$ (bonding), $1\pi_g$ (non-bonding) and $2\pi_u$ (anti-bonding), the first two orbitals being fully occupied and leading to the idea that the three atoms are bonded by two three-centre π bonds. The valence bond theory which depends upon two-electron two-centre bonding provides the CO_2 molecule with two C=O double bonds (with the two π C—O bonds at right angles to each other with regard to the molecular axis) and can cope with the delocalization of the π system only by the postulation of the participation of ionic structures, $^{-}O—C\equiv O^{+}$ and $^{+}O\equiv C—O^{-}$, in the resonance hybrid. This ensures in a clumsier manner than does m.o. theory that the doubly degenerate π bonding is delocalized over the three atom centres. The delocalization allows the electrons of the π system to occupy larger orbitals than they would occupy in a two-electron two-centre conventional bond. This allows a reduction in the interelectronic repulsion and leads to stronger bonding and stabilizes the molecule as a whole. This point can be demonstrated by a simple calculation. The standard enthalpy of formation of the CO_2 molecule is -394 kJ mol^{-1}. An estimate of what this would be if the molecule were to be constructed from two-electron two-centre bonds (two σ and two π bonds) is calculated by adding the enthalpy changes for (i) the sublimation of carbon (715 kJ mol^{-1}), (ii) the dissociation of dioxygen (496 kJ mol^{-1}), and (iii) the production of two C=O bonds from one carbon atom and two oxygen atoms (2×-743 kJ mol^{-1}). The calculated enthalpy of formation of the CO_2 molecule is thus -275 kJ mol^{-1} which indicates that the localized two double bonded molecule would be 119 kJ mol^{-1} less stable than the real delocalized arrangement. Delocalization occurs in molecules wherever it is possible by symmetry considerations and wherever an energetic advantage can be gained from its operation. Molecular orbital theory deals very satisfactorily with delocalization, but with valence bond theory the concept is somewhat clumsily incorporated as an addition to the conventional two-electron two-centre bonding.

The electronic configuration of the XeF_2 molecule is written as $(1\sigma_g^+)^2(1\sigma_u^+)^2$ $(2\sigma_g^+)^2(2\sigma_u^+)^2\,(1\pi_u)^4\,(1\pi_g)^4\,(3\sigma_g^+)^2(2\pi_u)^4$ (the 1s pairs of the fluorine atoms and the forty-six electron core of the xenon atom being ignored). It follows the same pattern as that of the CO_2 molecule with the additional occupancy of the $3\sigma_g^+$ and the doubly degenerate $2\pi_u$ orbitals, all three orbitals being anti-bonding. Their occupancy reduces the bonding in the XeF_2 molecule to the single electron pair in the $2\sigma_u^+$ bonding orbital. The bonding of the molecule is equivalent to each Xe-F 'bond' possessing a bond order of 0.5. The molecule is only marginally stable with respect to its formation; $\Delta_f H(XeF_2,g) = -82$ kJ mol^{-1}, compared with CO_2; $\Delta_f H^{\circ}(CO_2,g) = -394$ kJ mol^{-1}. A fairer comparison would be to use the standard heat of formation of CO_2 from carbon in the gaseous state which is -1109 kJ mol^{-1}. The XeF_2 molecule would be difficult to describe in terms of valence bond theory since conventional electron pair bonds are impossible with the filled shell Xe atom and in molecular orbital theory the effects of π-electron delocalization are not present because of the filled $2\pi_u$ anti-bonding orbitals.

The molecular orbital treatment of XeF_2 does not depend upon (and so does not over-emphasize) the inclusion of d orbital contributions from the xenon atom. Reference to the $D_{\infty h}$ character table shows that the d_{z^2} orbital of xenon transforms as σ_g^+ and that the d_{xz} and d_{yz} orbitals transform as π_g. Those orbitals have suitable symmetries to interact with the appropriate orbitals of the fluorine atoms. The other two d orbitals (xy and x^2-y^2) transform as the representation, δ_g, and cannot participate in the bonding of XeF_2. It is doubtful whether any significant d orbital participation occurs because of the relatively higher energies of such orbitals.

6.2 SOME POLYATOMIC MOLECULES

This section consists of the treatment by VSEPR and m.o. theories for some small polyatomic molecules. The shapes and bonding of the following molecules are discussed: i) NH_3, (ii) CH_4, (iii) BF_3, NF_3 and ClF_3, (iv) B_2H_6, (v) and the HF_2^- ion.

Ammonia, NH_3

The VSEPR treatment takes the nitrogen atom ($2s^22p^3$) and adds the 1s electrons from the three hydrogen atoms producing the $2s^22p^6$ configuration in the valence shell of the nitrogen atom. The four pairs of σ electrons adopt a tetrahedral distribution to minimize electron pair repulsions. One of the four tetrahedral positions is occupied by a lone pair, thc other three being those of the three bonding pairs. The molecular shape should be trigonally pyramidal with bond angles somewhat smaller than those of a regular tetrahedron. The molecule is a trigonal pyramid with HNH bond angles of 107°.

Molecular orbital treatment of the ammonia molecule
Molecular orbital theory may be used to establish stability preferences between extreme geometries for any molecular system. Those for the ammonia molecule are the trigonally planar, D_{3h}, the trigonally pyramidal, C_{3h}, and the T-shaped, C_{2v}, symmetries.

The 2s and 2p orbitals of the nitrogen atom and the 1s orbital group combinations of the three hydrogen atoms transform, with respect to the point groups, D_{3h}, C_{3v} and C_{2v}, as indicated in Table 6.3.

Table 6.3 Orbitals of the nitrogen and hydrogen atoms of the ammonia molecule

Orbital	Point group		
	D_{3h}	C_{3v}	C_{2v}
2s(N)	a_1'	a_1	a_1
$2p_x$(N)			b_1
	e'	e	
$2p_y$(N)			b_2
$2p_z$(N)	a_2''	a_1	a_1
3 × 1s(H)	$a_1' + e'$	$a_1 + e$	$a_1 + a_1 + b_2$

In the D_{3h} case interaction between the nitrogen and hydrogen orbitals is possible for those with a_1' and e' symmetry, with the a_2'' orbital being non-bonding. The a_1' bonding combination is only slightly bonding because of the mismatch of energies of the contributing 1s(H) (ionization energy = 1312 kJ mol^{-1}) and 2s(N) (ionization energy = 1400 + 1060 = 2460 kJ mol^{-1}) orbitals. The major interaction occurs with the energetically favourable e' orbitals giving low energy bonding and high energy anti-bonding sets of molecular orbitals. The molecular orbital diagram of the D_{3h} ammonia molecule is shown in the correlation diagram of Fig. 6.10. The orbital numbering takes into account the $1s^2$ electrons of the nitrogen atom which would be labelled as $1a_1'$ in the trigonally planar molecule and $1a_1$ in the C_{3v} and C_{2v} cases.

In the C_{3v} and C_{2v} symmetries the 2s and $2p_z$ orbitals of the nitrogen atom belong to identical representations a_1 indicating that mixing is possible. Because the 2p-2s energy gap is large, any mixing is minimal and is best considered after the molecular orbitals are formed.

In the C_{3v} case the main a_1 interaction is between the $2p_z$(N) orbital and the a_1 hydrogen group orbital, shown in Fig. 6.11, the 2s(N) participation being minimal because of the mismatch of energies mentioned above. As may be seen from Fig. 6.11 the interaction between the hydrogen group orbital (fully H-H bonding) and the nitrogen $2p_z$ orbital (zero in the D_{3h} case) becomes bonding as the group orbital leaves the *xy* plane. There is less + to − anti-bonding overlap and more + to + bonding overlap as the bond angle decreases. It is this interaction which is mainly responsible for the C_{3v} symmetry being more stable than the D_{3h}. The $3a_1$ orbital becomes more stable as the symmetry deviates from D_{3h}, its energy is reduced by 240 kJ mol^{-1} on going from the trigonally planar molecule to the 90° C_{3v} trigonally pyramidal molecue. The e-type interactions are not as effective as in the D_{3h} case since the hydrogen orbitals are no longer in the *xy* plane. Slight mixing of the $2a_1$ and $3a_1$ orbitals stabilizes the former at the expense of the latter. The m.o. diagram is shown in Fig. 6.10.

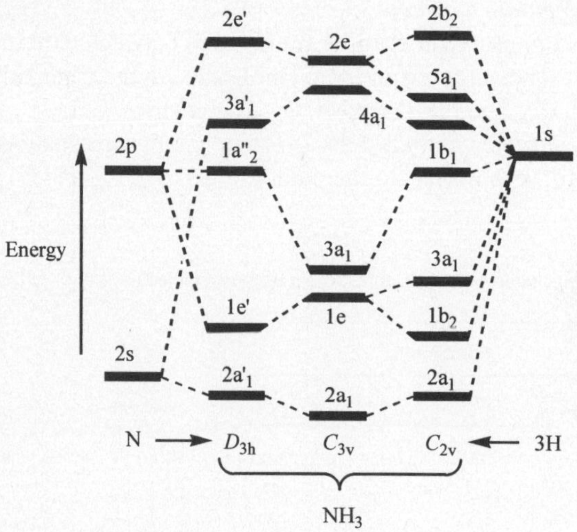

Fig. 6.10 Correlation diagram for the molecular orbitals of the ammonia molecule with trigonally planar (D_{3h}), trigonally pyramidal (C_{3v}) and T-shaped (C_{2v}) symmetries

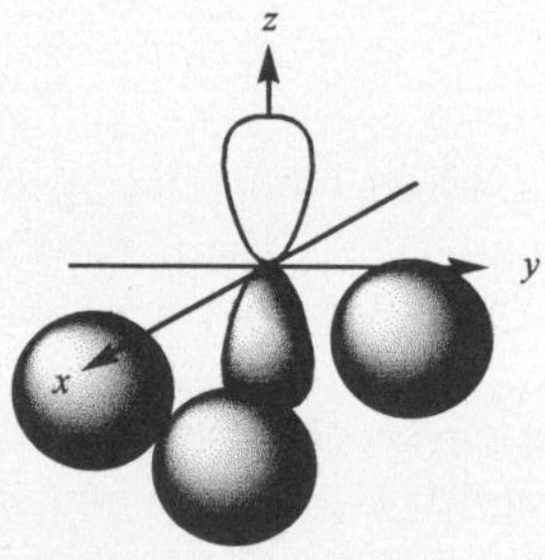

Fig. 6.11 The atomic orbital contributions to the $3a_1$ molecular orbital of the ammonia molecule

The third option for the ammonia molecule is the T-shape which belongs to the C_{2v} point group. The $3a_1$ interaction is mainly between the $2p_z$(N) orbital and the hydrogen orbital positioned along the *z* axis. The interaction between the 2s(N) orbital and the 1s(H)+1s(H) combination along the *y* axis produces the $2a_1$ (slightly bonding) and $4a_1$ (slightly anti-bonding) molecular orbitals. The $2p_x$(N) and $2p_y$(N) orbitals lose their degeneracy in C_{2v} symmetry. The $2p_x$(N) orbital becomes b_1 and reverts to being a non-bonding orbital (the $1a_2''$

orbital of the D_{3h}) and the $2p_y$(N) orbital interacts strongly with the appropriate hydrogen group orbital. The m.o. diagram is shown in Fig. 6.10.

The electronic configurations of the ammonia molecule in its three symmetries are shown in the Table 6.4.

Table 6.4 Electronic configurations of the ammonia molecule with different symmetries

NH_3 symmetry	Electronic configuration
D_{3h}	$(2a_1')^2(1e')^4(1a_2'')^2$
C_{3v}	$(2a_1)^2(1e)^4(3a_1)^2$
C_{2v}	$(2a_1)^2(1b_2)^2(3a_1)^2(1b_1)^2$

An *ab initio* estimation of the total energies of the D_{3h}, and various C_{3v} and C_{2v}, molecules indicates that the minimum total energy is that of the trigonally pyramidal molecule, C_{3v}, with a bond angle of 105° which is very close to the observed value.

The main stabilizing factor in the ammonia molecule is the bonding nature of the $3a_1$ orbital in C_{3v} symmetry. The two electrons in that orbital cause the molecule to assume C_{3v} symmetry in its ground state. The stabilization is opposed by the increase in interproton repulsion. The alternative D_{3h} and C_{2v} shapes are less stable because the one of the 2p(N) orbitals is non-bonding ($2p_z$(N) in D_{3h} and $2p_x$(N) in C_{2v}). The photoelectron spectrum of the ammonia molecule is consistent with the m.o. conclusions - there are three ionization energies at 1050, 1445 and 2605 kJ mol^{-1}, corresponding to the removal of electrons from the $3a_1$, 1e and $2a_1$ molecular orbitals respectively.

Methane, CH_4

The methane molecule is a very important molecule in organic chemistry, the geometry around the four-valent carbon atom being basic to the understanding of the structure, isomerism and optical activity of a very large number of organic compounds. It is a tetrahedral molecule belonging to the tetrahedral point group, T_d.

The VSEPR theory assumes that the four electrons from the valence shell of the carbon atom plus the valency electrons from the four hydrogen atoms form four identical electron pairs which, at minimum repulsion, give the observed tetrahedral shape.

Molecular orbital theory of tetrahedral and square planar methane

It is instructive to compare the stabilities of the tetrahedral and square planar forms of the methane molecule. The transformation properties of the orbitals of the carbon atom and the group orbitals of the four hydrogen atoms, with respect to the T_d and D_{4h} point groups are given in Table 6.5.

The molecular orbital diagrams for the two symmetries of the methane molecule are shown in Fig. 6.12. In T_d symmetry there is a total match between the two a_1 and two t_2 sets of orbitals (C and H). In D_{4h} symmetry the a_{1g} combinations are similar in energy to the a_1 in T_d. There is a smaller amount of H–H bonding character in the D_{4h} case which is the reason for placing the $1a_{1g}$ orbital slightly higher in energy than the $1a_1$ orbital.

Table 6.5 Symmetries of the valence orbitals of the carbon atom and the group orbitals of the hydrogen atoms of the methane molecule for tetrahedral, T_d, and square planar, D_{4h}, shapes of the molecule

Orbital	T_d	D_{4h}
2s(C)	a_1	a_{1g}
$2p_x$(C)		
		e_u
$2p_y$(C)	t_2	
$2p_z$(C)		a_{2u}
4 × 1s(H)	$a_1 + t_2$	$a_{1g} + e_u + b_{1g}$

The $1t_2$ orbitals lose their degeneracy in D_{4h} symmetry where they become the e_u bonding pair of orbitals and the non-bonding $1b_{1g}$ orbital. The latter orbital is placed at a fairly high level since it is considerably H–H anti-bonding. This character makes its energy higher than that of the carbon $2p_z$—$1a_{2u}$ orbital. The anti-bonding $2t_2$ orbitals lose their degeneracy in D_{4h} symmetry and become the anti-bonding $2e_u$ and non-bonding $1a_{2u}$.

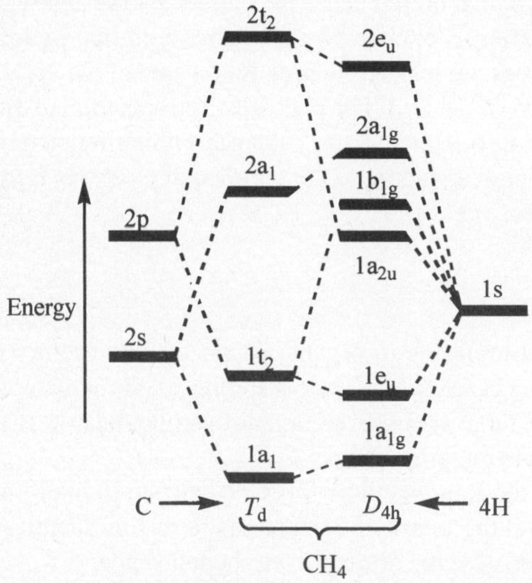

Fig. 6.12 Correlation diagram of the molecular orbitals of methane with tetrahedral (T_d) and square planar (D_{4h}) symmetries

The electronic configurations of the methane molecule with T_d and D_{4h} symmetries are $(1a_1)^2(1t_2)^6$ and $(1a_{1g})^2(1e_u)^4(1a_{2u})^2$ respectively. In the T_d case all eight electrons occupy bonding orbitals, but in D_{4h} symmetry two of the electrons occupy the non-bonding $1a_{2u}$ orbital which is the reason for methane being tetrahedral in its electronic ground state. As the symmetry deviates from D_{4h} the $2p_z$ orbital of the carbon atom is able to participate in the formation of a bonding orbital.

The VSEPR assumption that there are four identical localized electron pair bonds in the four C-H regions, made up from sp^3 hybrid carbon orbitals and the hydrogen 1s orbitals, are not consistent with the experimentally observed photoelectron spectrum. The m.o. theory is consistent with the two ionizations shown in the photoelectron spectrum of CH_4 and implies that the bonding consists of four electron pairs which occupy the $1a_1$ and $1t_2$ five-centre molecular orbitals.

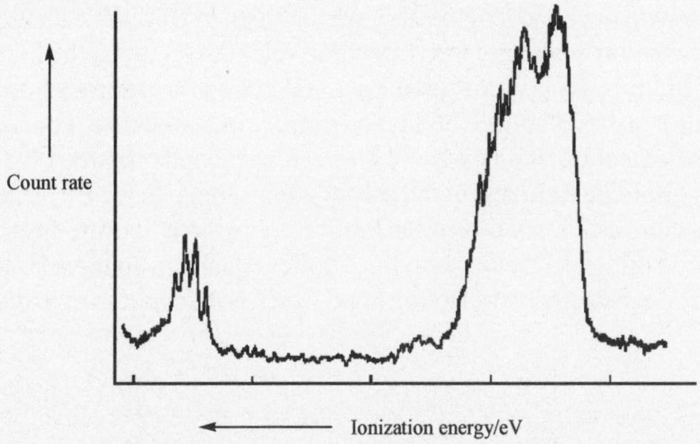

Fig. 6.13 A photoelectron spectrum of the methane molecule. Modified from the spectrum given by W. C. Price, Developments in photoelectron spectroscopy, a paper presented at The Institute of Petroleum Conference on Molecular Spectroscopy, 1968, published by Elsevier, p. 221–236

The photoelectron spectrum is shown in Fig. 6.13 and shows that there are two ionization energies at 1230 and 2160 kJ mol^{-1} corresponding to the removal of electrons from the 1t$_2$ and 1a$_1$ molecular orbitals respectively. The spectrum in the 1230 kJ mol^{-1} region consists of two main overlapping systems. This is consistent with the expectation that the 'odd' $(1t_2)^5$ configuration (the $^2T_2$ state of the CH_4^+) is subject to the Jahn-Teller effect (dealt with in Chapter 10) and should lose its triple degeneracy and realise two states, $^2B_2$ and ^2E, if the symmetry of the CH_4^+ is reduced to D_{2d} (a distorted tetrahedron) by the accompanying distortion. The valence electrons in CH_4 are accommodated in two levels of energy rather than the single level implied by the hybridization approach.

Boron, nitrogen and chlorine trifluorides

Most of the principles by which such AB_3 molecules as BF_3, NF_3 and ClF_3, are to be treated have been dealt with in previous sections. For this reason their treatments are presented in a relatively concise fashion with only additional points being stressed. As was the case with the ammonia molecule, the molecules of this section may adopt one of the three symmetries, D_{3h}, C_{3v} or C_{2v}.

Boron trifluoride

The VSEPR treatment of the shape of the BF_3 molecule depends upon the three σ electron pairs repelling each other to a position of minimum repulsion giving the molecule its trigonally planar configuration. The molecule belongs to the D_{3h} point group but experimental observation of the B-F internuclear distance (130 pm) shows it to be smaller than that in the BF_4^- ion (153 pm). Between the same two atoms a shorter distance implies a higher bond order. The VSEPR treatment ignores any possible contribution to the bonding which may be made by the other electrons of the ligand fluorine atoms. It assumes that the fluorine contributions are just the three σ electrons (one electron per fluorine atom).

In the molecular orbital treatment of the D_{3h} BF_3 molecule the 2p$_z$ orbital of the boron atom transforms as an a_2'' representation as does one of the linear combinations of fluorine 2p$_z$ group orbitals (all F–F π bonding). There are three σ bonding orbitals and the

π 1a″ orbital which are fully occupied and account for the bond order of each B-F linkage being $1^1/_3$, and is in accordance with the bond length being smaller than in BF_4^- which has a bond order of 1.0. Although the boron atom only provides three valence electrons, the four bonding orbitals are occupied additionally by three σ electrons and a pair of π electrons from the fluorine atoms. The bonding consists of three σ bonds and a *dative* or *coordinate* π bond. Dative bonds, alternatively called coordinate bonds are electron pair bonds between two atoms in which both electrons originate from one of the participating atoms. In the BF_3 case the two electrons forming the coordinate π bond between the boron and three fluorine atoms are derived from a fluorine group orbital so that the bonding applies equally to all three B–F linkages. This is an example of a four-centre two-electron bond which is *delocalized* over the four atoms of the molecule.

Nitrogen trifluoride, NF_3

The VSEPR theory of NF_3 takes the valence shell configuration of the nitrogen atom; $2s^22p^3$ and adds the three σ-type electrons from the fluorine atoms giving: $2s^22p^6$—four pairs of σ electrons which distribute themselves tetrahedrally around the nitrogen atom. The three identical bonding pairs are squeezed together somewhat by the lone pair to give a trigonally pyramidal molecule with bond angles slightly less than the regular tetrahedral angle. The observed angle is 104°.

If the NF_3 molecule possessed a trigonally planar structure its electronic configuration would be that of the BF_3 molecule with the extra pair of electrons occupying the anti-bonding $2a_2''$ (π) orbital. That would cancel out any bonding effect of the two electrons in the bonding $1a_2''$ (π) orbital and it would not have the extra stability of the π bonding possessed by the BF_3 molecule. For trigonally planar NF_3 the highest energy occupied molecular orbital (HOMO) would be the anti-bonding $2a_2''$ orbital, whereas with the actual C_{3v} trigonally pyramidal symmetry the HOMO is non-bonding.

Chlorine trifluoride, ClF_3

The VSEPR treatment of the ClF_3 molecule is of some interest in that it uses a chlorine 3d orbital to contain one of the electron pairs. The chlorine atom is normally: $3s^23p^5$, and as such is expected to be uni-valent. To arrange for it to be tri-valent one of the paired-up 3p electrons must be excited to a 3d level. Then the three electrons from the ligand fluorine atoms may be added to give the configuration: $3s^23p^63d^2$, making five σ electron pairs, three of which are bonding and two are non-bonding. The basic electron pair distribution is expected to be trigonally bipyramidal. There are three possible shapes for ClF_3 which may result from permuting the three fluorine atoms between the two different positions in the trigonal bipyramid; in the trigonal plane or in a position along the C_3 axis. There is a general rule which is based upon a detailed consideration of the repulsions between bonding-bonding and bonding-non-bonding and between two non-bonding pairs of electrons. This is that (for the case of five electron pairs) non-bonding pairs are more stable in the trigonal plane. Applied to ClF_3 this rule implies that the molecule should be T-shaped. The relatively greater repelling effects of the two in-plane non-bonding pairs cause the angle F–Cl–F to be slightly less than 90°. The observed angle is 87° as is shown in the diagram of Fig. 6.14. The VSEPR treatment gives the impression that there are three σ bonds of equal strength, but this does not seem to be consistent with the observed instability of the ClF_3 molecule.

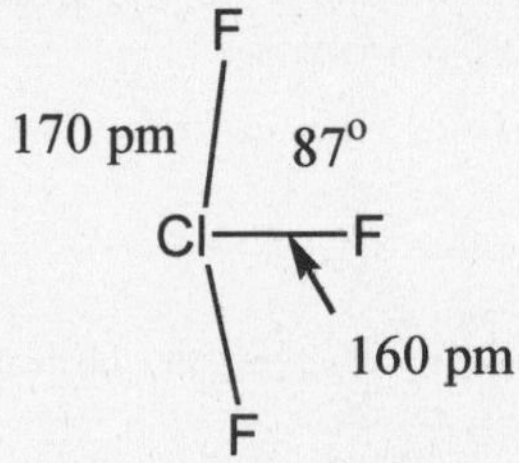

Fig. 6.14 The structure of the ClF_3 molecule

The molecule belongs to the C_{2v} point group. The possibility of ClF_3 trigonally planar is remote since the four electrons extra to those possessed by BF_3 would occupy two anti-bonding orbitals. With the C_{3v} trigonally pyramidal symmetry a σ anti-bonding orbital would be occupied which would rule out that shape for ClF_3. Molecular orbital theory indicates that there are only two bonding pairs of electrons, one participating in a four centre bond and the other in a three-centre arrangement along the F–Cl–F axis. The symmetries of the d orbitals of the chlorine atom allow some stabilization of the two bonding orbitals which is consistent with the assumptions made by the VSEPR approach. With an excess of only two bonding pairs of electrons it is not surprising that the molecule is not very stable.

Ethane, C_2H_6 and diborane, B_2H_6

The ethane molecule, C_2H_6, can be regarded as the dimer of the free radical, CH_3, which has an odd electron. The two odd electrons from two methyl radicals can pair up to make the conventional C-C σ bond to give the covalently saturated ethane molecule. The two methyl groups have rotational freedom with respect to rotation around the C-C axis, although it has been estimated that the difference in energy between the eclipsed and staggered forms of the molecule is 12.6 kJ mol^{-1}, an amount not sufficiently large to offer the possibility of their independent separate existence. The geometry around each carbon atom is expected to be tetrahedral as indicated from the VSEPR theory applied to four valence shell electron pairs.

The diborane molecule, B_2H_6, is the simplest of the boron hydrides and can be considered as the dimer of the BH_3 molecule which has only a transient existence. There are important differences between the diborane molecule and that of ethane. In addition to being a molecule with two 'central' atoms in the VSEPR sense, diborane poses an unusual problem in chemistry. The molecule is described as 'electron deficient' in the sense that there are only twelve valency electrons (three from each B plus one from each H) available for the bonding of the eight atoms. The structure of the molecule is shown in Fig. 6.15. The boron atoms are surrounded by four hydrogen atoms, two of which (the bridging atoms) are shared between them. The coordination of the boron atoms is irregularly tetrahedral with the HBH angles being 97° for the bridging hydrogen atoms, and 121.5° for the terminal angles. The molecule belongs to the D_{2h} point group. The terminal B–H distances are 119 pm and the bridge bonds are all 133 pm long.

The VSEPR theory can be applied to the geometry around either of the boron atoms. The boron atom has the configuration: $2s^22p^1$ and could be made 'trivalent' by the excitation of one of the 2s electrons to an otherwise vacant 2p orbital: $2s^12p^2$. Bearing in mind that the boron is involved in some form of bonding to four hydrogen atoms, the only sensible distribution is that in which the three electrons from the hydrogen atoms (being half the number of 1s(H) electrons available) are placed in the valence shell of the boron atom to give the configuration: $2s^2 2p^2 2p^1 2p^1$ with two singly occupied 2p orbitals.

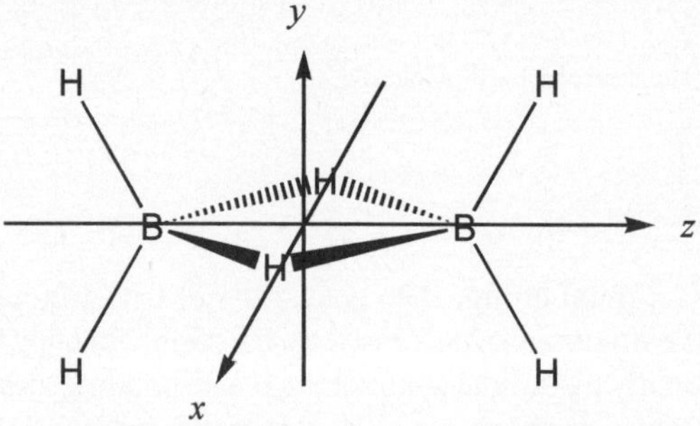

Fig. 6.15 The structure of the diborane molecule

The normal practice of making electron pairs has to be abandoned. The VSEPR argument continues by considering the minimization of the repulsions between the two pairs of electrons and the two unpaired electrons. This gives a distorted tetrahedral distribution with the electron pairs (those involved with the terminal B-H bonding) being further apart than the 'bonds' associated with the single electrons (which participate in the bridging B-H bonding). The two irregular tetrahedra, so formed, join up by sharing the bridging hydrogen atoms to give the observed D_{2h} symmetry of the molecule. Localized bonding could be inferred from the treatment and this leads to the concept of there being four localized one-electron bonds in the bridging between the two boron atoms. The one electron bond in the H_2^+ molecule-ion is reasonably strong (section 3.2). It is a general rule that when ever delocalization can occur it does. This is because the larger orbitals involved allow for a minimization of inter-electronic repulsion. The molecular orbital theory as applied so far would classify the orbitals of the boron atoms, and then those of the hydrogen atoms, with respect to their transformation properties in the D_{2h} point group. This may be done but is a very lengthy and complex procedure. With a relatively complex molecule, such as B_2H_6, there is a simpler way of dealing with the bonding known as the *molecules within molecule* method.

The 'molecules' within the B_2H_6 molecule, rather than two BH_3 groups, are chosen to be the two BH_2 (terminal) groups and a stretched version of the H_2 molecule. As a separate exercise the bonding in a BH_2 group may be dealt with in exactly the same way as was the H_2O molecule (see Fig. 6.4). It belongs to the C_{2v} point group and as such would have the electronic configuration:

$$BH_2\ (C_{2v}):\ (1b_2)^2(2a_1)^2(3a_1)^2$$

with the non-bonding $3a_1$ orbital (it is the $2p_x$(B) orbital) being singly occupied. There are two important differences from the H_2O case. The 2p–2s energy gap in the boron atom is relatively small (350 kJ mol^{-1}) and this allows the 2s(B) and $2p_z$(B) orbitals (both belonging to a_1) to mix. This has the consequence of making the higher energy $3a_1$ orbital non-bonding. The bonding $2a_1$ is of higher energy than the bonding $1b_2$ orbital because of the very large HBH bond angle which favours the latter orbital.

The two BH_2 groups may be set up so that their *z* axes are colinear and coincident with the cartesian z axis as in Fig. 6.16. Their molecular planes are contained within the cartesian yz plane. This ensures that the $1b_1(2p_x(B))$ orbitals of the two BH_2 groups are aligned in the x direction and in the xz plane. The two hydrogen atoms responsible for the bridging are situated along the x axis. The two hydrogen group orbitals are used in the construction of those molecular orbitals responsible for the cohesion of the three molecules in their formation of diborane. This arrangement of orbitals is shown in Fig. 6.17.

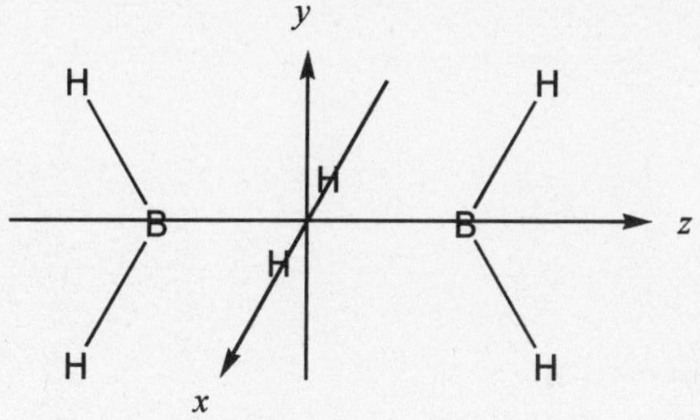

Fig. 6.16 Two BH_2 groups set up in the yz plane with their C_2 axes collinear with the z axis and with two hydrogen atoms symmetrically placed along the x axis

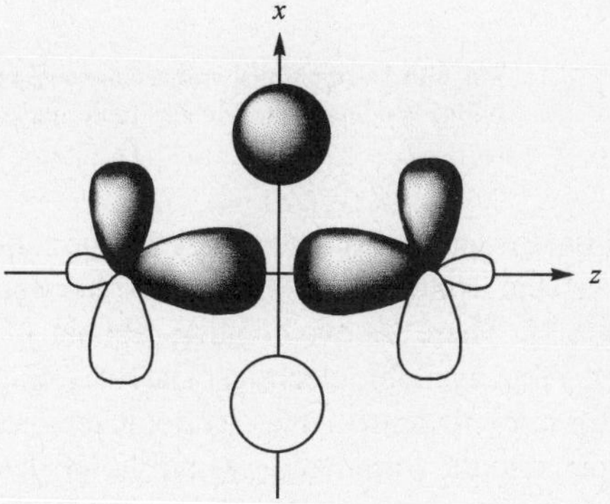

Fig. 6.17 The orbitals of two BH_2 groups and of two hydrogen atoms used in the formation of the molecular orbitals of the bridge bonding in diborane

In order to construct the m.o. diagram for the bridging between the two BH_2 groups it is essential to re-name the participating orbitals within the framework of the D_{2h} point group to which diborane belongs. This may be done by inspecting the D_{2h} character table and gives the following results:

The orbitals of BH_2, $3a_1$ and $1b_1$ form the following linear combinations:

$$\phi(a_g) = \phi(3a_1)_A + \phi(3a_1)_B \tag{6.5}$$

$$\phi(b_{1u}) = \phi(3a_1)_A - \phi(3a_1)_B \tag{6.6}$$

$$\phi(b_{3u}) = \phi(1b_1)_A + \phi(1b_1)_B \tag{6.7}$$

$$\phi(b_{2g}) = \phi(1b_1)_A - \phi(1b_1)_B \tag{6.8}$$

with their new D_{2h} names indicated.

The hydrogen 1s orbitals may be regarded in their usual group forms:

$$\phi(a_g) = \psi(1s)_A + \psi(1s)_B \quad (6.9)$$

$$\phi(b_{3u}) = \psi(1s)_A - \psi(1s)_B \quad (6.10)$$

also with their new D_{2h} names indicated.

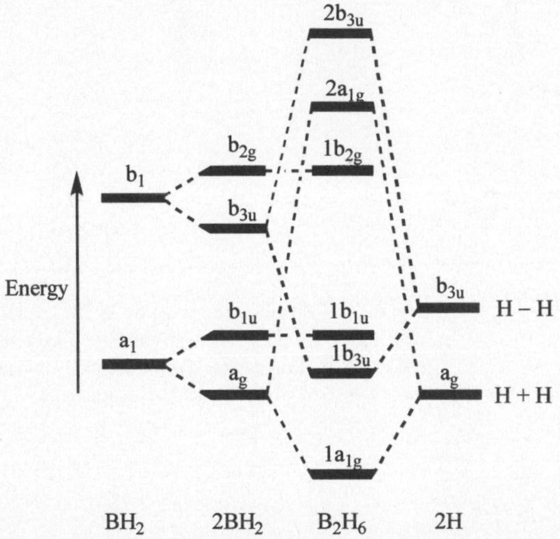

Fig. 6.18 The molecular orbital diagram for the bridge bonding in the diborane molecule

The m.o. diagram for the bridging in diborane may be constructed by the formation of bonding and anti-bonding combinations of the contributing orbitals of the same symmetries. The m.o. diagram is shown in Fig. 6.18. There are two bonding orbitals, of a_g and b_{3u} symmetries respectively, which contain the four available electrons. The two orbitals are four-centre molecular orbitals which allow the minimization of interelectronic repulsion and give the molecule some stabilization compared to the formulation with four localized one-electron bonds.

Dimers of some simple free radicals

The geometry of molecules which contain element-element bonds with other atoms or groups attached to those elements can be regarded as dimers of free radicals. For example, the molecule of hydrogen peroxide, H_2O_2, can be thought of as being the dimer of the hydroxyl free radical, OH. Free radicals usually contain at least one unpaired electron which can be used to form bonds and are usually highly reactive. The OH radical contains a σ bond between the two atoms and is an odd molecule with nine electrons, one 2p electron of the O atom being unpaired. The O−O bond in H_2O_2 is formed by the pairing up of the two 2p electrons to form a σ bond. The shape of the gas phase molecule is as shown in Fig. 6.19 with the dihedral angle between the two planes containing the two OH groups being 95° and the HOO angles being 111°. These angles are not far removed from the 90° values expected for OH bonds constructed from the 2p orbitals of the oxygen atoms. The third 2p orbitals of both the O atoms would contain lone pairs which would minimize their interactions in the observed geometry. The deviations from 90° of the two angles can be rationalized in terms of either some s orbital character or by the minimization of all the repulsions operating in the molecule. The O-O bond length in H_2O_2 is 149 pm and the bond dissociation enthalpy is 146 kJ mol^{-1}, these values being consistent with a somewhat weak single bond (c.f. F_2, bond length 143 pm, bond dissociation enthalpy 158 kJ mol^{-1}).

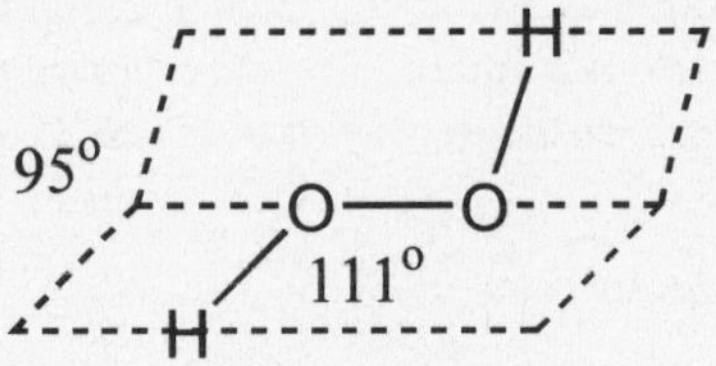

Fig. 6.19 The structure of the hydrogen peroxide molecule

The ethene molecule, C_2H_4, is planar and belongs to the point group, D_{2h}. It can be regarded as the dimer of the free radical CH_2 which has a transient existence and has a bond angle of 130°. It is known to possess two unpaired electrons. The bond angle value is consistent with its electronic structure (derived from Fig. 6.4), $2a_1^2 1b_2^2 3a_1^1 1b_1^1$, and this rationalizes the formation of the dimer in terms of a double bond between the two carbon atoms via one σ bond (overlap of the $3a_1$ orbitals) and one π bond (overlap of the $1b_1$ or $2p_x$ orbitals). The C-C bond has a bond dissociation term of 620 kJ mol^{-1} and a length of 134 pm consistent with the electronic description. The π bonding orbital formed by the interaction of the two 2p carbon orbitals perpendicular to the molecular plane ensures that the molecule is most stable when all six atoms are co-planar. Any rotation of one CH_2 group around the C=C bond would tend to destroy the π bonding, the strength of the π bond representing the barrier to such a rotation.

The NH_2 free radical has only a transient existence and has a bond angle of 103° consistent with its electronic configuration, $2a_1^2 1b_2^2 3a_1^2 1b_1^1$ (one electron fewer than water, bond angle 105°—the removal of a $1b_1$ ($2p_x$) electron having virtually no effect on the bond angle, a result not consistent with that expected from VSEPR theory), and can be considered to dimerize to give the molecule hydrazine, N_2H_4, which has the *gauche* structure shown in Fig. 6.20 in which presumably the repulsions between the hydrogen atoms and the lone pairs on the two nitrogen atoms are minimized. A single N-N σ bond is formed which has a characteristic length of 147 pm and a bond dissociation enthalpy of 276 kJ mol^{-1}.

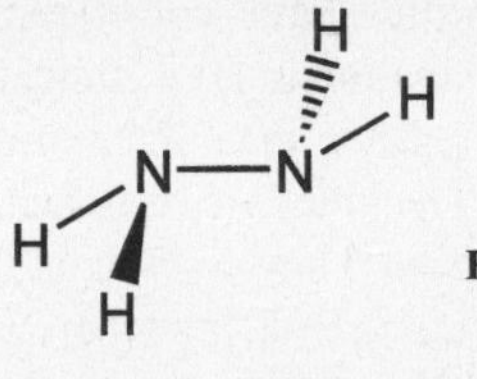

Fig. 6.20 The *gauche* structure of the hydrazine molecule

Some free radicals are reasonably stable, e.g. O_2, NO and NO_2 in that they have a permanent existence under certain circumstances. The odd molecule, NO_2, dimerizes at temperatures below 140° to give the planar dimer, N_2O_4, which has an N-N bond length of 178 pm with a dissociation enthalpy of only 57 kJ mol^{-1} (average values for N-N single bonds are 145 pm and 160 kJ mol^{-1}). The monomer is a brown gas, but dimerization to the colourless and diamagnetic (i.e. the electrons are paired up) N_2O_4 is complete in the solid state (below -11°C). The planarity of the dimeric molecule is possibly associated with some partial π bonding as the 2p orbitals of the six atoms perpendicular to the molecular plane interact to give delocalization over the whole molecule. It is also possible that the geometry of the dimer is determined by the efficiency of packing the molecules in the lattice rather than any electronic consideration. This example serves to indicate the complexities which affect the geometry of ostensibly simple molecules. The dithionite ion, $S_2O_4^{2-}$, which exists in solution and in the solid state, dissociates slightly in aqueous solution to give the monomer ion, SO_2^-. In the solid state the dithionite ion has the C_{2v} structure shown in Fig. 6.21 in which the two SO_2 units are only weakly bonded.

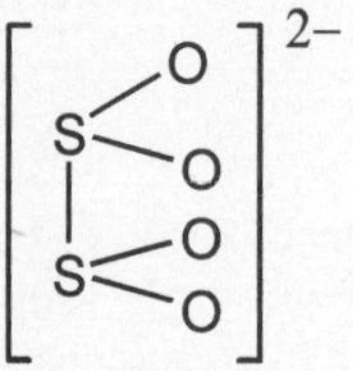

Fig. 6.21 The structure of the dithionite ion

The S–S distance is 239 pm whereas the normal distance for an S–S single bond is 208 pm. In solution there might be free rotation about the S–S bond, but in the solid state lattice forces are probably the main factor determining the geometry of the ion. The S–O bonds have a bond order of 1.5, a valence bond structure would have one of the two bonds to each sulfur atom drawn as a double bond. In molecular orbital theory the π bonding is delocalized over the O–S–O units.

Hydrogen bonding and the hydrogen difluoride ion, HF_2^-

The hydrogen difluoride ion has a central hydrogen atom and belongs to the $D_{\infty h}$ point group. It is important in explaining the weakness of hydrogen fluoride as an acid in aqueous solution, and as an example of hydrogen bonding. Hydrogen bonding occurs usually between molecules which contain a suitable hydrogen atom and one or more of the very electronegative atoms; N, O or F. It is to be considered, in general, as a particularly strong intermolecular force.

In the case of HF_2^- it may be considered that the ion is a combination of an HF molecule and a fluoride ion, F^-. Both of those chemical entities have considerable thermodynamic stability independently of each other. Their combination in HF_2^- indicates the operation of a force of attraction which is significantly greater than the normal intermolecular forces which operate in all systems of molecules. The latter are responsible for the stability of the liquid and solid phases of chemical substances. They counteract the tendency towards chaos. The majority of hydrogen bonding occurs in systems where the hydrogen atom, responsible for the bonding, is asymmetrically placed between the two electronegative atoms (one from each molecule). In those cases the bonding may be regarded as an extra-ordinary interaction between two dipolar molecules. The extraordinary nature of the effect originating in the high effectiveness of the almost unshielded hydrogen nucleus in attracting electrons from where they are concentrated—around the very electronegative atoms of neighbouring molecules. The hydrogen bond in the HF_2^- ion is capable of treatment by bonding theories.

The VSEPR treatment is best approached by considering the ion as being made up from three ions: $F^- + H^+ + F^-$. The central proton possesses no electrons until the ligand fluoride ions supply two each. The two pairs of electrons repel each other to give the observed linear configuration of the three atoms. The two pairs of electrons would occupy the 1s and 2s orbitals of the hydrogen atom and, what with a considerable amount of interelectronic repulsion, would not lead to stability.

The molecular orbital theory is straightforward. The hydrogen 1s orbital transforms as σ_g^+ within the $D_{\infty h}$ point group. The $2p_z$ orbitals of the fluorine atoms (since the ion is set-up with the molecular axis coincident with the cartesian z axis) may be arranged in the linear combinations:

$$\phi(\sigma_g^+) = \psi(2p_z)_A + \psi(2p_z)_A \tag{6.11}$$

and

$$\phi(u_g^+) = \psi(2p_z)_A - \psi(2p_z)_A \tag{6.12}$$

Molecular orbitals may be constructed from the two orbitals of σ_g^+ symmetry to give bonding and anti-bonding combinations. The orbital, localized on the fluorine atoms, of σ_u^+ symmetry remains as a non-bonding orbital. The m.o. diagram is shown in Fig. 6.22. The four electrons, which may be regarded to have been supplied by two fluoride ions, occupy the bonding and non-bonding orbitals and the F-H bond order is 0.5. This is consistent with the observed weakness of the hydrogen bond of 126 kJ mol^{-1} (the $\Delta H^\ominus$ of the reaction of HF_2^- to give $H^+ + HF$) although that figure is a very large amount relative to 'normal' hydrogen bond strengths of 10–30 kJ mol^{-1}.

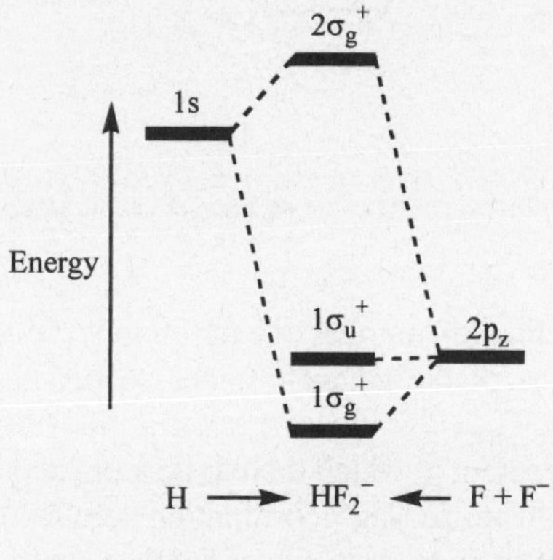

Fig. 6.22 The molecular orbital diagram for the HF_2^- ion

The HF_2^- ion is isoelectronic with the unknown compound, helium difluoride, HeF_2. The latter compound would have a very similar electronic configuration to that of HF_2^-. The reason for its non-existence is indicated by the values of the ionization energies of H^- and He (which are isoelectronic) of 73 and 2372 kJ mol^{-1} respectively. The attraction for electrons represented by the large ionization energy of the helium atom is greater than that exerted by the two fluorine atoms, making the formation of a stable compound impossible. It is of interest to note that the molecule XeF_2 has a stable existence (even though the ionization energy of Xe is 1170 kJ mol^{-1}) although it is very weakly bonded.

6.3 ELEMENTARY FORMS

The majority (~80%) of the elements are solid metals in their standard states at 298 K. Of the metallic elements only mercury is a liquid at 298 K—caesium melts at 302 K, gallium at

302.9 K). The non-metallic elements exist as either discrete small molecules, in the solid (S_8), liquid (Br_2) or gaseous (H_2) states, or as extended atomic arrays in the solid state (C as graphite or diamond), the elements of Group 18 being monatomic gases at 298K.

Metallic character decreases across any period and increases down any group, the non-metals being situated towards the top right hand region of the periodic table. The trends in ionization energies and electronegativity coefficients are of an opposite nature to that of metallic character. A large ionization energy is associated with a large effective nuclear charge and the atoms in this class are those which can participate in the formation of strong covalent bonds. Metal atoms, on the contrary, can only form weak covalent bonds with each other and usually exist in the solid state with lattices in which the coordination number of each atom is relatively high (between eight and fourteen). Diagrams of the three most common lattices are shown in Fig. 6.23.

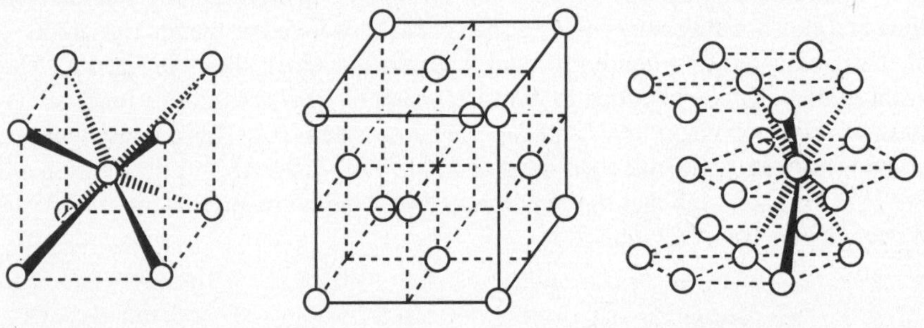

(a) (b) (c)

Fig. 6.23 Three common crystal lattices adopted by elements; (a) body-centred cubic packing, (b) cubic closest packed (or face centred cubic) and (c) hexagonal closest packed

In the body centred cubic lattice (Fig. 6.23a) the coordination number of each atom is eight. There are six next nearest neighbours in the centres of the adjacent cubes so that the coordination number can be regarded as being fourteen. The cubic closest packed lattice (Fig. 6.23b) may be regarded as a face centred cubic arrangement in which there is an atom at the centre of each of the six faces of the basic cubic of eight atoms. The coordination number of any particular atom is twelve, consisting of (considering an atom in one of the face centres of the diagram in Fig. 6.23a) four atoms at the corners of that face and the eight atoms in the centres of adjacent faces of the two cubes sharing the atom in question. At the centre of the cubic arrangement there is an octahedral hole—a possible location for an interstitial atom. In the hexagonally closest packed arrangement (Fig. 6.23c) the coordination consists of six atoms in the same plane as the atom under consideration plus three atoms from both of the adjacent planes, making a total of twelve.

A discussion of the elementary forms of lithium and hydrogen serves to illustrate the factors which determine their metallic or non-metallic nature.

Elementary Lithium

The atom of lithium has the outer electronic configuration, $2s^1$, and might be expected to form a diatomic molecule, Li_2. The molecule, Li_2, does exist in lithium vapour and has a dissociation energy of only 108 kJ mol^{-1}. The enthalpy of atomization of the element is 161 kJ mol^{-1} which is a measurement of the strength of the bonding in the solid state. To produce two moles of lithium atoms (as would be produced by the dissociation of one mole of dilithium molecules) would require 2 x 161 = 322 kJ of energy. From such figures it can be seen that solid metallic lithium is more stable than the dilithium molecular form by (322 – 108)/2 = 107 kJ mol^{-1}. The weakness of the covalent bond in dilithium is understandable in terms of the low effective nuclear charge which allows the 2s orbital to be very diffuse. If the effectiveness of the nuclear charge is low, a 2s electron would not be expected to show much interaction with an electron from another atom. The diffuseness of the 2s orbital of lithium is indicated by the large bond length (267 pm) in the dilithium molecule. The metal exists in the form of a body centred cubic lattice in which the radius of the lithium atoms is 152 pm—again a very high value indicative of the low cohesiveness of the structure.

The high electrical conductivity of lithium (and metals in general) indicates considerable electron mobility. This is consistent with the molecular orbital treatment of an infinite three dimensional array of atoms in which the 2s orbitals are completely delocalized over the system with the formation of a band of *n*/2 bonding orbitals and *n*/2 anti-bonding orbitals for the *n* atoms concerned. Fig. 6.24 shows a diagrammatic representation of the 2s band of lithium metal. The successive levels (molecular orbitals) are so close together as to be regarded as almost a continuum of energy. The formation of a band of delocalized molecular orbitals allows for the minimization of interelectronic repulsion. The small gaps in energy between adjacent levels in a band of molecular orbitals allows a considerable number of the higher ones to be singly occupied. Such single occupation is important in explaining the electrical conduction typical of the metallic state.

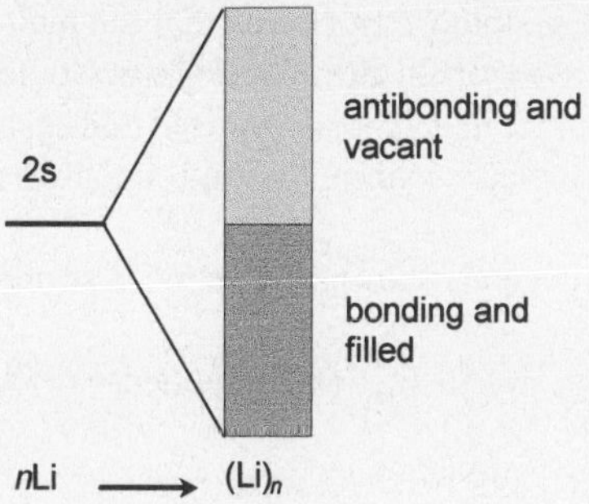

Fig. 6.24 The 2s band of lithium metal; the darker shading implies that the lower half is occupied by electrons, the upper half being vacant

The existence of an incompletely filled infinite band of molecular orbitals leads to one general definition of the metallic state. The highest filled level in a band is known as the Fermi level (equivalent to the HOMO level in small molecules) and when the Fermi level is not at the top of a filled band (as it would be in an insulator or some semi-conductors) it leads to the concept of a Fermi surface for the infinite array of metal atoms. A metal, then, is that state which has a Fermi surface, one where electron promotion to the nearest higher level is very facile.

Elementary Hydrogen

The simplest element exists normally as the dihydrogen molecule, H_2. The high value of the ionization energy of the hydrogen atom is indicative of the electron attracting power of the proton which contributes to the large single bond energy in the dihydrogen molecule. It is instructive to consider the energetics of formation of larger molecules such as H_4.

The hypothetical H_4 molecule could have the possible structures, (i) tetrahedral, (ii) square planar or (iii) linear. The results of some molecular orbital calculations on these molecules are shown in Table 6.6, the nearest neighbour distances being assumed to be 100 pm.

Table 6.6 The calculated energies of some H_4 molecules

Point group	Energy/kJ mol^{-1}		
	Electronic	Nuclear	Total
T_d	-12138	8344	-3794
D_{4h}	-12333	7518	-4815
$D_{\infty h}$	-11524	6017	-5507

The most stable symmetry for H_4 is the linear $D_{\infty h}$ form, the main factor being the low value of the internuclear repulsion energy. The calculated values for the internuclear repulsion, electronic and total energies for the H_2 molecule (r_{eq} = 74 pm) are 1876, -4807 and -2930 kJ mol^{-1} respectively. The reaction:

$$H_4 \text{ (linear, H-H, 100 pm)} \rightarrow 2H_2 \text{ (74 pm)}$$

would have a ΔU° value (change in standard internal energy) of $(2 \times -2930) - (-5507) = -353$ kJ mol^{-1}. There would be a positive change in the entropy of the system which would also contribute to the instability of the H_4 molecule. These and other calculations show that any form of hydrogen which is more than diatomic is less stable than H_2 and that the main reason for this is the nuclear repulsion term. For hydrogen to exist in the metallic form would require tremendous pressure to overcome the internuclear repulsion. It is supposed that such conditions exist in the core of Jupiter and the presence of metallic hydrogen could explain the magnetic properties of that planet.

The non-metallic elements whose cohesiveness depends upon participation in strong covalent bond formation exist in the following forms.

(i) Diatomic molecules which may be singly bonded, e.g. H_2 and F_2, doubly bonded, e.g. O_2 or triply bonded, e.g. N_2.

(ii) Polyatomic small molecules involving single covalent bonds between adjacent atoms, e.g. P_4 and S_8.

(iii) Three-dimensional arrays, e.g. graphite, diamond, boron and silicon. A capacity to be at least three-valent is necessary for a three-dimensional array to be formed.

The individual molecules form crystals in which the cohesive forces are intermolecular, the form of a particular crystal being determined usually by the economy of packing the units together.

Of the elements which participate in catenation (chain formation) in their elementary states only carbon retains the property in its compounds to any great extent. The relatively high strength of the single covalent bond between two carbon atoms gives rise to such a large number, and wide variety, of compounds which form the basis of the branch of the subject known as organic chemistry. Boron and silicon chemistry contain some examples of compounds in which catenation is exhibited.

6.4 THE METALLIC BOND

The band of molecular orbitals formed by the 2s orbitals of the lithium atoms, described above, is half filled by the available electrons. Metallic beryllium, with twice the number of electrons, might be expected to have a full '2s band.' If that were so the material would not exist since the 'anti-bonding' half of the band would be fully occupied and the element would not have a Fermi surface. Metallic beryllium exists because the band of molecular orbitals produced from the 2p atomic orbitals overlaps (in terms of energy) the 2s band. This makes possible the partial filling of both the 2s and the 2p bands, giving metallic beryllium a greater cohesiveness and a larger electrical conductivity than lithium. The overlapping of the 2s and 2p bands of beryllium, and their partial occupancies, are shown diagrammatically in Fig. 6.25.

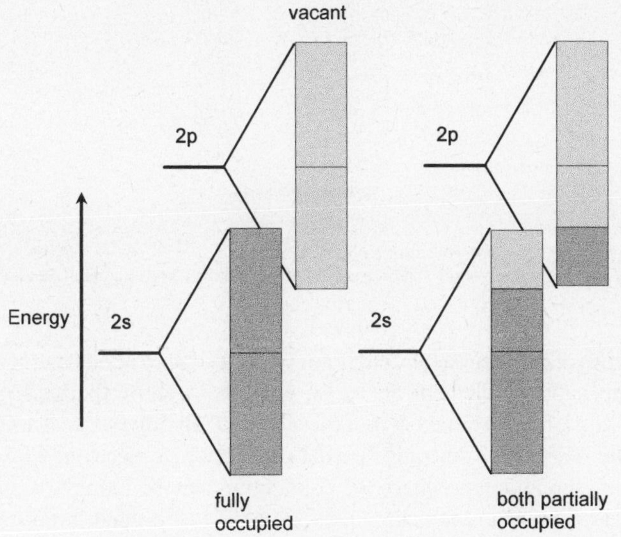

Fig. 6.25 The 2s and 2p bands of beryllium metal before (left-hand side) and after (right-hand side) overlap is allowed to occur; darker shading indicates filled levels and lighter shading vacant levels

If overlap between bands does not occur, the size of the band gap (between the lowest level of the vacant band and the highest level of the filled band) determines whether the element exhibits insulator properties (large band gap), or is a semi-conductor (small band gap). The band gaps in the diamond form of carbon and in silicon (which has the diamond, four coordinate structure) are 521 kJ mol^{-1} and 106 kJ mol^{-1} respectively. Diamond is an insulator but silicon is a semi-conductor.

Electrical conduction in a semi-conductor may occur because of (i) thermal excitation of electrons from the *highest filled band* (HFB) to the *lowest vacant band* (LVB), (ii) photo-excitation from the HFB to the LVB, or (iii) by the presence of an impurity element. If such an element is introduced deliberately this is known as doping. Ultra-pure silicon is an *intrinsic* semi-conductor but would be of no practical use because the band gap is too high. Most commercial semi-conductors are based on doped silicon, examples of extrinsic semi-conductors. The difference between an insulator and the various types of semi-conductors is shown in Fig. 6.26.

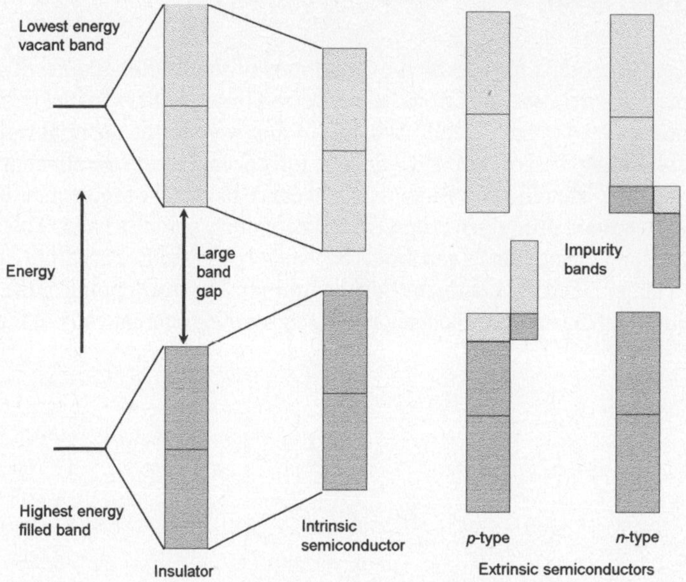

Fig. 6.26 A representation of band gaps in insulators and intrinsic semiconductors, and of the production of semiconduction in *p*-type and *n*-type materials; horizontal shading indicates filled levels and diagonal shading vacant levels

There are two main types of extrinsic semi-conductor which are based upon whether a suitable energy levels of the impurity element are close enough to allow thermal or photo-excitation to occur from the HFB or to the LVB of the silicon. If an otherwise *vacant* impurity band overlaps with the HFB of the silicon its partial occupancy by electrons from the silicon allows for *p*-type semi-conduction, where the conduction may be thought of in terms of the movement of positive holes in the main conduction band. If a filled impurity band overlaps with the LVB of the silicon and causes the latter to be partially populated this gives rise to *n*-type (normal or negative) semi-conduction by electrons in the previously vacant upper band of the silicon. Even if there is no overlap in either of the cases, semi-conduction may occur by either thermal or photo-excitation. Thermal semi-conduction depends upon the extent to which a conduction band is populated, that being governed by the magnitude of the Boltzmann factor:

$$e^{-\left(\frac{bandgap}{kT}\right)}$$

Such conductivity increases as the temperature increases. Photoconduction is initiated by photon-induced transitions of electrons from the lower energy populated band to the normally vacant upper energy band.

With regard to the above considerations, silicon doped with arsenic behaves as an *n*-type semiconductor—the extra electrons from the arsenic populating the otherwise vacant conduction band of the silicon. Silicon doped with boron behaves as a *p*-type semiconductor because some of the electrons from the otherwise filled band of the silicon enter the lower levels of the boron band.

Metallic conduction exhibits a reduction in conductivity with an increase in temperature, in spite of the population of more of the band with unpaired electrons, there being a very small difference between adjacent levels in any one band. This inverse relationship is thought to be due to higher temperatures causing greater amplitudes of atomic vibrations or even discontinuities in the metal lattice (less extensive delocalization of the electrons) which would lead to an increase in resistance at their boundaries. At very low temperatures (such as that of liquid helium, 4K) some metals, alloys and compounds become superconductive. This means that their resistance to the passage of an electrical current becomes zero.

Recently, interest has developed into the superconductive properties of a range of mixed oxides. The latest developments have resulted in the production of a compound of the composition, $HgBa_2Ca_2Cu_3O_{8+\delta}$, which is superconductive at temperatures up to 133K. The δ in the formula represents the incorporation of some oxygen atoms in sites in the lattice which are only partially filled. Further progress towards the production of materials which show superconductivity at even higher temperatures will depend upon further experimentation, and the development of a theory with predictive value. It is possible that under suitable circumstances (interionic distances and temperature) population of an otherwise vacant '3s' band of the oxide ions allows superconductivity to occur. The subjection of the superconducting compounds to high pressures (e.g. 15 GPa) increases the superconductive temperature limit to as much as 153K.

Yellow gold and liquid mercury

The colour of gold adds to the attractiveness of the metal and the liquid state of mercury allows the metal to be used over a wide range of temperatures in thermometers and switches. The relativistic effects on the 6s orbital are at a maximum in gold and are considerable in mercury.

In gold the 6s band is stabilized making the Fermi level lower than expected and the 5d level is destabilized and split by spin-orbit coupling. These effects combine to make the electronic transitions from the higher d band into the 'Fermi band' have energies that correspond to the blue/violet region of the visible spectrum so that the metal reflects the red/yellow frequencies. The majority of metals have transitions into their 'Fermi bands' which lie in the ultra-violet region of the spectrum and consequently reflect white light with very little selective absorption to produce sensations of colour.

The relativistic stabilization of the 6s band in mercury with respect to the 6p band reduces the extent of overlap between the two bands so that the mercury atoms behave almost as though they had a pseudo-inert gas configuration $5d^{10}6s^2$. The poor metallic cohesiveness between mercury atoms makes covalency a more probable bonding mode and leads to the existence of the unusual Hg_2^{2+} dication.

6.5 PROBLEMS

1. Give a reasoned explanation of the following observations.

(a) The molecules $N(CH_3)_3$ and $N(SiH_3)_3$ have different shapes and different bond angles at the nitrogen atom.
(b) The molecules NH_3 and NF_3 have dipole moments which are of opposed direction.
(c) The shape of gaseous PCl_5 is different from the two P environments found in the solid compound.
(d) The NO_2^- and NO_2^+ have different shapes and bond angles. Discuss their bonding and compare them with the neutral molecule, NO_2.

2. Comment on the bond angles (in degrees) given for the following listed compounds.

Compound				
POF_3	FPF	101.3	FPO	117.7
PSF_3	FPF	99.6	FPS	122.7
NSF_3	FSF	94.0	FSN	125
F_2SO_2	FSF	96.1	OSO	124.0
Cl_2SO_2	ClSCl	100.3	OSO	123.5
Me_2SO_2	CSC	102.6	OSO	119.7

7

Ions in Solids and Solutions

Ionic compounds have structures which are related to those of metals and crystal lattices are discussed from that standpoint. The energetics of ionic bond formation are considered and the thermodynamics of ion stabilities and reactivities as indicated by values of standard reduction potentials, E°, are also discussed.

7.1 THE IONIC BOND

The interaction between any two atomic orbitals is at a maximum if the energy difference between them is zero and that they have appropriate symmetry properties. A difference in energy between the two participating orbitals causes the bonding orbital to have a greater than 50% contribution from the lower of the two atomic orbitals. Likewise the anti-bonding orbital contains a greater than 50% contribution from the higher of the two atomic orbitals. In a molecule such as HF the bonding orbital has a major contribution from the fluorine 2p orbital and in consequence the two bonding electrons are unequally shared between the two nuclei which results in the molecule being dipolar. In HF the energies of the participating orbitals are indicated by the first ionization energies of the atoms H (1312 kJ mol^{-1}) and F (1681 kJ mol^{-1}). The valence electron of the sodium atom has an ionization energy of only 496 kJ mol^{-1} and a diatomic molecule, NaF, would have a difference of 1681 - 496 – 1185 kJ mol^{-1} between the valence electrons 3s(Na) and 2p(F). That difference is so great as to make the bonding level virtually the same as the 2p(F) orbital so that if the bond were formed it would be equivalent to the transfer of an electron from the sodium atom to the fluorine atom causing the production of the two ions, Na^+ and F^-.

The standard enthalpy change for the reaction:

$$Na(s) + \tfrac{1}{2}F_2(g) \rightarrow Na^+(g) + F^-(g) \qquad (7.1)$$

is estimated as the sum of the enthalpy changes of the following four component reactions.

(i) $Na(s) \rightarrow Na^+(g)$, the atomization of sodium, $\Delta_{atom}H^{\circ}(Na) = 108$ kJ mol^{-1}

(ii) $\tfrac{1}{2}F_2(g) \rightarrow F(g)$, the atomization of fluorine, $\tfrac{1}{2}D(F_2,g)$, 79 kJ mol^{-1}, where $D(F_2,g)$ is the dissociation energy of the fluorine molecule.

(iii) $Na(g) \rightarrow Na^+(g)$, the ionization of sodium, $I_1(Na)$, 496 kJ mol^{-1}, where I_1 is the first ionization energy of the sodium atom.

(iv) $F(g) \rightarrow F^-(g)$, the attachment of an electron to the fluorine atom, $E_{ea}(F)$, −333 kJ mol^{-1}, where $E_{ea}(F)$ is the electron attachment energy or the negative value of the electron affinity of the fluorine atom. (The *electron affinity* of an atom is defined as the energy *released* when an atom accepts an electron to become a negative ion, i.e. 333 kJ mol^{-1} for fluorine. The *electron attachment energy* is the enthalpy change for the same process, i.e. −333 kJ mol^{-1} for fluorine.)

The standard enthalpy change for equation (7.1) is given by:

$$\Delta H^\circ(7.1) = \Delta_{atom}H^\circ(Na) + \tfrac{1}{2}D(F_2,g) + I_1(Na) + E_{ea}(F)$$

$$= 108 + \tfrac{1}{2} \times 158 + 496 - 333 = 350 \text{ kJ mol}^{-1}.$$

The reaction is considerably endothermic and would not be feasible. The above calculation does not take into account the electrostatic attraction between the oppositely charged ions, nor the repulsive force which operates at small interionic distances. In the crystal of NaF the smallest distance apart of the sodium and fluoride ions is 231 pm and Coulomb's law may be used to calculate the energy of stabilization due to electrostatic attraction between such ion pairs:

$$E(Na^+F^-) = -\frac{N_A e^2}{4\pi\epsilon_0 r} \qquad (7.2)$$

where N_A is the Avogadro constant, e the electronic charge, ϵ_0 the vacuum permittivity and r the interionic distance. Putting the values for the terms in equation (7.2) gives a value of $E(Na^+F^-)$ of −601 kJ mol^{-1}.

The standard enthalpy of formation of the substance Na^+F^- with discrete ion pairs, but with no interaction between pairs, is then calculated to be:

$$\Delta_f H^\circ(Na^+F^-,g) = 350 - 601 = -251 \text{ kJ mol}^{-1}$$

The actual substance formed when sodium metal reacts with difluorine is solid sodium fluoride and the observed standard enthalpy of its formation is −569 kJ mol^{-1}. The actual substance is 318 kJ mol^{-1} more stable than the hypothetical substance consisting of ion pairs, $Na^+F^-(g)$ described above. The added stability of the observed solid compound arises from the long range interactions of all the positive Na^+ ions and negative F^- ions in the solid lattice which forms the structure of sodium fluoride. The ionic arrangement is shown in Fig. 7.1. Each Na^+ ion is octahedrally surrounded (coordinated) by six fluoride ions, and the fluoride ions are similarly coordinated by six sodium ions.

The overall stability of the NaF lattice is represented by the resultant of the many stabilizing attractions (Na^+−F^-) and destabilizing repulsions (Na^+−Na^+ and F^-−F^-) which amount to a stabilization which is 1.74756 times that of the interaction between the individual

Na^+-F^- ion pairs. The factor, 1.74756, is the so-called Madelung constant, *M*, for the particular lattice arrangement and arises from the forces experienced by each ion. These are composed of six attractions at a distance *r*, twelve repulsions at a distance $2^{1/2}r$ eight attractions at a distance $3^{1/2}r$ six repulsions at a distance of $4^{1/2}r$ twenty four attractions at a distance $5^{1/2}r$ and so on to infinity.

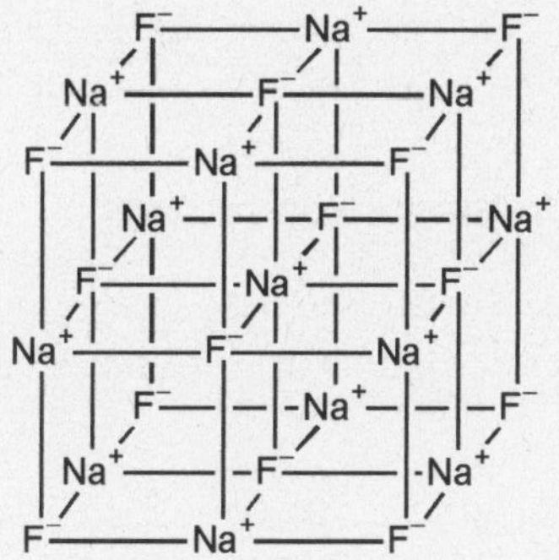

Fig. 7.1 The structure of the sodium fluoride crystal

The series: $6 - 12/2^{1/2} + 8/3^{1/2} - 6/4^{1/2} + 24/5^{1/2} - ....$ eventually becomes convergent and gives the value for the Madelung constant for the sodium chloride type of lattice (that which is adopted by sodium fluoride). The values of Madelung constants for some common crystal lattices are given in Table 7.1.

Table 7.1 Values of some Madelung constants

Lattice type	Madelung constant
NaCl	1.74756
CsCl	1.76267
CaF_2 (fluorite)	2.51939
TiO_2 (rutile)	2.408
ZnS (blende, sphalerite)	1.63806
ZnS (wurtzite)	1.64132
Al_2O_3 (corundum)	4.17186

The electrostatic contribution to the *lattice energy*, *L*, for the sodium fluoride arrangement—the energy required to convert a solid lattice into its constituent ions in the gas phase—is the negative value of the standard enthalpy change for the reaction:

$$Na^+(g) + F^-(g) \rightarrow Na^+F^-(s) \tag{7.3}$$

is given by the equation:

$$L = \frac{MN_A e^2}{4\pi\epsilon_0 r} \tag{7.4}$$

where M is the Madelung constant, N_A is the Avogadro number, ϵ_0 is the vacuum permittivity and r is the distance of closest approach between the oppositely charged ions.

In addition to the electrostatic forces there is a repulsive force which operates at short distances between ions as a result of the overlapping of filled orbitals (this is potentially a violation of the Pauli exclusion principle). This repulsive force may be represented by the equation:

$$E_{\mathrm{rep}} = \frac{B}{r^n} \tag{7.5}$$

B being a proportionality factor. The full expression for the lattice energy becomes:

$$L = \frac{MN_{\mathrm{A}}e^2}{4\pi\epsilon_0 r} - \frac{B}{r^n} \tag{7.6}$$

When r is the equilibrium interionic distance r_{eq}, the differential: $\mathrm{d}L/\mathrm{d}r$ has a value of zero, so that:

$$\frac{\mathrm{d}L}{\mathrm{d}r} = 0 = -\frac{MN_{\mathrm{A}}e^2}{4\pi\epsilon_0 r_{\mathrm{eq}}^2} + \frac{nB}{r_{\mathrm{eq}}^{n+1}} \tag{7.7}$$

which gives for B:

$$B = \frac{MN_{\mathrm{A}}e^2 r^{n-1}}{4\pi\epsilon_0 n} \tag{7.8}$$

and substituting this value into equation (7.6) with $r = r_{\mathrm{eq}}$ gives:

$$L = \frac{MN_{\mathrm{A}}e^2}{4\pi\epsilon_0 r_{\mathrm{eq}}}\left[1 - \frac{1}{n}\right] \tag{7.9}$$

which is known as the Born-Landé equation. The equation may be applied to any lattice provided that the appropriate value of the Madelung constant is used. The *positive value* of the product of the ionic charges, z^+z^-, should be included in the equation to take into account any charges which differ from ± 1, the final equation being:

$$L = \frac{MN_{\mathrm{A}}z^+z^-e^2}{4\pi\epsilon_0 r_{\mathrm{eq}}}\left[1 - \frac{1}{n}\right] \tag{7.10}$$

The values of the Born exponent, n, for various crystal structures are estimated from compressibility data. The values recommended for use with various ion configurations are shown in Table 7.2.

Table 7.2 Some values of the Born exponent, n

Ion configuration	Value of n
He	5
Ne	7
Ar, Cu^+	9
Kr, Ag^+	10
Xe, Au^+	12

For a crystal with mixed ion-types the average value of n should be used in the calculation of lattice energy. Using the value $n = 7$ for sodium fluoride with r_{eq} equal to 231 pm gives a value of L of 901 kJ mol^{-1}.

It is now possible to produce a theoretical value for the standard enthalpy of formation of sodium fluoride based upon the ionic model described above. The necessary equation is produced by adding together equations (7.1) and (7.3) to give:

$$Na(s) + \tfrac{1}{2}F_2 \rightarrow Na^+F^-(s) \qquad (7.11)$$

and the calculated value for $\Delta_f H^\ominus$ (Na^+F^-) is $350 - 901 = -551$ kJ mol^{-1} which is fairly close (~97%) to the observed value of -569 kJ mol^{-1} and gives considerable respectability to the theory. Other evidence for the ionic nature of sodium fluoride is that the molten substance is a good conductor of electricity as are aqueous solutions of the compound. The ionic substances which are soluble in water dissociate to give their component hydrated ions which cause the solution to exhibit good electrical (i.e. electrolytic) conductance.

Born-Haber thermochemical cycles can be used to summarize the various steps which are used to calculate the standard enthalpy of formation of a compound. The cycle for the formation of sodium fluoride is shown in Fig. 7.2.

$$Na(g) + F(g) \xrightarrow{\Delta H^\ominus = I_1(Na) + E_{ea}(F)} Na^+(g) + F^-(g)$$

$$\Delta_a H^\ominus(F) = \tfrac{1}{2}D(F_2, g)$$

$$\Delta_{lat} H^\ominus = -L(Na^+F^-)$$

$$\Delta_a H^\ominus(Na) = S(Na)$$

$$Na(s) + \tfrac{1}{2}F_2(g) \xrightarrow{\Delta_f H^\ominus(Na^+F^-)} Na^+F^-(s)$$

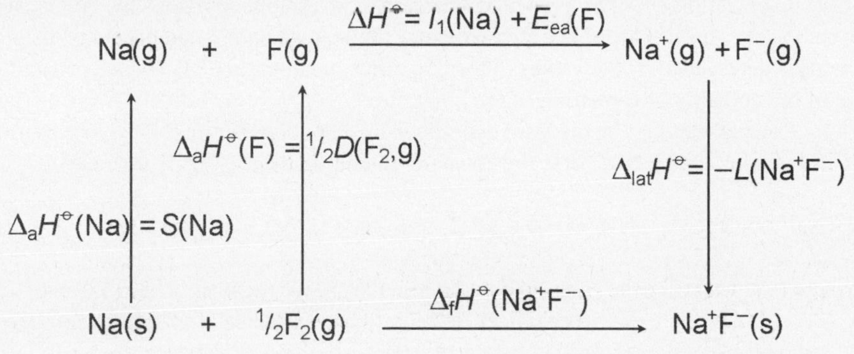

Fig. 7.2 A Born-Haber thermochemical cycle for the formation of sodium fluoride from its elements in their standard states

The basis of the cycle is the first law of thermodynamics, one statement of which is that the enthalpy changes which occur in the change from state *A* (e.g. the elements in their standard states) to state *B* (e.g. the solid ionic compound produced) are independent of the path by which *A* reaches *B*. It is an alternative to Hess's law of heat summation as used in the derivations of some of the above equations, but both methods have the same basis and give good results—providing due attention is given to the signs of the quantities used.

In general terms the feasibility of ionic bond production depends upon the $\Delta_f H^\circ$ value for the compound, MX_m, the entropy of formation being of minor importance. The value of $\Delta_f H^\circ$ may be estimated from known thermodynamic quantities and the calculated value for the lattice energy:

$$\Delta_f H^\circ = \Delta_{atom} H^\circ(M) + (n/2)\Delta_{atom} H^\circ(X_2) + \Sigma I(M) + nE_{ea}(X) - L(MX) \qquad (7.12)$$

the enthalpy of atomization of the metal, $\Delta_{atom} H^\circ(M)$, sometimes being called the *sublimation energy*, and that of the non-metal, $\Delta_{atom} H^\circ(X_2)$, being a combination of the energy required to produce the gas phase molecule plus the dissociation energy of the molecule. The ionization energy term is the appropriate sum of the first *n* ionization energies to produce the required *n*-positive ion. The *n* electrons are used to produce either *n* singly negative ions or *n*/2 doubly negative ions and the $E_{ea}(X)$ term has to be modified accordingly. The $L(MX)$ value is calculated for the appropriate ionic arrangement (if known, otherwise appropriate guesses have to be made).

The first three terms of equation (7.12) are positive and can only contribute to the feasibility of ionic bond production by being minimized. The last two terms are negative for uni-negative ions (E_{ea} represents energy released) and for ionic bond feasibility should be maximized. From such considerations it is clear that ionic bond formation will be satisfactory for very electropositive and easily atomized metals with very electronegative and easily atomized non-metals.

The lattice energy term may be increased, for ions with charges, z^+ and z^-, by the positive value of the product of the charges, z^+z^-. To increase the cation charge involves the expense of an increase in $\Sigma I(M)$ and, although there is some pay-back in the production of more anions (a larger E_{ea}) there are severe limits of such activity. This is because the successive ionization energies of an atom increase more rapidly than does the lattice energy with increasing cation charge. The stoichiometry of the ionic compound produced from two elements is determined by the values of the five terms of equation (7.12) which minimize the value of the enthalpy of formation of the compound.

Calculations may be made for hypothetical compounds such as NaF_2. Assuming that the compound contains Na^{2+} ions, the standard enthalpy of formation is given by:

$$\Delta_f H^\circ(NaF_2,\text{ionic}) = \Delta_{atom} H^\circ(Na) + \Delta_{atom} H^\circ(F_2) + I_1(Na) + I_2(Na) + 2E_{ea}(X) - L(NaF_2) \qquad (7.13)$$

The numerical values of the terms in equation (7.13) are respectively: 108, 158, 496, 4562, 2×-333 and (assuming the hypothetical Na^{2+} ion to have a radius of 65 pm, as does the real Mg^{2+} ion, and using a Madelung constant of 2.381, that appropriate for the MgF_2 lattice) 2821 kJ mol^{-1}, which gives a value of +1837 kJ mol^{-1} for the standard enthalpy of formation of ionic NaF_2 — a compound which would not be expected to exist. The reason for that is the very large value of the second ionization energy of the sodium atom, the electron being removed from one of the very stable 2p orbitals.

Exercise Born-Haber thermochemical cycles are a useful aid to understanding problems of energetics of reactions. Construct a cycle for the formation of NaF_2.

The relation of ionic structures to the packing of anion/cation lattices

An instructive way of thinking about ionic structures is to consider that the anions (usually the larger of the two participating ions in a binary compound) are arranged in either cubic or hexagonal closest packing. Such arrangements of spherical objects may be studied at the orange seller's stall! With a hard sphere model in mind the interstices (or holes) in, for instance, the cubic closest packed lattice would have a size dependent upon the size of the spheres from which it was composed. Fig. 7.3 shows a diagram of two layers of closest packed spheres which shows that there are two kinds of holes. One is bounded by four spheres and is termed a *tetrahedral hole*, the other being bounded by six spheres and is termed an *octahedral hole*. One tetrahedral hole in Fig. 7.3 is that directly beneath the white atom—that atom and the three atoms below it form the tetrahedron surrounding the hole. Another tetrahedral hole is that beneath the black atom. An octahedral hole exists in the centre of the octahedron formed by three atoms in one layer and another three in the next layer as indicated in Fig. 7.4. The radius, r^+, of an entering sphere which is to fit exactly into either a tetrahedral or an octahedral hole is dependent upon the radius, r^-, of the cubic closest packed spheres.

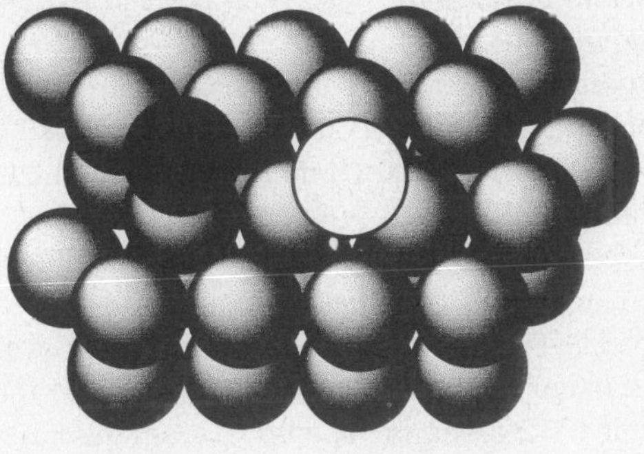

Fig. 7.3 Two layers of closest packed atoms (viewed from above) with two ions of possible third layers shown; the white atom is directly above a similar atom in the bottom layer, the black atom is in an alternative position—not directly above an ion in the bottom layer

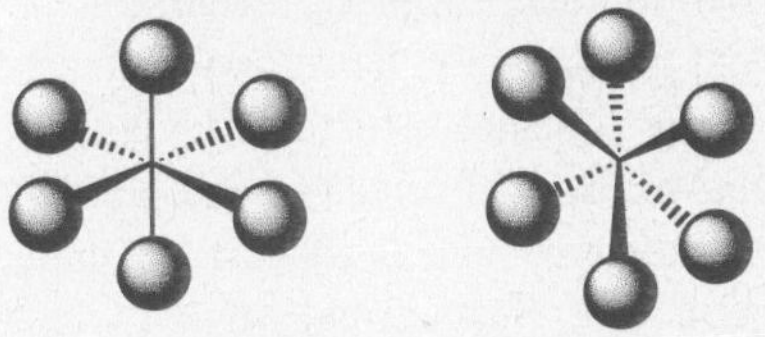

Fig. 7.4 The left-hand diagram is the conventional illustration of an octahedral arrangement of six atoms, the octahedral hole being at the origin of the steric projections. The right-hand diagram is produced by rotating the left-hand diagram through 45° to show the upper three atoms as part of one layer of an infinite array and the lower three atoms as part of the next lower layer

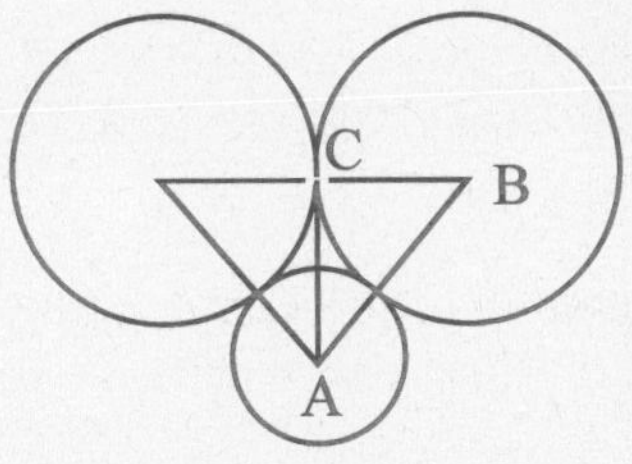

The calculations for the two types of hole are illustrated by the diagram in Fig. 7.5.

Fig. 7.5 The two large circles represent two adjacent atoms in a closest packed structure, the small circle represents the largest atom which can fit into a hole in the structure. The distance B–C is the radius of the larger atom and the distance A–B is the sum of the radii of the two sizes of atom.

The diagram shows the small (cation) sphere in contact with two large (anion) spheres with the two large spheres in contact with each other. The right-angled triangle, ABC, is constructed so that the side, BC, is equal to the anion radius, r^-, and the side, AB, is equal to the sum of the two radii, $r^+ + r^-$. For a tetrahedral hole the anion-cation-anion angle is the tetrahedral angle of 109°28', so the angle BAC is 109°28'/2 = 54°44'. For an octahedral hole angle BAC is 90°/2 = 45°. The sines of these angles are given by the equation:

$$\sin(\text{BAC}) = \frac{r^-}{r^- + r^+} \tag{7.14}$$

Since the triangle, ABC, in the case of the octahedral hole contains a right-angle the theorem of Pythagoras may be used as an alternative to equation (7.14) to achieve the desired result. In that case the equation to be solved is: $2(r^-)^2 = (r^+ + r^-)^2$.

The solutions of the equation (7.14) for the two angles show that for a tetrahedral hole the ratio, r^+/r^-, is equal to 0.225, and for an octahedral hole has the value, 0.414. The calculations indicate that an octahedral hole could accommodate an ion which was around 84% larger in radius than one which would fit into a tetrahedral hole.

Ions behave only very roughly like hard spheres but the calculations above do indicate that only relatively very small ions would be expected to be able to occupy tetrahedral holes. The ratio of the sodium/fluoride ionic radii is 95/136 = 0.698, but the sodium fluoride lattice is that which would be produced if the octahedral holes in a cubic close packed fluoride lattice were filled by sodium ions.

The crystal structure which is produced when the cation/anion radius ratio approaches unity is that adopted by CsCl. The structure is based on a simple cubic packing of anions with the cations in the body-centre positions. The structure is that of Fig. 6.23a with eight anions at the cube corners and a cation in the body-centre.

Fig. 7.3 offers an insight into the difference between the cubic and hexagonal closest packing arrangements which are adopted by many metals. The first layer of closest packed spheres in the diagram is overlain by the second layer in the manner shown. The third layer of spheres may be arranged so that the position indicated by the white atom is occupied The atoms in the third layer, in this case, would be directly above corresponding atoms in the first layer. This ABABAB.... arrangement is that of the hexagonally closest packed lattice (see Fig. 6.23c for a diagram of three layers). If, alternatively, the third layer makes use of the position indicated by the black atom in Fig. 7.3 the arrangement is that of the cubic closest packed lattice, ABCABC.... a conventional view of which is shown in Fig. 6.23b.

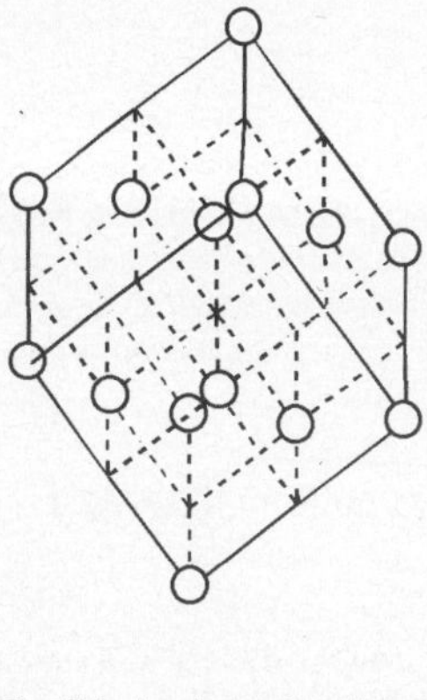

Fig. 7.6 An alternative view of the diagram shown in Fig. 6.23b in which the members of the four ABCA layers are made more obvious

The relationship between the two views of the cubic closest packing, as described by Figs 7.3 and 6.23b, is hopefully made clear by the view in Fig. 7.6 which consists of a C_8 rotation (45°) of the diagram of Fig. 6.23b around an axis perpendicular to the paper. It is more obvious from the diagram of Fig. 7.5 that the atoms belong to four layers of the kind exhibited by Fig. 7.3 of the ABCA... type. It should be realized that, atom for atom, the second and third layers are not directly above each other.

Consideration of an extended version of Fig. 7.3, leads to the conclusion that, for a host lattice of N atoms arranged in either cubic or hexagonal closest packing, there are N octahedral holes and $2N$ tetrahedral holes. If all the octahedral holes in a ccp host lattice of chloride ions are occupied by sodium ions, the sodium chloride structure is produced. Variations on this theme allow the rationalization of many crystal structures, some of which are summarized in Table 7.3.

Table 7.3 Hole occupancy in some crystal structures

Holes occupied	ccp Host	hcp Host
All octahedral	NaCl	NiAs
Half octahedral-alternate layers	$CdCl_2$	CdI_2
Half tetrahedral	ZnS (blende)	ZnS (wurtzite)
All tetrahedral	CaF_2 (fluorite)	Li_2O (anti-fluorite)

In the nickel arsenide structure the arsenic atoms may be considered to have the hcp arrangement (as in Fig. 6.23c ABAB...), the nickel atoms occupying all the octahedral holes. The nickel atoms form a trigonally prismatic coordination around each arsenic atom as shown in Fig. 7.7. In NiAs the Ni–Ni distance of closest approach is 253 pm which is very close to twice the value of the metallic radius of the nickel atom (248 pm). It is likely, therefore, that there is considerable Ni–Ni metal-metal bonding in the compound. Nickel and arsenic do not differ greatly in electronegativity and their combination would not be expected to be ionic. Binary transition metal sulfides adopt the NiAs structure because of the stabilizing facility of metal-metal bonding. They are covalently bound semiconductors.

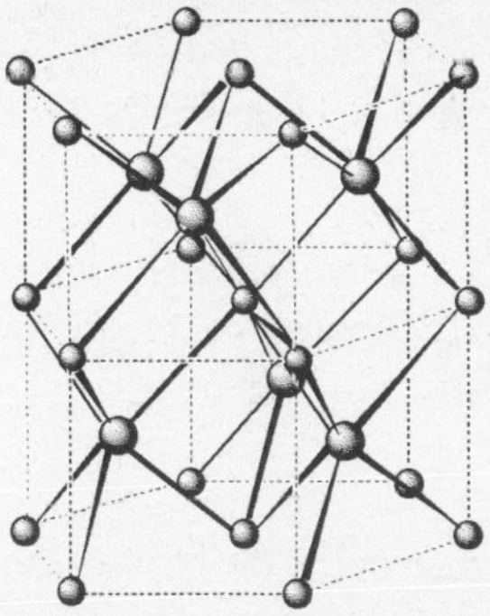

Fig. 7.7 The nickel arsenide structure; the small nickel atoms are in the hcp arrangement with the arsenic atoms occupying the octahedral holes

The half-filling of the octahedral holes in a host lattice leads to compounds with AB_2 stoichiometry. In the examples quoted in Table 7.3, the two layer-lattices ($CdCl_2$ and CdI_2) are formed when the half-filling is achieved by filling the octahedral holes in alternate layers in the ccp and hcp host halide lattices respectively.

In the zinc blende (sphalerite) form of zinc sulfide, ZnS, half of the tetrahedral holes in a sulfide cubic close packed lattice are occupied by zinc ions (Fig. 7.8, left-hand diagram). A similar filling of an hcp host sulfide lattice produces the wurtzite form of zinc sulfide (Fig. 7.8, right-hand diagram).

In calcium fluoride (fluorite), the calcium ions may be considered to have a cubic close packed arrangement with all the tetrahedral holes filled by the fluoride ions to give the CaF_2 stoichiometry (Fig. 7.9). A ccp host lattice of oxide ions with all the tetrahedral holes occupied by lithium ions produces the anti-fluorite lattice arrangement.

The above approach is very useful in the discussion of the structures of so-called interstitial compounds. For example, the majority of the transition metal hydrides may be thought of as slightly expanded metallic lattices with some or all of the tetrahedral and octahedral interstices occupied by hydrogen atoms. Thus all stoichiometries from the metal, M, to MH_3, are understandable. In general such hydrides are 'non-stoichiometric' in that the hydrogen/metal ratios are usually non-integral, their formulae depending upon a balance between the number of holes filled with hydrogen atoms and the higher entropies associated with non-stoichiometry.

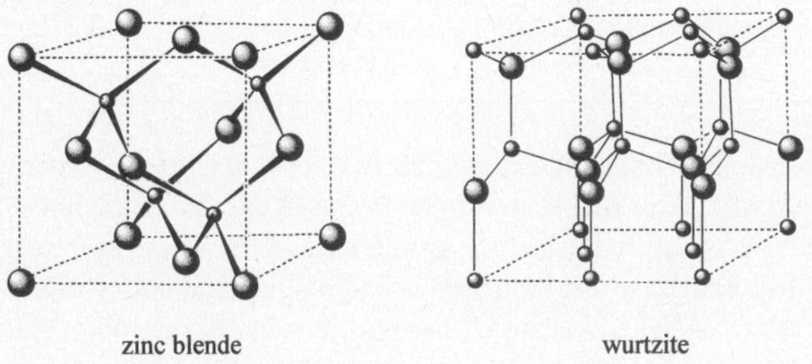

Fig. 7.8 The structures of the zinc blende (sphalerite) and wurtzite forms of zinc sulfide

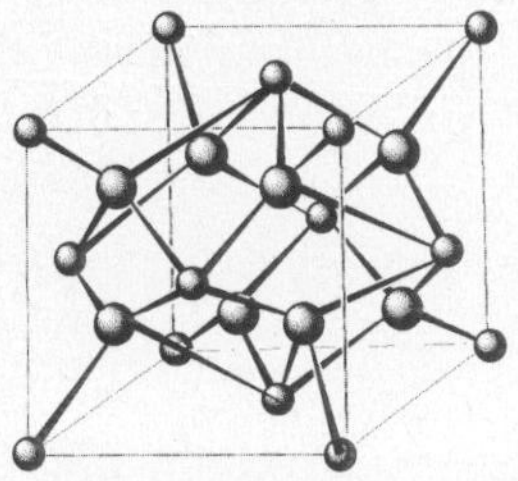

Fig. 7.9 The structure of fluorite, calcium fluoride

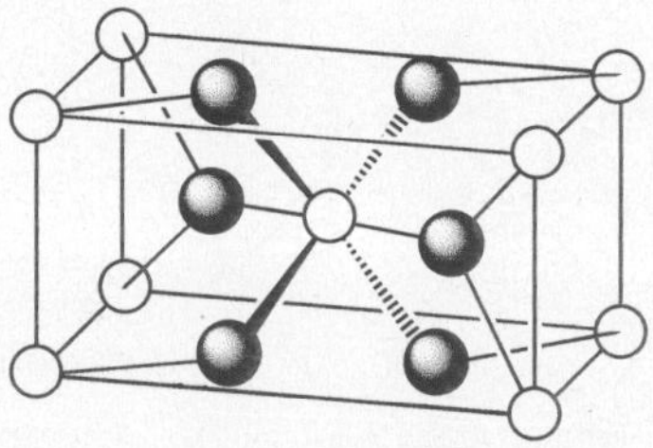

Fig 7.10 The structure of rutile

If half of the nickel atoms in the NiAs structure are removed systematically, the basis of the rutile structure is produced. This structure is shown in Fig. 7.10. Rutile is one form of TiO_2 in which the Ti atoms have a coordination number of six, the Ti atom coordination being octahedral. The oxygen atoms are surrounded by three Ti atoms in a triangular coplanar fashion. The structure is best regarded as chains of TiO_6 octahedra in which each octahedron shares opposite edges and where the chains are linked together by sharing vertices to give a 3D arrangement. The rutile structure is adopted theoretically by MX_2 compounds where the r^+/r^- ratio is in the range 0.414–0.732 and in practice by the transition metal fluorides and many oxides of the formulae MO_2.

Although this section has referred mainly to ionic compounds the lattices described apply also to crystalline compounds which are principally covalent.

The relationship between lattice diagrams and compound stoichiometry

There are simple rules which allow the stoichiometry of a compound to be derived from its lattice diagram. Fig. 7.1 serves to demonstrate the application of the rules. The diagram of the NaF structure in Fig. 7.1 contains representations of thirteen sodium ions and fourteen fluoride ions. Only one ion - a sodium ion - is wholly contained within the cubic arrangement. There are twelve sodium ions half way along each cube edge. In the actual crystalline material the ions along such edges will each be shared by four cubes so that each ion on an edge counts only as a quarter for purposes of determining the stoichiometry of the compound. Therefore, the twelve edge-type sodium ions of Fig. 7.1 count as 12/4 = three whole sodium ions. There is a total of four whole sodium ions represented by the diagram of Fig. 7.1. There are six fluoride ions in the cube faces which count as three whole ions, since each face is shared by two cubes. The eight fluoride ions at the cube corners count as a single whole ion since any ion at a corner is shared by eight cubes. There is a total of four whole fluoride ions represented by the Fig. 7.1 diagram. Thus the stoichiometry of NaF as given by applying these rules to the lattice diagram is 4:4 = 1:1, so all is well! Similar rules apply to non-cubic systems.

7.2 IONS IN AQUEOUS SOLUTION

This section deals with the factors which influence the chemical form and stability of ions in solution with respect to their formation and with respect to their reactivities with either water or other ions. Stability with respect to formation is discussed in terms of $\Delta G^\ominus$ and $E^\ominus$, the appropriate standard reduction potential. The reactivity of ions is discussed in terms of $E^\ominus$ only.

Thermodynamics and electrode potentials

The basis of the discussion of thermodynamic stability and reactivity of ions depends upon the equation:

$$\Delta G^\ominus = -RT \ln K \tag{7.15}$$

which relates the value of $\Delta G^\ominus$, the change in standard Gibbs energy, to that of the equilibrium constant, K, for any process. It is convenient to express the change in standard Gibbs energy, for any reaction, as the difference between the two standard reduction potentials, for the two 'half-reactions' which, when added together, produce the overall equation. For example, the overall equation:

$$Cu^{2+}(aq) + Zn(s) \rightleftharpoons Cu(s) + Zn^{2+}(aq) \tag{7.16}$$

may be considered to be the sum of the two half-reactions:

$$Cu^{2+}(aq) + 2e^- \rightleftharpoons Cu(s) \tag{7.17}$$

and

$$Zn(s) \rightleftharpoons Zn^{2+}(aq) + 2e^- \tag{7.18}$$

where the 'state' of the electrons is not defined —this is deliberate and immaterial because they cancel out when the two half-reactions are summed.

The equation which expresses the relationship between the standard electrode potential, E°, and the change in standard Gibbs energy change for any reaction is:

$$\Delta G^\circ = -nFE^\circ \tag{7.19}$$

where n is the number of electrons involved in the process, and F is the Faraday constant - the conversion factor with the value: 96485 C mol^{-1}, (one Volt–Coulomb = one Joule).

The value of ΔG° for reaction (7.16) is −212.3 kJ mol^{-1} corresponding to the observed maximum voltage for the Daniell cell (copper and zinc electrodes in contact with solutions of their +2 sulfates) of 1.1 V. The standard electrode potential for the reaction may be considered to be made up from the *difference* between the values of the standard *reduction* potentials of the two half-reactions, (7.17) and (7.18):

$$E^\circ = E^\circ(Cu^{2+}/Cu) - E^\circ(Zn^{2+}/Zn) \tag{7.20}$$

the appropriate reductions being indicated for the two half-reaction potentials.

Consider the meaning of equation (7.20) in terms of ΔG° values. The value of ΔG° for the reduction of Cu^{2+} to Cu (7.17) is:

$$\Delta G^\circ\ (Cu^{2+}/Cu) = -2F\,E^\circ(Cu^{2+}/Cu) \tag{7.21}$$

and the value for the reduction of Zn^{2+} to Zn (the reverse of equation (7.18)) is:

$$\Delta G^\circ\ (Zn^{2+}/Zn) = -2F\,E^\circ(Zn^{2+}/Zn) \tag{7.22}$$

the values of $n = 2$ being included because the half-reactions are two-electron reductions.

Equation (7.16) may be constructed from its constituent half-reactions (7.17) and (7.18), the appropriate value for its ΔG° being the sum of the ΔG° values for the half-reactions:

$$\begin{aligned}\Delta G^\circ(7.16) &= \Delta G^\circ(7.17) + \Delta G^\circ(7.18)\\ &= -2F\,E^\circ(Cu^{2+}/Cu) + 2F\,E^\circ(Zn^{2+}/Zn)\\ &= -2F\,(E^\circ(Cu^{2+}/Cu) - E^\circ(Zn^{2+}/Zn))\end{aligned} \tag{7.23}$$

It is most important to notice that the **sign** of $E^\circ(Zn^{2+}/Zn)$ in equation (7.23) is the reverse of that in equation (7.22) since equation (7.23) is relevant to the oxidation of Zn to Zn^{2+}.

The standard potential for the overall reaction (7.16) is calculated from equation (7.23) (applying equation (7.19)) as:

$$-\Delta G^\circ(7.16)/2F = (E^\circ(Cu^{2+}/Cu) - E^\circ(Zn^{2+}/Zn) \tag{7.24}$$

—the difference between the two standard reduction potentials.

Since neither of the two half-reaction potentials is known it is necessary to propose an arbitrary zero relative to which all half-reaction potentials may be quoted. The half-reaction chosen to represent the arbitrary zero is the reduction of aqueous hydrogen ion to gaseous dihydrogen:

$$H^+(aq) + e^- \rightleftharpoons \tfrac{1}{2}H_2(g) \qquad (7.25)$$

where the state of the electron is not defined. The standard reduction potential for the reaction is taken to be zero: $E^\circ(H^+/H_2) = 0$. This applies strictly to the standard conditions where the activity of the hydrogen ion is unity and the pressure of the hydrogen gas is 1 atmosphere at 298 K. It is conventional to quote *reduction* potentials for the general half-reaction:

$$\text{Oxidized form} + ne^- \rightleftharpoons \text{Reduced form} \qquad (7.26)$$

The E° value for the half-reaction:

$$H^+(aq) + e^-(aq) \rightleftharpoons \tfrac{1}{2}H_2(g) \qquad (7.27)$$

with the state of the electron being defined as aqueous can be estimated to be 2.6 V.

The hydrated electron, $e^-(aq)$, is a well characterized chemical entity produced in aqueous solutions by irradiation with γ-rays or by an electron beam from an accelerator. It reacts with water according to the equation:

$$e^-(aq) + H_2O(l) \rightarrow H(aq) + OH^-(aq) \qquad (7.28)$$

The pseudo-first order rate constant for reaction (7.28) has a value of 16 s^{-1} at 298 K, the second order rate constant for the reverse reaction being 2×10^7 L mol^{-1} s^{-1}. The second-order rate constant for reaction (7.28) is obtained by dividing the pseudo-first order constant by the molarity of liquid water (55.5 *M*) giving a value of 0.288 L mol^{-1} s^{-1}. The equilibrium constant for the reaction is the ratio of the two second order rate constants which is 1.44×10^{-8} for the reaction written as (7.28). This corresponds to a ΔG° value of 44.7 kJ mol^{-1}. The equilibrium may be combined with that governing the self-ionization of water:

$$H^+(aq) + OH^-(aq) \rightleftharpoons H_2O(l) \qquad (7.29)$$

which has the value, 1×10^{14} at 298 K, ($\Delta G^\circ = -79.7$ kJ mol^{-1}) and the change in standard Gibbs energy for the formation of dihydrogen from hydrogen atoms of −218 kJ mol^{-1} (of H atoms) to give a value of $\Delta G^\circ = -253$ kJ mol^{-1} for the reaction (7.27) which is equivalent to an E° value of 2.62 V, indicative of the great reducing power of the hydrated electron.

If reaction (7.27) is reversed and added to reaction (7.25) the result is:

$$e^- \rightleftharpoons e^-(aq) \qquad (7.30)$$

the appropriate value of E° being −2.6 V, so allowing the hydrated electron to be compared to any other reducing agent on the arbitrary scale.

The accepted values of the standard reduction potentials for the Cu^{2+}/Cu and Zn^{2+}/Zn couples are 0.34 V and −0.76 V respectively. The standard potential for reaction (7.16)is thus:

$$E^{\ominus}(7.16) = 0.34 - (-0.76) = 1.1\ \text{V}$$

The overall standard potential for the reaction:

$$2Cr^{2+}(aq) + Cu^{2+}(aq) \rightarrow Cu(s) + 2Cr^{3+}(aq) \tag{7.31}$$

is calculated from the standard reduction potentials for the constituent half-reactions, i.e. equation (7.17), and:

$$Cr^{3+}(aq) + e^- \rightleftharpoons Cr^{2+}(aq) \qquad E^{\ominus} = -0.41\ \text{V} \tag{7.32}$$

as:

$$E^{\ominus}(7.31) = 0.34 - (-0.41) = 0.85\ \text{V}$$

the chromium half-reaction potential having its sign reversed because the half-reaction itself is reversed to make up the overall equation (7.31). Notice that the 2:1 stoichiometry of equation (7.31) has no effect upon the calculation of the overall $E^{\ominus}$ value. This is because $E^{\ominus}$ values are values of the changes in standard Gibbs energy *per electron*, and since the number of electrons cancels out in an overall reaction it has no influence upon the calculation. The overall change in the standard Gibbs energy for reaction (7.31) is given by:

$$\Delta G^{\ominus} = -nF\,E^{\ominus} = -2F \times 0.85\ \text{V} = -164\ \text{kJ mol}^{-1}$$

the negative value indicating that the reaction is thermodynamically feasible. Thermodynamic feasibility for any process is indicated by either a negative $\Delta G^{\ominus}$ value or a positive $E^{\ominus}$ value.

It is often necessary to calculate the $E^{\ominus}$ value of a half-reaction from two or more values for other half-reactions. In such cases the above considerations do not apply and care must be taken to include the number of electrons associated with each half-reaction. For example, the two half-reactions:

$$Cr^{VI}(aq) + 3e^- \rightleftharpoons Cr^{3+}(aq) \tag{7.33}$$

and equation (7.32) have $E^{\ominus}$ values of 1.38 V and −0.41 V respectively. When added together they give the half-reaction:

$$Cr^{VI}(aq) + 4e^- \rightleftharpoons Cr^{2+}(aq) \tag{7.34}$$

The calculation of $E^{\ominus}(7.34)$ via the value of $\Delta G^{\ominus}$ illustrates why the answer is not the arithmetic sum of the two half-reaction potentials.

$$\begin{aligned}\Delta G^{\ominus}(7.34) &= \Delta G^{\ominus}(7.33) + \Delta G^{\ominus}(7.18)\\ &= -3F\,E^{\ominus}(7.33) - F\,E^{\ominus}(7.18)\\ &= -F\,(3 \times E^{\ominus}(7.33) + E^{\ominus}(7.18))\end{aligned} \tag{7.35}$$

The value of $E^{\ominus}(7.34)$ is given by:

$$E^{\ominus}(7.34) = -\Delta G^{\ominus}(7.34)/4F = (3 \times E^{\ominus}(7.33) - E^{\ominus}(7.18))/4 = (3 \times 1.38 - 0.41)/4 = 0.93\ \text{V}$$

The final $E^{\ominus}$ value is the sum of the individual $E^{\ominus}$ values appropriately weighted by their respective numbers of electrons, the sum then being divided by the total number of electrons in the final equation. This is because the electrons do not cancel out in such cases.

On the hydrogen scale the range of values for standard reduction potentials varies from about 3.0 to −3.0 V for solutions in which the activity of the hydrogen ion is 1 *M*. It is also conventional to quote values for standard reduction potentials in solutions with unit activity of hydroxide ion; $E_B^{\ominus}$. Typical values range from about 2.0 to −3.0 V.

Examples of half-reactions at each end of the $E^{\ominus}$ and $E_B^{\ominus}$ ranges are:

$$\tfrac{1}{2}F_2(g) + e^- \rightleftharpoons F^-(aq), \qquad E^{\ominus} = 2.65\ \text{V}$$

$$Li^+(aq) + e^- \rightleftharpoons Li(s), \qquad E^{\ominus} = -3.04\ \text{V}$$

$$O_3(g) + H_2O(l) + 2e^- \rightleftharpoons O_2 + 2OH^-(aq), \qquad E_B^{\ominus} = 1.24\ \text{V}$$

$$Ca(OH)_2(s) + 2e^- \rightleftharpoons Ca(s) + 2OH^-(aq), \qquad E_B^{\ominus} = -3.03\ \text{V}$$

The following generalizations are helpful in the interpretation of $E^{\ominus}$ values.
1. A highly negative value of $E^{\ominus}$ implies that the reduced form of the couple is a good reducing agent.
2. A large positive value of $E^{\ominus}$ implies that the oxidized form of the couple is a good oxidizing agent.

Forms of ions in aqueous solution

The form of any ion in aqueous solution depends largely upon the size and charge (oxidation state) of the central atom. These two properties influence the extent of any interaction with the
solvent molecules. A large singly charged ion is likely to be simply hydrated — surrounded by a small number (4 to 10) of water molecules some of which may be formally bonded to it (as in the case of a transition metal ion). The interaction of gaseous ions with water to give their hydrated versions causes the liberation of enthalpy of hydration, $\Delta_{hyd}H^{\ominus}$, the negative value, $-\Delta_{hyd}H^{\ominus}$, being quoted as the hydration energy (a positive quantity).

Forms of ions other than the simply hydrated ones may be understood in terms of the increasing interaction between the central ion and its hydration sphere as its oxidation state increases and its size decreases. The first stage of hydrolysis may be written as:

$$M(H_2O)^{n+} \rightleftharpoons MOH^{(n-1)+} + H^+ \qquad (7.36)$$

implying that a metal−oxygen bond is formed with the release of a proton into the solution. As the value of *n* (the oxidation state of M) increases the next stage of hydrolysis would be expected to occur:

$$M(H_2O)OH^{(n-1)+} \rightleftharpoons M(OH)_2^{(n-2)+} + H^+ \qquad (7.37)$$

There is then the possibility that water could be eliminated from the two OH^- groups:

$$M(OH)_2^{(n-2)+} \rightleftharpoons MO^{(n-2)+} + H_2O \tag{7.38}$$

Further hydrolyses involving more pairs of water molecules would yield the oxo-ions, $MO^{(n-4)+}$, $MO^{(n-6)+}$ and $MO^{(n-8)+}$.

It may be envisaged that water molecules could be eliminated between two hydrolysed ions to give a dimeric product:

$$2MOH^{(n-1)+} \rightleftharpoons MOM^{2(n-1)+} + H_2O \tag{7.39}$$

Further hydrolysis of the dimeric product could yield ions such as: $O_3MOMO_3^{2(n-7)+}$. In some cases further condensations occur to give polymeric ions.

Examples of these forms of ions are to be found in the transition elements and in those main group elements which can exist in higher oxidation states. As may be inferred from equations (7.36) and (7.37) alkaline conditions encourage hydrolysis, so that the form an ion takes is dependent upon the acidity of the solution. Highly acid conditions tend to depress the tendency of an ion to undergo hydrolysis. Table 7.4 contains some examples of ions of different form, the form depending upon the oxidation state of the central element.

Table 7.4 Some ionic forms of vanadium in solution

Oxidation state of V	Form in acid solution	Form in basic solution
II	V^{2+}	VO (insoluble)
III	V^{3+}	V_2O_3 (insoluble)
IV	VO^{2+}	$V_2O_5^{2-}$
V	VO_2^+	VO_4^{3-}

In acidic solution the (II) and (III) oxidation states are simple hydrated ions, the (IV) and (V) states being oxocations. In basic solution the (II) and (III) states form neutral insoluble oxides (the lattice energies of the oxides giving even more stability than any soluble form in these cases) and the (IV) and (V) states exist as oxyanions (dimeric in the (IV) case).

Simple cations such as $Fe(H_2O)_6^{3+}$ undergo a certain amount of primary hydroysis depending upon the pH of the solution. The ion is in the hexaaqua-form only at pH values lower than 2.0. Above that value the hydroxypentaaquairon(III) ion is prevalent. Further increase in the pH of the solution causes more hydroysis until a complex solid material (sometimes described erroneously as iron(III) hydroxide, $Fe(OH)_3$ is precipitated. The solid contains iron(III) oxohydroxide (FeOOH) and iron(III) oxide in various states of hydration, $Fe_2O_3.xH_2O$.

With some ions the extent of polymer formation is high, even in acidic solution, e.g. vanadium(V) can exist as a polyvanadate ion, $V_{10}O_{28}^{6-}$.

The above ideas are not limited to species with central metal ions. They apply to the higher oxidation states of non-metallic elements. Many simple anions do exist with primary solvation spheres in which the positive ends of dipoles are attracted to the central negative charge. Examples of ions which may be thought about in terms of the hydrolysis of the parent hydrated ions are: S^{IV}: SO_3^{2-}, S^{VI}: SO_4^{2-} and $S_2O_7^{2-}$, Cl^I: ClO^-, Cl^{III}: ClO_2^-, Cl^V: ClO_3^-, Cl^{VII}:

ClO_4^-, and the various polymeric silicate(IV) ions.

Factors affecting the magnitude of $E^\ominus$

This section contains considerations of the trends in $E^\ominus$ values for:
(i) the reduction of Group 1 unipositive cations to the metal, and a comparison with the reduction of silver(I) ion to silver(0).
(ii) the reduction of Group 17 elements to their uninegative anions, and
(iii) the reduction of Na^+, Mg^{2+} and Al^{3+} cations to their metals.

Group 1 – trend in $E^\ominus$ down a typical group

The accepted values for the standard reduction potentials for the Group 1 unipositive cations being reduced to the solid metal are given in Table 7.5, together with some relevant thermochemical data and the appropriate ionic radii. The units for all the energies are kJ mol^{-1}.
The $E^\ominus$ values imply that all the elements are very powerful reducing agents (M^+ being difficult to reduce to M) and that lithium is the most powerful with sodium being the least powerful. The reducing powers of the elements K, Rb and Cs are very similar and intermediate between those of Li and Na.

Table 7.5 Data for Group 1 elements and silver, M, and their cations, M^+

	Li	Na	K	Rb	Cs	Ag
$E^\ominus(M^+/M)/V$	-3.04	-2.71	-2.92	-2.92	-2.92	0.8
$\Delta_{atom}H^\ominus(M)$	161	108	90	86	79	289
$I_1(M)$	519	496	418	402	376	732
$\Delta_{hyd}H^\ominus(M^+)$	-523	-419	-331	-314	-285	-464
$r_{ionic}(M^+)/pm$	76	102	138	152	167	115

The calculations which follow are carried out in terms of enthalpy changes with entropy changes being ignored. This is because the entropy changes are (a) difficult to estimate, (b) expected to be fairly similar for the reactions of the elements of the same periodic group, and (c) relatively small compared to the enthalpy changes.

The standard enthalpy change for the reduction of $M^+(aq)$ to M(s) may be estimated from the above data by applying Hess's Law (first law of thermodynamics) to the reactions:

$$M^+(aq) \rightarrow M^+(g)$$
$$M^+(g) + e^- \rightarrow M(g)$$
$$M(g) \rightarrow M(s)$$

their sum being:

$$M^+(aq) + e^- \rightarrow M(s) \qquad (7.40)$$

The general expression for the standard enthalpy change for reaction (7.40) is:

$$\Delta H^{\circ}(7.40) = -\Delta_{\text{hyd}}H^{\circ}(\text{M}^{+}) - I_1(\text{M}) - \Delta_{\text{atom}}H^{\circ}(\text{M}) \qquad (7.41)$$

The reduction enthalpy as calculated from equation (7.41) may be compared with that calculated for the reference reduction of the hydrated proton to molecular dihydrogen. The latter enthalpy change is calculated from the ΔH° values for the reactions:

$$\text{H}^{+}(\text{aq}) \rightarrow \text{H}^{+}(\text{g})$$
$$\text{H}^{+}(\text{g}) + \text{e}^{-} \rightarrow \text{H}(\text{g})$$
$$\text{H}(\text{g}) \rightarrow \tfrac{1}{2}\text{H}_2(\text{g})$$

their sum being:

$$\text{H}^{+}(\text{aq}) + \text{e}^{-} \rightarrow \tfrac{1}{2}\text{H}_2(\text{g}) \qquad (7.42)$$

The standard enthalpy change for reaction (7.42) is given by:

$$\begin{aligned}\Delta H^{\circ}(7.42) &= -\Delta_{\text{hyd}}H^{\circ}(\text{H}^{+}) - I_1(\text{H}) - \Delta_{\text{atom}}H^{\circ}(\text{H}) \\ &= 1091 - 1312 - 218 = -439 \text{ kJ mol}^{-1}\end{aligned} \qquad (7.43)$$

The reduction enthalpies for the Group 1 cations are obtained by subtracting -439 kJ mol^{-1} from the results of substituting the appropriate data into equation (7.41), and are given in Table 7.6, together with their conversion into volts (division by $-F$). The overall equation to which the reduction enthalpies are relevant is that obtained by adding the reverse of reaction (7.42) to reaction (7.40):

$$\text{M}^{+}(\text{aq}) + \tfrac{1}{2}\text{H}_2(\text{g}) \rightleftharpoons \text{H}^{+}(\text{aq}) + \text{M}(\text{s}) \qquad (7.44)$$

Table 7.6 Calculated reduction enthalpies and E° values for the Group 1 cations and for the Ag^{+} cation

Cation/element	**Reduction enthalpy /kJ mol^{-1}**	**E°/V**
Li^{+}/Li	282	−2.92
Na^{+}/Na	254	−2.63
K^{+}/K	262	−2.72
Rb^{+}/Rb	265	−2.75
Cs^{+}/Cs	269	−2.79
Ag^{+}/Ag	−118	1.22

The calculated values for the Group 1 elements are similar to those observed, the small differences being attributable to the ignored entropy terms. The calculation of reduction enthalpy shows that the value is the resultant, relatively small quantity, of the interaction between two large quantities; the ionization energy of the gaseous metal atom and the hydration energy of the cation, the heat of atomization being a relatively small contribution. The trends in the values of the three contributing quantities are understandable in terms of the changes in size of the ions (hydration enthalpy decreases with increasing ionic radius, see data in Table 7.5), the electronic configuration of the atoms (ionization energy decreases as Z increases) and the strength of the metallic bonding (bonding becomes weaker as atomic size increases).

The aqueous silver(I) ion, with its $3d^{10}$ outer electronic configuration, is more easily reduced to its metallic state than are the unipositive ions of the elements of Group 1, the Ag^+/Ag standard reduction potential being 0.8 V. The calculated value for this quantity, using the above approach is 1.22 V, the discrepancy between theory and calculation being somewhat greater than those of the Group 1 cases. The calculated value arises from the greater standard enthalpy of atomization of silver due to the stronger metallic bonding (silver metal bonding is enhanced by overlap of the 3d and 4s bands), and a higher first ionization energy. The hydration enthalpy of the silver(I) ion is a relatively minor factor in determining the differences between the above potentials.

Group 17

The Group 17 elements are the most electronegative of their respective periods and the reductions of the elements to their uninegative ions are thermodynamically feasible as can be seen from the values for E° given in Table 7.7. The table contains the data for the calculation of the appropriate reduction enthalpies, the results of the calculations (in kJ mol^{-1}) and the calculated E° values.

Table 7.7 Data and calculated E° values for the reduction potentials of the Group 17 elements/uninegative anions, X_2/X^- (enthalpies in kJ mol^{-1})

	F	Cl	Br	I
E°(observed)/V	2.87	1.36	1.07	0.54
$\Delta_{atom}H^\circ(X_2)$	158	244	224	212
$E_{ea}(X)$	-333	-348	-324	-295
$\Delta_{hyd}H^\circ(X^-)$	-490	-356	-310	-255
$r_{ion}(X^-)$/pm	133	181	196	220
Reduction enthalpy	-305	-143	-83	-5
E°(calculated)/V	3.16	1.48	0.86	0.05

The appropriate equation for the calculation of the enthalpy change for the reaction:

$$\tfrac{1}{2}X_2(g, l \text{ or } s) + \tfrac{1}{2}H_2(g) \rightleftharpoons H^+(aq) + X^-(aq)$$

is:

$$\text{Reduction enthalpy} = \tfrac{1}{2}\Delta_{atom}H^\circ(X_2) + E_{ea}(X) + \Delta_{hyd}H^\circ(X^-) + 439 \qquad (7.45)$$

where $E_{ea}(X)$ is the electron attachment enthalpy (the energy released upon ion formation) of the gaseous X atom, the 439 amount being the 'oxidation enthalpy' for the standard reference reaction (the reverse of reaction (7.42)).

Exercise. Check the calculations for the reduction enthalpies and E° values.

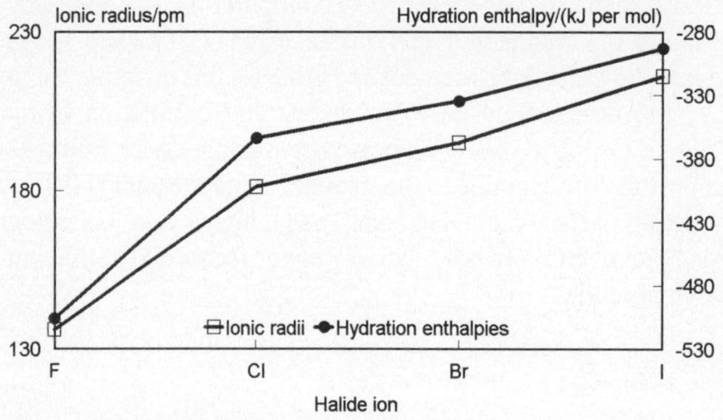

Fig. 7.11 Variations in the ionic radii and hydration enthalpies of the halide ions

The discrepancies between the observed and calculated values for $E^{\ominus}$ are attributable to the missing entropy term, but are not large. Fig. 7.11 shows the trends in the hydration energies and in the ionic radii, the former depending on the latter.

$E^{\ominus}$ values of sodium, magnesium and aluminium

As the third short period is traversed the reducing powers of the elements decrease. This is shown for the elements Na, Mg and Al, appropriate data being presented in Table 7.8. The reduction enthalpies are calculated as in the above cases with suitable modification for the varying number of electrons involved. The general equation for the calculation of the reduction enthalpies is:

$$\text{Reduction enthalpy} = -\Delta_{\text{hyd}}H^{\ominus}(\text{M}^{n+}) - \Sigma I_{1-n}(\text{M}) - \Delta_{\text{atom}}H^{\ominus}(\text{M}) + 439n \qquad (7.46)$$

where $n = 1$ for Na, 2 for Mg and 3 for Al.

The calculations are consistent with the observed trends and it should be noticed that the major factor responsible for the lessening reducing power of the elements is the increasing total ionization energy which is not offset sufficiently by the relatively smaller increase in hydration enthalpy.

Table 7.8 Data and calculated $E^{\ominus}$ values for the reduction potentials of Na, Mg and Al (enthalpies in kJ mol^{-1})

	Na	Mg	Al
$E^{\ominus}$(observed)/V	-2.71	-2.37	-1.66
$\Delta_{\text{atom}}H^{\ominus}$(M)	108	150	314
I_1	494	736	577
I_2		1450	1820
I_3			2740
$\Delta_{\text{hyd}}H^{\ominus}(\text{M}^{n+})$	-419	-1942	-4697
$E^{\ominus}$ (calculated)/V	-2.65	-2.5	-1.95

It is clear from the above examples that quite simple calculations can identify the major factors which govern the values of $E^{\ominus}$ for any couple and that major trends may be rationalized.

Exercise. Check the calculations for Mg and Al.

7.3 THE STABILITY OF IONS IN AQUEOUS SOLUTION

This section is not concerned with the thermodynamic stability of ions with respect to their formation. It is concerned with whether or not a given ion is capable of existing in aqueous solution without reacting with the solvent. Hydrolysis reactions are dealt with above, the only reactions discussed in this section are those in which either water is oxidized to dioxygen or reduced to dihydrogen. The limits of stability for water at different pH values may be defined by the Nernst equations for the hydrogen reference half-reaction and that for the reduction of dioxygen to water.
The equation:

$$\Delta G = \Delta G^{\circ} + RT\ln Q \tag{7.47}$$

represents the change in Gibbs energy for non-standard conditions of a system. The quotient, Q, is the product of the activities of the products divided by the product of the reactant activities, taking into account the stoichiometry of the overall reaction. It may be converted into the *Nernst equation* by using equation (7.19), so that:

$$E = E^{\circ} - (RT/nF)\ln Q \tag{7.48}$$

The Nernst equation may be applied to the hydrogen standard reference half-reaction and gives:

$$E(\mathrm{H^+/H_2}) = -\frac{RT}{F}\ln\left(\frac{1}{a_{\mathrm{H^+}}}\right) \tag{7.49}$$

considering the activity of dihydrogen to be unity and $E^{\circ}(\mathrm{H^+/H_2}) = 0$. In order to obtain an equation which relates E to pH it is convenient to convert equation (7.49) to one in which decadic logarithms are used:

$$E(\mathrm{H^+/H_2}) = -\frac{2.303RT}{F}\log_{10}\left(\frac{1}{a_{\mathrm{H^+}}}\right) \tag{7.50}$$

which, at 298 K, becomes:

$$E(\mathrm{H^+/H_2}) = -0.059\log\left(\frac{1}{a_{\mathrm{H^+}}}\right) \tag{7.51}$$

The theoretical definition of pH is given by the equation:

$$pH = -\log a_{H^+} \tag{7.52}$$

and substitution into equation (7.51) produces:

$$E(H^+/H_2) = -0.059pH \tag{7.53}$$

The equation for the half-reaction expressing the reduction of dioxygen to water in acid solution is:

$$O_2(g) + 4H^+(aq) + 4e^- \rightleftharpoons 2H_2O(l) \tag{7.54}$$

the $E^{\ominus}$ value being 1.23 V when the activity of the hydrogen ion is unity. The Nernst equation for the dioxygen/water couple is:

$$E = 1.23 - 0.059pH \tag{7.55}$$

It has been assumed that the activities of dioxygen and water are unity.

Exercise. Check that you can derive equation (7.55). Take into account the $4H^+$ and the four electrons in equation (7.54).

Equations (7.53) and (7.55) are plotted in Fig. 7.12 with pH values varying from 0 to 14. Theoretically the oxidized form of any couple whose E value lies above the oxygen line should be unstable with respect of its potential for oxidizing water to dioxygen. Likewise the reduced form of any couple whose E value lies below the hydrogen line should be unstable and should be capable of reducing water to dihydrogen.

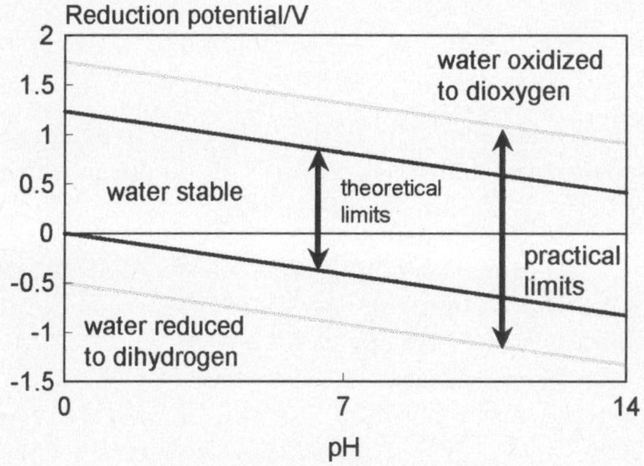

Fig. 7.12 A diagram of the Nernst equations for the reduction of dioxygen to water (upper black line) and for the reduction of water to dihydrogen (lower black line). The grey lines represent the practical limits for the two processes

In practice it is found that wherever gases are evolved (in particular the necessity for atoms to come together to form diatomic molecules) there is a barrier to the process (sometimes known as overpotential) so that the E value needs to be around 0.4 V lower than the hydrogen line or above the oxygen line for water to be reduced or oxidized, respectively. These practical limits are shown in Fig. 7.12. The oxidized and reduced forms of couples with

E values anywhere between the two practical limits are stable in aqueous systems with respect to either the reduction or the oxidation of the solvent.

Couples with E values outside of the practical limits of stability do not necessarily cause the destruction of the solvent. Some reactions, although they may be thermodynamically feasible, are kinetically very slow. The Co^{3+}/Co^{2+} couple has an E° value of 1.92 V and Co^{3+} should not exist in aqueous solution but its oxidation of water to oxygen is very slow and solutions containing $[Co(H_2O)_6]^{3+}$ evolve dioxygen slowly. The E° value for the Al^{3+}/Al couple is −1.66 V and indicates that aluminium should dissolve in acidic aqueous solutions, but the reaction is normally prevented by a stable protective oxide layer.

7.4 LATIMER, VOLT-EQUIVALENT AND POURBAIX DIAGRAMS

There are three main methods of summarizing the thermodynamic stabilities of the oxidation states of elements in aqueous solution. Latimer and volt-equivalent (sometimes known as Frost —**Fr**ee energy **o**xidation **st**ate—or Frost–Ebsworth diagrams) diagrams are usually restricted to the two extremes of standard 1 M hydrogen ion (pH = 0) or 1 M hydroxide ion (pH = 14) solutions. Pourbaix diagrams express the variations in stabilities of oxidation states as a function of pH between pH values of zero to fourteen and are more comprehensive than the previously mentioned diagrams.

Latimer diagrams

A Latimer diagram is a list of the various oxidation states of an element arranged in descending order from left to right, with the appropriate standard reduction potentials (in volts) placed between each pair of states. The diagram for chromium in acid solution is written below:

$$Cr_2O_7^{2-} \xrightarrow{1.33} Cr^{3+} \xrightarrow{-0.41} Cr^{2+} \xrightarrow{-0.91} Cr$$

The more exact forms of the aqueous cations, with their primary hydration shells, are normally omitted from the diagrams. As is the case with Cr^{VI}, which in acid solution exists as the dichromate ion, $Cr_2O_7^{2-}$, the forms of any oxo-anions are indicated by their formulae in the diagrams. The diagram for chromium summarizes the following important properties:
(i) $Cr_2O_7^{2-}$ is a powerful oxidizing agent (it is used as such in volumetric analysis) as indicated by the high positive value of the E° for its reduction to Cr^{III}.
(ii) Chromium metal should dissolve in molar acidic solutions to give the III state. Written as oxidation reactions (i.e. with the signs of the reduction potentials reversed) the positive potentials for the two stages indicate that the reactions would be thermodynamically feasible. In practice chromium does dissolve in hydrochloric and sulfuric acids, but is rendered passive by concentrated nitric acid. This latter effect is because of the production of a surface layer of Cr^{III} oxide which is impervious to further acidic attack. It is thought that this is also the reason for chromium (together with nickel) giving stainless steel its non-corrosive property.
(iii) The Cr^{2+} should be, and is, unstable in aqueous solution in the presence of dissolved dioxygen. Dioxygen (see equation (7.53) has the potential to oxidize Cr^{2+} to Cr^{3+}. The (II) state is a good reducing agent. It has almost the 'practical' potential to reduce water to dihydrogen (see Fig. 7.12).

(iv) The (III) state, placed between the positive and negative $E^{\ominus}$ values, is the most thermodynamically stable state. The Latimer diagram for the states of chromium in basic solution is:

$$CrO_4^{2-} \xrightarrow{-0.13} Cr(OH)_3 \xrightarrow{-1.1} Cr(OH)_2 \xrightarrow{-1.4} Cr$$

The information contained by the diagram is:
(i) Chromium(VI) is the monomeric tetroxochromate(VI) ion. It does not have any oxidant property and is the most stable state of the element under basic conditions.
(ii) There is no other solution chemistry as the (II) and (III) states exist as solid hydroxides which are easily oxidized to the (VI) state. It should be noted that, although the thermodynamic data are presented as $E_B^{\ominus}$ values, none of the numerical data has been (or could have been) derived by measurements of potentials. Such data are derived from thermochemical experiments.

Latimer diagrams are available for all the elements which exhibit more than one oxidation state and form an excellent and concise summary of the aqueous chemistry of such elements.

Volt-equivalent diagrams (Ebsworth-Frost diagrams)

Equation (7.19) relates the standard reduction potential (volts) to the standard Gibbs energy (in J mol^{-1}). This equation can be rearranged to give $-n\,E^{\ominus} = \Delta G^{\ominus}/F$. The value of the standard reduction potential is multiplied by n (the number of electrons participating in the reduction—the change in the oxidation state of the element) and the product represents, with its sign changed, the volt-equivalent of the change in standard Gibbs energy for the reduction process.

The plot of the volt-equivalent ($\Delta G^{\ominus}/F$) against oxidation state for chromium in acid solution is shown in Fig. 7.13, with the oxidation state decreasing from left to right. In this way the volt-equivalent or Ebsworth-Frost diagrams (Frost is a composite word derived from **Fr**ee energy **o**xidation **st**ate) can be associated easily with the corresponding Latimer diagram. The origin of such a graph is the volt-equivalent of zero of the zero oxidation state of the element. The point for Cr^{2+} is placed at the oxidation state 2 and with a volt-equivalent value of -1.82 V (2×-0.91 V) indicating that Cr^{2+} is so much more stable than the neutral element. The point for Cr^{3+} is placed a further 0.41 V down from -1.82 V at -2.23 V, indicating the further stability of the (III) state. The point for $Cr_2O_7^{2-}$ is placed $1.33 \times 3 = 3.99$ V higher than that for the (III) state, indicating the instability of the former state with respect to the latter. The factor of three is used because the (VI) to (III) conversion is a three-electron change ($n = 3$).

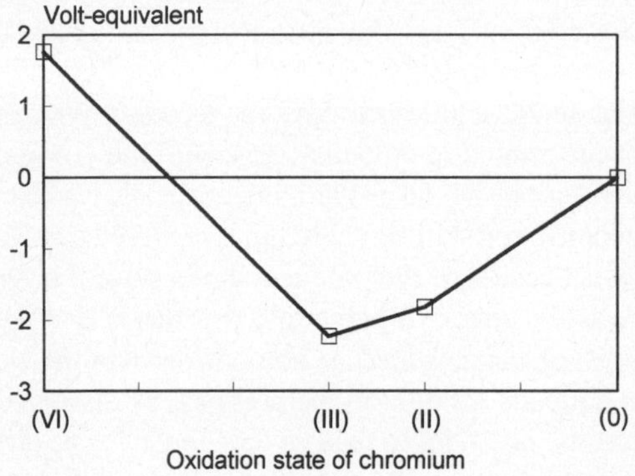

Fig. 7.13 The volt-equivalent diagram for the oxidation states of chromium in 1 *M* acidic solution

From Fig. 7.13 it can be seen that the reduction of Cr^{VI} to Cr^{III} has a change in volt-equivalent of $-2.23 - 1.76 = -3.99$ V. The corresponding reduction potential is thus: $-(-3.99)/3 = 1.33$ V. Volt-equivalent diagrams convey the same amount of information as do Latimer diagrams about the relative stabilities of the oxidation states of an element and their oxidation/reduction properties, but do it in a graphical manner. More examples are given in Chapters 9 and 11.

Pourbaix diagrams

Pourbaix diagrams are plots of E versus pH for the various couples in the oxidation of an element. They are useful in defining 'areas' of stability for any particular oxidation state of that element, and may include the Nernst equations for the reduction and oxidation of water to dihydrogen and dioxygen respectively.

The Pourbaix diagram for the $Fe/Fe^{II}/Fe^{III}$ system is shown in Fig. 7.14. Pourbaix diagrams correlate the Latimer diagrams of the two extremes of the pH scale and take into account the speciation of the oxidation states of the element. Speciation takes into account the molecular form of each species which contributes to the system at any value of pH.

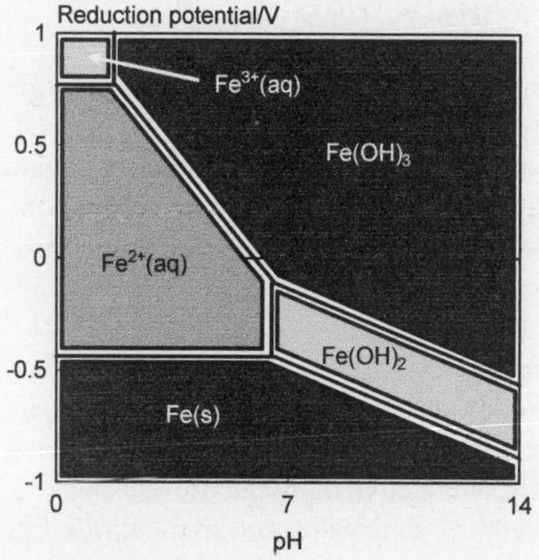

Fig. 7.14 A Pourbaix diagram for the 0, II, and III oxidation states of iron; the shaded areas indicate the conditions where the species of iron predominate

Horizontal lines on Pourbaix diagrams arise when no protons are needed in the half-reaction as in the cases:

$$Fe^{2+}(aq) + 2e^- \rightleftharpoons Fe(s)$$

and

$$Fe^{3+}(aq) + e^- \rightleftharpoons Fe^{2+}(aq)$$

Areas where different forms of an element in a particular oxidation state exist are separated by a vertical line. Examples are the phase changes when the pH changes so that iron(III) and iron(II) become insoluble in water.

An excellent article for further reading about Pourbaix diagrams is by Campbell and Whiteker in the Journal of Chemical Education, **46**, 92 (1969), entitled 'A Periodic Table Based on Potential-pH Diagrams'.

7.5 ACIDS AND BASES

The subject of acids and bases is very extensive. The discussion in this book is restricted to the main definitions of acids and bases and their applications to a small range of chemical problems. Some definitions of acids and bases are so comprehensive as to encompass all chemical reactions. They invite the sub-division of chemistry to make the subject relevant to specific areas and are not referred to in this book. The three main definitions are those accredited to (i) Brønsted and Lowry, (ii) G.N. Lewis and (iii) Pearson. They are the ones that are currently in most use by inorganic chemists.

Autoionization of solvents

Water, liquid ammonia (anhydrous liquid used at its b.p. of −33°C, 240 K), and sulfuric acid (pure liquid) are examples of liquids in which autoionization occurs. The process is one of proton transfer between two molecules in the liquid phase as indicated by the equilibria below.

$$2H_2O(l) \rightleftharpoons H_3O^+(aq) + OH^-(aq) \qquad (7.56)$$

$$2NH_3(l) \rightleftharpoons NH_4^+(solvated) + NH_2^-(solvated)$$

$$2H_2SO_4(l) \rightleftharpoons H_3SO_4^+(solvated) + HSO_4^-(solvated)$$

In the three cases the positive ion is a solvated hydrogen ion, the negative ions being the solvent molecules minus one hydrogen ion. The extent of the autoionization differs in the three cases and in none is very extensive. The autoprotolysis constants (the equilibrium constants for the processes described by equations (7.55-7.57) and which are sometimes referred to as *ionic products*) are 1.0×10^{-14} (298 K), $\sim 10^{-30}$ (223 K, -50°C) and 1.7×10^{-4} (283 K) respectively. In the sulfuric acid case there are complications from other equilibria in which the $HS_2O_7^-$ ion participates. Autoprotolysis determines the conductance of the liquid substances, sulfuric acid being a better electrical conductor than water, ammonia being poorer. The magnitude of the autoprotolysis constant is related to the dielectric constant (i.e. the permittivity) of the liquid. The permittivities of sulfuric acid, water and ammonia are 101 (298 K), 78.5 (298 K) and 22 (-33°C, 240 K) F m^{-1}. It is to be expected that the greater the value of the permittivity of a solvent is, the larger will be the value of the solvation energy of the ions produced in that solvent by autoionization. A larger value of solvation energy of an ion will result in a greater stabilization of the ion and lead to a higher equilibrium concentration of it in the solvent.

Brønsted and Lowry acids and bases

A Brønsted-Lowry acid is defined as a proton donor, a Brønsted-Lowry base as a proton acceptor. The definitions apply to protic systems; those in which proton transfers can occur. A general equation expressing proton transfer in aqueous solution is:

$$AH(aq) + B(aq) \rightleftharpoons A^-(aq) + BH^+(aq)$$

where AH is a general acid with a proton which is dissociable and B is a general base which can accept a proton. In the reverse process where the proton is donated by the BH^+ ion to the anion A^-, BH^+ is called the *conjugate acid* of the base B and the anion A^- is called the *conjugate base* of the acid AH.

The strength of a Brønsted-Lowry acid is quantified in terms of the magnitude of the equilibrium constant for the ionization reaction in which the solvent acts as the base, e.g. for the aqueous system the general reaction is:

$$AH(aq) + H_2O(l) \rightleftharpoons H_3O^+(aq) + A^-(aq)$$

the equilibrium constant (the acid dissociation constant), K_a, being given by the equation:

$$K_a = \frac{[H_3O^+][A^-]}{[AH]} \tag{7.57}$$

There is a very large range of values of K_a so that it is more convenient to express acid strengths as pK_a values ($pK_a = -\log_{10} K_a$).

Some binary hydrides (e.g. those of Groups 16 and 17) behave as acids in aqueous solution, but the majority of acids are oxo-acids derived from acidic oxides. An oxide can be either acidic, amphoteric (i.e. acidic or basic depending upon conditions) or basic. A general understanding is achieved by a consideration of the M–O–H grouping in the hydroxide (produced by the reaction of an oxide with water). The compound behaves as an acid if the O–H bond is broken in a heterolytic manner, i.e. the oxygen atom retains both of the bonding electrons. If the M–O bond breaks in a similar heterolytic manner the compound exhibits basic behaviour. Amphoteric behaviour is adopted by intermediate compounds. If M is a relatively large and electropositive, e.g. Na, the M–O–H arrangement is likely to be ionic, M^+OH^-, so that the aquated hydroxide ion is produced when the compound dissolves in water. If M is relatively small, electronegative and in a high oxidation state, e.g. as the chlorine(VII) atom in O_3Cl–O–H, the M–O bond is likely to be the stronger bond thus favouring heterolysis to give the ions, $MO^-(aq)$ and $H^+(aq)$. In $Al(OH)_3$, which is an intermediate case with a central element in the (III) state, both heterolytic processes can occur.

Sodium hydroxide dissolves in water to give $OH^-(aq)$ ions, the strongest base which can exist in an aqueous system. Chloric(VII) acid (perchloric acid) dissociates practically completely in water to give $H^+(aq)$ ions which represent the strongest acid which can exist in an aqueous system. The amphoteric behaviour of aluminium is noticed in the series of hydrated salts containing the Al^{3+} ion and in compounds such as $NaAlO_2$ containing the aluminate(III) ion, AlO_2^-.

Lewis acids and bases

The reaction of a hydrated hydrogen ion with a hydrated hydroxide ion:

$$H_3O^+(aq) + OH^-(aq) \rightleftharpoons 2H_2O(l)$$

is the reverse of the autoprotolysis reaction (7.56) and occurs when Brønsted-Lowry acids and bases react with each other in aqueous solution, a reaction known as *neutralization*. In Brønsted-Lowry terms the hydrated hydrogen ion (the acid) donates a proton to the hydroxide ion (the base). The reaction may also be regarded as the donation of a pair of electrons (a non-bonding pair) by the hydroxide ion to the hydrogen ion, the electrons becoming hydrogen-oxygen bonding. The definitions of an acid as an electron-pair acceptor and a base as an electron-pair donor are the basis of the Lewis acid-base theory. The definitions allow a broad range of reactions to be regarded in terms of acid/base behaviour, in particular those systems in which protons do not necessarily participate.

The Lewis definitions apply to the following disparate examples.

(i) The reaction of BF_3 (electron-pair acceptor) with ammonia (electron-pair donor) to give the adduct, $H_3N{\rightarrow}BF_3$.

(ii) The autoionization of aprotic solvents, e.g. BrF_3 and SO_2 which ionize slightly according to the equations:

$$2BrF_3 \rightleftharpoons BrF_2^+ + BrF_4^-$$

$$2SO_2 \rightleftharpoons SO^+ + SO_3^-$$

The neutral solvent molecules act both as acids and bases, the positive ions being electron-pair acceptors (from F^- and O^{2-} respectively) and the negative ions being electron pair donors.

(iii) The formation of metal-ligand bonds in complexes are reactions between metal ions (electron-pair acceptors) and ligands (electron-pair donors). The bonding of ligands such as the cyanide ion and the ethene and carbon monoxide molecules to transition elements in various states of oxidation is strengthened by metal to ligand 'π back-bonding' which, in Lewis terms, can be described as π acid behaviour by the ligands.

Pearson's hard and soft acids and bases (HSAB) theory

In an attempt to generalize the behaviour of ligands with metal ions, Ahrland, Chatt and Davies suggested the a, b and borderline approach. This was dependent upon the general observations that some metal ions (a-type) form particularly stable complexes with the smaller donor ligand ions/atoms F^-, O and N. Other metal ions (b-type) form particularly stable complexes with the larger donor ligand ions/atoms I^-, S and P. There are some intermediate cases known as borderline.

The approach was developed by Pearson who generalized complex formation in terms of hard and soft acids and bases. A hard acid is a metal ion in a higher oxidation state and is not easily polarized. A hard base is an ion or donor atom which is small and not easily polarized. Hard acids tend to react preferentially with hard bases to give particularly stable complexes in which the bonding is more ionic than covalent. A soft acid is a metal atom or a metal ion in a low oxidation state and is easily polarized. A soft base is a large ion or donor atom which is easily polarized. Soft acids tend to react preferentially with soft bases to give particularly stable complexes in which the bonding is predominantly covalent. Some typical hard and soft acids and bases, together with borderline cases, are given in Table 7.9.

Table 7.9 HSAB classification of some Lewis acids and bases

Hard	**Borderline**	**Soft**
ACIDS		
H^+, Na^+, Ca^{2+}	Fe^{2+}, Co^{2+}, Ni^{2+}	Cu^+, Ag^+, Hg^{2+}
Al^{3+}, La^{3+}, Fe^{3+}	Cu^{2+}, Zn^{2+}, Pb^{2+}	Hg_2^{2+}, Pd^{2+}, Pt^{2+}
Ti^{4+}, Pb^{4+}, U^{4+}	Sb^{3+}, Bi^{3+}, Rh^{3+}	Pt^{4+}, metal atoms
BASES		
H_2O, OH^-, F^-, Cl^-	Br^-, NO_2^-, SO_3^{2-}, N_2	I^-, CN^-, CO, C_2H_4
PO_4^{3-}, SO_4^{2-}, ClO_4^-		R_2S, R_3P, $C_5H_5^-$
NO_3^-, NH_3, CO_3^{2-}		C_6H_6, H^-

Attempts to quantify hardness and softness of metals, metal ions and ligands based upon experimentally observed formation constants have been only partially successful, and only over localized groupings of metal-ligand combinations. Nevertheless, the concepts of hardness and softness aid the experimental chemist as a practical guide in inorganic syntheses.

7.6 PROBLEMS

1. In the crystal of KCl, the constituent ions being isoelectronic, the internuclear distance is 314 pm. Pauling suggested that, in such cases, the radius of each ion is inversely proportional to its effective nuclear charge, Z_{eff}, i.e. $Z - S$, where S is the Slater screening constant. Calculate S for the two constituent ions and calculate their respective ionic radii.

2. The internuclear distance for CsI is 395 pm. Calculate Z_{eff} for the constituent ions and estimate their ionic radii.

3. Calculate the radius ratios, r^+/r^-, for the Group 2 binary (MO/MS) oxides and sulfides from the data for the radii of the ions given (in pm): Mg^{2+} 65, Ca^{2+} 99, Sr^{2+} 113, Ba^{2+} 135, O^{2-} 140, S^{2-} 184 and predict their structures. Compare your results with the known structures.

4. Using the following data (all in units of kJ mol^{-1}), calculate the lattice energies of the four oxides and comment on the chemical significance of the results.

Metal	Pb	Sn		
$\Delta_a H^\circ$	196	301		
I_{1+2}	2166	2117		
I_{3+4}	7160	6870		
Compound	PbO	PbO_2	SnO	SnO_2
$\Delta_f H^\circ$	-218	-277	-286	-581

ΔH° (O(g) → O^{2-}(g)) = 704; $D(O_2$,g) = 496

5. Calculate the lattice energies of AuCl and $AuCl_3$ from the following thermochemical data (all in units of kJ mol^{-1}):

$\Delta_a H^\circ$(Au) = 369; I_1(Au) = 891; I_{1+2+3}(Au) = 5811; $D(Cl_2,g)$ = 242; E_{ea}(Cl) = −364; $\Delta_f H^\circ$(AuCl) = −35; $\Delta_f H^\circ(AuCl_3)$ = −118

Calculate the value of ΔH° for the disproportionation of $3AuCl \rightarrow AuCl_3 + 2Au$ and comment upon the relative stability of the two oxidation states of gold.

6. (a) From the following data (all in units of kJ mol^{-1}) calculate ΔH° for the process $E^-(g) + e^- \rightarrow E^{2-}(g)$ (i.e. the second electron attachment energy $E_{ea}(2)$), for O and S and comment upon their magnitudes and compare them with the first electron attachment energy of the two elements.

	$\Delta_a H^\circ$	I_{1+2}
Ba	157	1468
		$E_{ea}(1)$
S	223	−200
O	248	−140
	$\Delta_f H^\circ$	L
BaO	−560	3130
BaS	−444	2624

(b) Comment upon the magnitudes of the lattice energies for the two barium compounds and compare them with those of the magnesium–Group 16 binary compounds:

Compound	MgO	MgS	MgSe	MgTe
L	3925	3297	3167	2925

7. The electrical resistance of V^{IV} oxide varies with temperature, T, as shown:

T/K	297	309	325	336	344	352	355
	105	60.8	24.3	10.5	0.3	0.1	0.08

What conclusions can be drawn from the data about the nature of the compound at temperatures < 340 K and > 340 K given that the V...V distances are 312 pm and 265 pm in the former range and are all equal at 288 pm in the latter range?

8. A dark compound formulated as $YBa_2Cu_3O_{7-x}$ was cooled in liquid nitrogen when its electrical resistance dropped sharply to zero. 0.15 g of the solid were dissolved in 10 cm^3 of 1 M HCl and the solution was boiled. After dilution, 1.5 g KI were added and the iodine liberated was titrated with 0.032 M sodium thiosulfate solution of which 20.85 cm^3 were required. An identical mass was dissolved, under an inert atmosphere, in 10 cm^3 of 1 M HCl/0.7 M KI and then diluted and titrated as above when 16.10 cm^3 were required. Calculate the percentage of Cu^{III} (or positive holes) in the sample.

Under less rigorous preparative conditions, a green compound was obtained whose electrical resistance increased on cooling. What are the likely differences between the two compounds? (In the first experiment Cu^{III} is assumed to be reduced to Cu^{II} and the total Cu^{II} liberates I_2, in the second experiment: $Cu^{3+} + 3I^- \rightarrow CuI + I_2$).

8

Chemistry of Hydrogen and the s Block Metals

8.1 INTRODUCTION

Hydrogen is located in the periodic table as the first member of Group 1. Some advocates would have it placed as the first member of Group 17 since it readily accepts an electron to become the hydride ion and is similar in this respect to the other members of that group (the halogens). Likewise, some advocates would have helium placed as the first member of Group 2, rather than Group 18, since it has a filled s orbital as its valence shell. If both alternatives were to be adopted there would be a peculiar situation with helium appearing before hydrogen and neither option can be recommended.

The electronic configurations of hydrogen and the Group 1 metals (the Alkali metals) are $1s^1$ and [noble gas] ns^1 respectively and for Group 2 metals (the Alkaline Earth metals) they are [noble gas] ns^2. The ionization energy of the hydrogen atom (1312 kJ mol^{-1}) is considerably larger than those of the Group 1 metals and is the factor which is mainly responsible for the differences in chemistry between H and the other members of the group.

8.2 HYDROGEN

The reasons for hydrogen ($1s^1$) occurring as a diatomic molecule are dealt with in detail in Chapter 5. The element has an extensive chemistry because of the simplicity of the atom and its consequent ability to enter into combination as a positive hydrogen ion, a negative hydride ion, a participant in covalency and, as the diatomic molecule, as a ligand in complex formation. The bare proton, H^+, does not have a permanent role in the chemistry of hydrogen, but as the solvated hydroxonium ion, H_3O^+, has an extensive role in the aqueous chemistry of acidic substances. The hydrogen atom is produced transiently when metals dissolve in aqueous acids (nascent hydrogen) and when hydrated electrons (themselves produced by the interaction of X- and γ-ray quanta with liquid water) react with water:

$$e^-(aq) + H_2O(l) \rightarrow H + OH^-(aq)$$

The H atom is a powerful reducing agent and tends to react with organic materials by hydrogen atom abstraction reactions:

$$H + RH \rightarrow H_2 + R$$

with the R free radicals either dimerizing (to give R_2) or undergoing dismutation reactions to give stable products (usually one saturated and one with a double C=C bond). With unsaturated molecules the H atom undergoes addition, producing a free radical which can either dimerize or dismute.

The many types of covalent hydrides are dealt with element by element in the remainder of the text. Compounds of the type, EOH, where E is either an element or an atom with other attachments, behave in aqueous solution as either acids or alkalies. If E is a relatively large atom with a low oxidation state, e.g. Na, the E–O bond tends to be weaker than thc O–H bond and, aided by the evolution of hydration energy, the compound acts as an alkali giving E^+ and OH^- ion in aqueous solution. If E is a relatively small atom with a high oxidation state, e.g. S^{VI} as in H_2SO_4, the E–O bond is stronger than the O–H bond and, again with the assistance from the evolution of hydration energy, the acidic dissociation into EO^- and H_3O^+ occurs. Some intermediate types, e.g. AlOH as in $Al(OH)_3$, can act as either acids or bases and are termed amphoteric.

Hydrogen forms three types of binary hydrides, E^+H^-, ionic or saline which contain the hydride ion and react with water to yield dihydrogen, covalent hydrides and interstitial hydrides. In the latter the hydrogen atoms enter the interstitial positions of a metal lattice to give mainly non-stoichiometric compounds ranging from EH (equivalent to the filling of the tetrahedral holes in a ccp or hcp lattice) to EH_3 (all tetrahedral and octahedral holes filled).

The one outstanding property of hydrogen in combination is that of participation in hydrogen bonding in which it allows two or more molecules to interacts because of its unshielded proton which can attract particularly electronegative atoms, e.g. F, O and N. The formation of hydrogen bonds influences the properties of many molecules, e.g. from the small HF and water molecules, whose physical properties are intimately concerned with intermolecular hydrogen bonding, to macromolecules such as proteins, enzymes and DNA.

8.3 THE s BLOCK METALS

The s electrons of the Group 1 and 2 metals are well screened from their nuclear charges, resulting in the first ionization energy being relatively low for Group 1, but somewhat higher for Group 2 (due to higher Z_{eff} values). The second ionization energy of the latter metals is even higher since the second electron is removed from a smaller and positively charged ion. The sum of the two ionization energies is considerable and is relevant to the chemistry of the Group 2 metals. For both groups, ionization energies decrease down the group (Z_{eff} does not change much but the atomic radius increases). Hence, Cs has the lowest first ionization energy of any non-radioactive element and each Group 1 metal has the lowest value in its period.

The loosely held single s electron of the Group 1 metals results in a larger atom compared to its Group 2 neighbour.

Fig. 8.1 shows the metallic radii for the two groups together with the radii of the Group 1 metal M^+ cations and the Group 2 M^{2+} cations. The preference of the latter metals to form divalent rather than univalent ions is based on thermodynamic considerations. The energy required to form M^{2+} ions (rather than M^+ ions) in solution or in solid compounds, is more than compensated by the energy released when the M^{2+} ions are hydrated or M^{2+} ionic solids are formed.

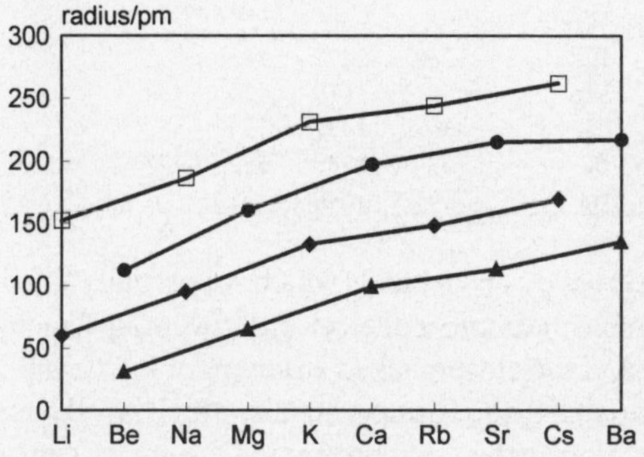

Fig. 8.1 The metallic and cationic radii of the Group 1 (open squares and diamonds respectively) and Group 2 elements (circles and triangles respectively)

Problem 8.1

The hypothetical CaCl crystal would be expected to have an NaCl structure and an internuclear distance of 299 pm. Estimate its lattice energy, L, taking L for NaCl(s) as 768 kJ mol^{-1} and its internuclear distance r_{eq} as 276 pm (assume that the Born exponent n is the same for the CaCl and NaCl.) Then calculate $\Delta_f H^\circ$ for CaCl(s) using the following ΔH° values at 298K (all in kJ mol^{-1}):

	$\Delta_{atom} H^\circ$	I_1		$D(X_2)$	E_{ea}
Ca	172	589	Cl	248	-364

and calculate ΔH° for the reaction: 2CaCl(s) = $CaCl_2$(s) + Ca(s), given that $\Delta_f H^\circ$ for $CaCl_2$(s) is -795 kJ mol^{-1}.

Answer: L (CaCl) = 768 × 276/299 = 709 kJ mol^{-1}

$\Delta_f H^\circ$(CaCl) = 172 + 589 + 124 − 364 − 709 = −188 kJ mol^{-1}

ΔH°(2CaCl → $CaCl_2$ + Ca) = −795 + 2 × 188 = −419 kJ mol^{-1}

i.e. more exothermic than $\Delta_f H^\circ$(CaCl) and therefore favoured thermodynamically (its L = 2268 kJ mol^{-1})

8.4 STRUCTURES AND PHYSICAL PROPERTIES

The solid Group 1 metals and barium have body-centred cubic structures (bcc, see Fig. 6.23a) unlike the close-packed structures (ccp and hcp, see Fig. 6.23b and c) of the other Group 2 metals. These latter elements have either (Be and Mg) a hexagonal close-packing (hcp) of atoms or (Ca and Sr) cubic close-packing (ccp). The melting and boiling points and the densities of the elements are given in Table 8.1.

Table 8.1 The melting and boiling points (°C) and densities (kg m^{-3}) of the metallic elements of Groups 1 and 2

Group 1	m.p.	b. p.	Density	Group 2	m.p.	b. p.	Density
Li	180	1330	534	Be	1280	2477	1850
Na	97.8	890	970	Mg	650	1110	1740
K	63.7	774	860	Ca	850	1487	1540
Rb	38.9	688	1530	Sr	768	1380	2620
Cs	28.7	690	1879	Ba	714	1640	3510

In both Groups of elements, the solids have s and p bands which are partially filled, i.e. conduction bands. Such bands contain double the number of electrons in a Group 2 metal than its Group 1 metal neighbour. This fact and the close-packed structure of the Group 2 metals leads to stronger metallic bonding compared to the Group 1 metals, which is reflected in the lower melting points and boiling points of the latter compared to the former, as can be seen from the data in Table 8.1. The melting point of sodium is close to the boiling point of water. However, the melting points and boiling points of the Group 2 elements do not vary regularly down the group, unlike the Group 1 metals where both melting points and the respective boiling points decrease down the group as the metallic radii increase. The irregularities in the trends of the densities of the elements down their respective groups are connected with the combination of the trends in the R. A. M. values and the metallic radii.

Exercise. Plot the function, R. A. M./r_{met}^3, against the densities of the Group 1 and 2 elements respectively and note the correlations.

A possibly more significant parameter is the atomization enthalpy, $\Delta_a H^\ominus$, values of which are shown below (kJ mol^{-1}), and which exhibit a trend that is roughly parallel to the boiling points.

Li	Na	K	Rb	Cs
161	109	90	86	79
Be	Mg	Ca	Sr	Ba
321	150	193	164	176

Both the atomization enthalpy and the boiling point of any metal are indicative of the extent of cohesive forces operating in the liquid state, greater cohesive forces leading to higher values of the two parameters, although the atomization enthalpy refers to a change of phase from the solid to gaseous state.

8.5 STANDARD REDUCTION POTENTIALS AND THE ELECTROLYTIC EXTRACTION OF METALS

All the metals of Groups 1 and 2 are very electropositive elements as can be seen from the values of their standard reduction potentials, $E^\ominus(M^{n+}/M)$, given in Table 8.2. This indicates that the reduction of their compounds to produce the metals is difficult and requires an electrolytic process.

Table 8.2 Standard reduction potentials for the Group 1 and 2 metals

Group 1 metal	$E^{\ominus}(M^+/M)$	Group 2 metal	$E^{\ominus}(M^{2+}/M)$	$E^{\ominus}(M^{2+}/M)$ calculated
Li	-3.04	Be	-1.97	-2.03
Na	-2.71	Mg	-2.36	-2.39
K	-2.92	Ca	-2.84	-2.81
Rb	-2.92	Sr	-2.89	-3.04
Cs	-2.92	Ba	-2.92	-3.08
Fr	~-2.92	Ra	-2.92	

The factors which mainly affect the values of the reduction potentials of the cations of the Group 1 elements are discussed in Section 7.4 and the results of the calculations of their values are given in Table 7.6. Section 7.4 also contains a comparison of the reduction potentials of the cations of Na and Mg (as well as Al). The same approach can be used to compare the trend of reduction potentials of the Group 2 cations with a modification of equation (7.41) to take into account the two electron reduction process. The equation which expresses the values for the enthalpies of reduction of the $M^{2+}(aq)$ ions of the Group 2 elements is:

$$\Delta H^{\ominus}(M^{2+}/M) = -\Delta_{hyd}H^{\ominus}(M^{2+}) - I_2 - I_1 - \Delta_a H^{\ominus}(M) + 2 \times 439 \quad (8.1)$$

The difference between equations (7.41) and (8.1) arises because of the two electron reduction process, there being two ionization energies to be incorporated and two amounts of 439 kJ mol^{-1} to be added to account for the oxidation of two moles of hydrated protons. The data used in the calculations are given in Table. 8.3 and the results for the calculated reduction potentials for the Group 2 cations are shown in Table 8.2. The particularly less negative value for Be arises from the large hydration enthalpy associated with the small radius of the Be^{2+} cation and the relatively large value of the enthalpy of atomization of the metal. The magnesium cation is more easily reduced than the cations of the heavier members of the group because its smaller size leads to a relatively high negative value for its hydration enthalpy.

Table 8.3 Thermodynamic data for the Group II elements (kJ mol^{-1})

Element	$\Delta_{hyd}H^{\ominus}(M^{2+})$	I_2	I_1	$\Delta_a H^{\ominus}(M,g)$
Be	-2494	1760	900	321
Mg	-1920	1450	736	150
Ca	-1598	1150	590	193
Sr	-1480	1060	548	164
Ba	-1360	966	502	176

The s block metals occur in nature either as chlorides or oxosalts which can be converted to chlorides. When an electric current is passed through a fused chloride, the M^{n+} ions are discharged at an inert cathode, where the metal is deposited. To minimize the energy required for melting the chloride, it is mixed with another metal chloride to lower the melting point. Metals other than the s block metals can be also extracted by electrolysis of fused compounds, e.g. fused Al_2O_3, mixed with $Na_3[AlCl_6]$ is used for the production of Al.

Because of the powerful reducing power of the Group 1 and Group 2 metals, they are not usually obtained by thermal reduction. However, K is produced by the thermal reduction of KCl using Na. Mg is usually obtained from $MgCa(CO_3)_2$ using ferrosilicon as a reductant.

In the electrolysis of brine (NaCl solution), using Hg cathodes, Na is discharged at the cathode forming an amalgam. This takes place rather than liberation of H_2 (expected from the electrochemical series) due to the high H_2 overvoltage at the Hg electrode. The amalgam reacts with water forming NaOH and H_2.

8.6 REACTIVITY AND REACTIONS

The s block metals are the most reactive metals; the Group 1 metals being more reactive than the Group 2 metals. In both groups, reactivity increases down the Group. Although this trend is parallel to the ionization energy, reactivity is a kinetic property which should not be related to thermodynamic predictions. The reactivity towards water is a clear example. Although Li has the most negative $E^\ominus$, its reaction with water is considerably less vigorous than that of Na which has the least negative $E^\ominus$ among the Group 1 metals. Its reaction with water produces sufficient heat to melt it and the hydrogen produced ignites in air. The other Group 1 metals with lower melting points react explosively with water. Among the Group 2 metals reactivity with water runs parallel to the $E^\ominus$ values. Although these values suggest reduction of water, Be does not react with water or steam whereas Mg does react with steam. Even in the case of Ba whose $E^\ominus$ is close to K, Rb and Cs, its reaction with water is less vigorous than that of Na. In all cases reaction with water forms $M(OH)_n$ and H_2. It has been argued, on the basis of thermodynamic considerations, that all the s block metals, except Be, are capable of detaching hydrated electrons [e^-(aq) or H_2O^-(aq)] from water, which are very rapidly converted to hydrogen:

$$M(s) + H_2O(l) \rightarrow M^+(aq) + H_2O^-(aq)$$
$$H_2O^-(aq) \rightarrow H(aq) + OH^-(aq)$$
$$2H(aq) \rightarrow H_2(g)$$

If the first reaction is rate-determining the first ionization energy of the metal atom might be expected to give a good guide to the rate trend.

Exercise. Read around equation (7.41) and decide whether the above statement regarding Be should not also apply to Mg.

The s block metals liberate H_2 from acids forming the salts of the acid, the reaction being more vigorous than the reaction with water. Thus Be reacts with dilute acids especially when amalgamated with Hg, although it is not affected by water. Be is also unique in dissolving in alkaline solutions forming $[Be(OH)_4]^{2-}$ and H_2. All the s block metals, except Be and Mg, react when heated in H_2 to form MH_n (n = 1 or 2). The metals are slowly attacked by O_2 when a layer of oxide is formed. When heated in O_2, an oxide is formed. Li and the Group 2 metals form nitrides when heated in N_2. The other members of Group 16

(S, Se and Te) also form binary compounds by direct reaction with the Groups 1 and 2 metals.

8.7 IONIC COMPOUNDS OF THE s BLOCK METALS

In spite of the limitations of the concept of ionic bonding, it is generally accepted that most of the solids formed by the s block metals with non-metals are ionic. It is convenient to consider the cations M^+ and M^{2+} as spheres with an assigned radius, although the choice of a set of ionic radii is not unequivocal. Related to the ionic radius, the concept of *ionic potential*, ϕ, is helpful in explaining the departure from purely ionic character. ϕ can be defined as the ratio of the cation charge to its radius. ϕ can be taken as a measure of the *polarizing power of cations*, i.e. their ability to distort the electron cloud of the larger anions. Distortion of the electron arrangements around anions may be thought of as promoting covalency. ϕ can also be correlated with the lattice energy, L, which is proportional to the product of the ionic charges and inversely proportional to the sum of the ionic radii:

$$L \propto \frac{z^+z^-}{r^+ + r^-} \tag{8.2}$$

where z refers to the charge and r stands for the radius, + denoting cation and – denoting anion. L and $\Delta_f H^{\circ}$ are related in a Born-Haber thermochemical cycle.

There is a general *inverse* correlation between the value of the ionic potential and the ionic nature of the compound, although from an empirical standpoint the correlation is better with the square root of the ionic potential. For example, the anhydrous halides for which $\phi^{1/2}$ values are less than 2.2 are electrolytic conductors when molten whereas those with values greater than 2.2 are non-conductors in the liquid state.

ϕ values are evidently higher for Group 2 than for Group 1, whose larger M^+ ions have only half the charge of the Group 2 cations. ϕ decreases down each group so it is lowest for Cs and highest for Be, whose compounds are expected to exhibit the greatest departure from ionic character.

The least polarizing cations, i.e. of the lowest ϕ, are able to stabilize di- or poly-atomic anions, where non-metal atoms are covalently bound such as O_2^{n-} ($n = 1$ or 2), O_3^-, S_x^{2-}, N_3^- and various polyhalide ions. The departure from 'purely' ionic character, which is related to ϕ, is also reflected in the structure of solid compounds.

Fajans proposed that the usually smaller cations polarize the anions, which are mostly larger than the cations. The polarizing power of a cation increases as its charge increases and as its radius decreases. The electron cloud of an anion is distorted as a result and can be interpreted as leading to a greater covalent contribution (more electron sharing) to the bond. The polarizability of an anion increases with the radius and with the charge. Cations with a noble gas configuration are less polarizing than those with other configurations, e.g. the Cu^+ ion is more likely to induce a degree of covalency in a compound than the K^+ ion. Fajans' rules summarize the ideas about polarization of anions by cationic charge as four factors which favour covalency in a compound, assuming them initially to be 100% ionic. The four factors are (i) a small cation, (ii) a large anion, (iii) large charges on either or both participating ions and (iv) a cation with a non-inert gas configuration.

8.8 HYDRIDES AND BINARY COMPOUNDS WITH OXYGEN

Hydrides

Be and Mg are unique in not combining directly with H_2. The compounds BeH_2 and MgH_2 are better described as covalent, rather than as ionic as are the other hydrides of group 2 and those of group 1. Amorphous or crystalline BeH_2 is polymeric with Be–H–Be 3-centre 2-electron bonds (i.e. two electrons occupying a three-centre bonding molecular orbital), similar to boranes. For the other s block metals, $\Delta_f H^\circ$ and L are given in Table 8.4 (kJ mol^{-1}).

Table 8.4 Data for ionic hydrides

	Li	Na	K	Rb	Cs	Ca	Sr	Ba
L	858	782	699	674	648	2410	2250	2121
$\Delta_f H^\circ(MH_n)$	-90.5	-56.3	-57.7	-52.3	-54.2	-181.5	-180.1	-178.7

L is much higher for Group 2 hydrides compared to MH, as expected. For LiH, $\Delta_f H^\circ$ is more negative than for the other MH compounds. A better indication of thermal stability is the temperature at which the solid starts to decompose which decreases from 550°C for LiH to 170°C for RbH and CsH, this trend reflects the decrease in L. The MH_2 solids are thermally more stable and the corresponding temperatures: 885°C for CaH_2 and 230°C for BaH_2 exhibit the same trend. LiH melts before dissociating and on electrolysis H^- ions are discharged at the anode liberating H_2. The ionic character of the hydrides is reflected in labelling them as saline (salt-like) hydrides. In fact MH solids have the NaCl structure. All the hydrides are powerful reductants, e.g. NaH reduces BF_3 to B_2H_6 or to $NaBH_4$ in ether. $NaAlH_4$ is similarly prepared using $AlBr_3$. CO_2 is reduced to HCO_2Na and $TiCl_4$ to Ti.

Oxides, peroxides, superoxides and ozonides

Lithium burns in O_2 to give Li_2O whereas Na gives mainly Na_2O_2 (peroxide) and the other three group 1 metals give MO_2 (superoxides). Equations representing the appropriate Born-Haber cycles for the standard enthalpies of formation of M_2O, M_2O_2 and MO_2 are given below.

$$\Delta_f H^\circ(M_2O) = 2\Delta_a H^\circ(M) + \Delta_a H^\circ(O_2) + 2I_1(M) + E_{ea(1)}(O) + E_{ea(2)}(O) - L(M_2O) \quad (8.3)$$

$$\Delta_f H^\circ(M_2O_2) = 2\Delta_a H^\circ(M) + 2I_1(M) + E_{ea(1)}(O_2) + E_{ea(2)}(O_2) - L(M_2O_2) \quad (8.4)$$

$$\Delta_f H^\circ(MO_2) = \Delta_a H^\circ(M) + I_1(M) + E_{ea(1)}(O_2) - L(MO_2) \quad (8.5)$$

For a proper comparison of relative stabilities of the three types of metal–oxygen compound the terms in equation (8.5) should be doubled to take into account the stoichiometries and to compare mole-for-mole compounds with the same number of metal ions. The standard enthalpies of atomization are relatively small quantities and would not be expected to exert a major effect on the comparative stabilities of the three types of compound as the group 1 metal varies. The formation of the oxide anion (O^{2-}) requires the dissociation of the dioxygen molecule and its enthalpy of formation is given by $\Delta_a H^\circ(O_2) + E_{ea(1)}(O) + E_{ea(2)}(O) = 248 - 142 + 844 = 950$ kJ mol^{-1}. By comparison the peroxide and superoxide ions do not require the breaking of the dioxygen bond and their enthalpies of formation (see Problem 8.2) are given by:

$$\Delta_f H^\circ(O_2^{2-}) = E_{ea(1)}(O_2) + E_{ea(2)}(O_2) = -74 + 784 = 710 \text{ kJ mol}^{-1}$$
$$\text{and } \Delta_f H^\circ(O_2^{-}) = E_{ea(1)}(O_2) = -74 \text{ kJ mol}^{-1}$$

giving a clear advantage to the formation of superoxides. Superoxide formation loses out with the lattice energy terms of the above equations and with Li and Na the smaller cations confer a relatively greater stability on the oxides and peroxides.

It is usual to ignore entropy terms in general comparative thermochemical calculations, but in this case that would be a serious omission. In the formation of oxides, peroxides and superoxides there is a loss of entropy of the gaseous dioxygen molecule to the extent of one-half, one and two moles respectively for a comparison of equal moles of metal atoms reacting. These losses of entropy amount to contributions of 30, 61 and 122 kJ mol^{-1} respectively to the $\Delta_f G^\circ$ values via the $T\,\Delta_f S^\circ$ terms for oxide, peroxide and superoxide formation. The lattice energies of the oxides, peroxides and superoxides of the group I metals are given in Table 8.5 and their enthalpies of formation are given in Table 8.6.

Table 8.5 The lattice energies (kJ mol^{-1}) of the oxides, peroxides and superoxides of the group 1 metals

	$L(M_2O)$	$L(M_2O_2)$	$L(MO_2)$
Li	2907	2592	878
Na	2518	2309	799
K	2229	2114	741
Rb	2146	2025	706
Cs	2016	1948	679

The L and $\Delta_f H^\circ$ values demonstrate the expected higher L for oxides and peroxides which contain doubly charged ions. Since the ionic radius increases from O^{2-} (140 pm) to O_2^- (163 pm) to O_2^{2-} (180 pm), the increase of L in the group 1 metal series: $MO_2 \ll M_2O_2 < M_2O$ is expected A similar effect on L for the Group 2 oxides: $MO_2 < MO$ may be demonstrated.

Table 8.6 The standard enthalpies of formation (kJ mol^{-1}) of the oxides, peroxides and superoxides of the group 1 metals

	$\Delta_f H^\circ(M_2O)$	$\Delta_f H^\circ(M_2O_2)$	$\Delta_f H^\circ(MO_2)$
Li	-596	-640	
Na	-416	-505	-270
K	-362	-494	-280
Rb	-330	-426	-310
Cs	-318	-402	-315

For the group 1 metals $\Delta_f H^\circ(M_2O_2)$ is more negative than for M_2O, in each case becoming less negative down the group. However $\Delta_f H^\circ(MO_2)$ becomes more negative from NaO_2 to CsO_2. For Group 2, the trends in $\Delta_f H^\circ$ are less regular.

Problem 8.2 Use the following standard enthalpy changes, ΔH° at 298K (data in kJ mol^{-1}) to calculate:

(a) ΔH° for processes: $O_2(g)+e^-(g) \rightarrow O_2^-(g)$ and $O_2(g) + 2e^-(g) \rightarrow O_2^{2-}(g)$ and comment on their relative values;

(b) L for Li_2O_2, Rb_2O_2, NaO_2, BaO_2 and CaO_2 and comment on their relative magnitudes;

(c) ΔH° for the reactions:

(i) $MO_2(s) \rightarrow MO(s) + \frac{1}{2}O_2(g)$ for Ca and Ba;

(ii) $MO_2(s) \rightarrow \frac{1}{2}M_2O_2(g) + \frac{1}{2}O_2$ for Na and Rb;

(iii) $M_2O_2(s) \rightarrow M_2O(s) + \frac{1}{2}O_2(g)$ for Li and Na;

(d)What general conclusions can be drawn about the relative stability of oxides, peroxides and superoxides of the s block metals?

Element	H	Cl	Na	Ca	Ba
$\Delta_a H^{\circ}$	218	121	108	172	157
I_1			494	590	502
I_2				1150	966
E_{ea}		-364			

Compound	BaH_2	CaH_2	NaH	NaCl
$\Delta_f H^{\circ}$	-171	-189	-57	-411
L		2414		

Answer (a) ΔH° for $O_2(g)+e^- \rightarrow O_2^-(g) = -86 - 289 + 70\ 3 - 480 = -74$ kJ mol^{-1}

ΔH° for $O_2(g)+2e^- \rightarrow O_2^{2-}(g) = -508 + 2425 - 218 - 988 = 710$ kJ mol^{-1}

The negative ΔH° for the first process indicates that O_2 has an affinity to attract an electron. The positive ΔH° for the second process shows that a second electron added to O_2^- is not favoured due to repulsions between the singly negative ion and the incoming electron.

(b) $L(Li_2O_2) = 636 + 322 + 1038 + 710 = 2706$ kJ mol^{-1}

$L(Rb_2O_2) = 427 + (2 \times 86) + (2 \times 402) + 710 = 2113$ kJ mol^{-1}

The higher L for Li_2O_2 is due to the smaller radius of Li^+

$L(NaO_2) = 259 + 494 - 74 = 788$ kJ mol^{-1}

The low value is due to the size of O_2^- and the presence of one cation (unlike the two in peroxides)

$L(BaO_2) = 632 + 157 + 1462 + 710 = 2961$ kJ mol^{-1}

$L(CaO_2) = 661 + 193 + 1740 + 710 = 3304$ kJ mol^{-1}

The higher L for CaO_2 compared to BaO_2 is due to the smaller Ca^{2+}. Both are higher than L for the peroxides of Group 1 due to the charge (+2).

(c) (i) For CaO_2, $\Delta H^{\circ} = -690 - (-661) = -29$ kJ mol^{-1} and for BaO_2, $\Delta H^{\circ} = -569 - (-632) = 63$ kJ mol^{-1}, therefore, BaO_2 is more stable than CaO_2.

(ii) For NaO_2, $\Delta H^{\circ} = -253 - (-259) = 6$ kJ mol^{-1} and for RbO_2, $\Delta H^{\circ} = -213.5 - (-289) = 75.5$ kJ mol^{-1}. RbO_2 is more stable than NaO_2.

(iii) For Li_2O_2, $\Delta H^{\circ} = -594 - (-636) = 42$ kJ mol^{-1} and for Na_2O_2, $\Delta H^{\circ} = -414 - (-506) = 92$ kJ mol^{-1}. Na_2O_2 is more stable than Li_2O_2.

(d) Normal oxides of Group 2 are more stable than M_2O of Group 1.

The ability to form stable peroxides is greater for M_2O_2 than for MO_2.

The largest Group 1 metal ions can stabilize the superoxide O_2^- (less polarised than O_2^{2-}).

These conclusions follow from the trends in ϕ in the s block cations.

Reactions

All the normal s block oxides, except BeO and MgO, react with water forming hydroxides. BeO is insoluble, but the sparingly soluble MgO gives an alkaline reaction. The heat evolved from the reaction of water with MO increases from Ca to Ba. The peroxides react with water to form H_2O_2 besides the hydroxide except on warming (H_2O_2 is unstable). The superoxides react forming O_2 in addition to H_2O_2. As expected the oxides can combine with CO_2 or other acidic oxides and dissolve in acids to form salts.

Ozonides, containing the angular O_3^- ions, are known for the heavier Group 1 metal ions. Both $Ca(O_3)_2$ and $Ba(O_3)_2$ have been reported. Thus, stabilization by the less polarizing M^+ or M^{2+} ions is again demonstrated.

Suboxides; Group 1 metal clusters

Under certain conditions, Rb and Cs form suboxides containing metal clusters. Rb_9O_2 and $Cs_{11}O_3$ are composed of two and three face-sharing M_6O octahedra respectively. In the former, the Rb–O bonds are estimated to be 85% ionic, therefore it can be formulated as $(Rb^+)_9(O^{2-})_2(e^-)_5$, weaker Rb–Rb bonds are assumed to be delocalized over the metal system.

8.9 CHALCOGENIDES AND POLYSULFIDES

The elements in Group 16 other than oxygen (i.e. S, Se and Te, the chalcogens) form binary compounds with the s block metals, the sulfides being the most important. The compounds can be considered as the salts of H_2E, where E = S, Se or Te. Because these elements are less electronegative than oxygen, the Group 1 M_2E or Group 2 ME are less ionic than the oxides. Reaction with acids liberates the gaseous H_2E. The tendency to form polysulfides is parallel to the tendency to form peroxides and is more pronounced in Group 1 than in Group 2, especially down the two groups. Because of the strength of the S-S bonds, catenated S_x^{2-} ions can combine with the s block metals, the value of x increasing from Li_2S_4 to Na_2S_5 to M_2S_6 (M = K, Rb or Cs). Lower values of x are observed in the heavier Group 2 compounds eg BaS_4 and SrS_3. The SSS angles are not far from the regular tetrahedral value of 109°28′.

8.10 HYDROXIDES

All the group 1 MOH solids are soluble in water giving strongly alkaline solutions. Of the group 2 $M(OH)_2$ compounds, the Be hydroxide is only sparingly soluble in water, Be salts precipitate polymeric species on addition of hydroxide ions to Be^{2+}(aq) and eventually soluble $[Be(OH)_4]^{2-}$ is formed. This unique amphoteric behaviour, which resembles Al is related to the similarity in their ionic potentials, leading to departure from ionic behaviour, an example of diagonal similarity. The other group 2 hydroxides are somewhat soluble in water, but by no means as soluble as their equivalent group 1 compounds are.

8.11 HALIDES, POLYHALIDES AND PSEUDOHALIDES

The s block metals form halides MX and MX_2 (X = F, Cl, Br or I) which are mostly ionic, with some departure in BeX_2. The tendency to form polyhalide solids or in solution follows the same pattern as observed above ie the greatest tendency is found in the heavier Group 1 metals and to a less extent within the less polarizing M^{2+} ions. Besides the linear X_3^- ions, numerous anions with a mixture of I, Br and Cl are known, e.g. linear IBr_2^-, square planar $[XY_4]^-$ or $[X_mY_nI_p]^-$ with $m + n + p$ being an odd number. The larger the central atom and the more symmetrical the ion, the more stable are the compounds.

In X_3^-, a 3-centre 4-electron bond has been proposed. Alternatively, promoting an electron to a vacant d orbital is envisaged, although the promotion energy is high. Molecular orbital theory (see Fig. 6.9) indicates that the three atoms are held together by a pair of electrons in a σ_u^+ bonding orbital, the orbital having contributions from the p atomic orbitals of the participating atoms which are directed along the molecular axis. An unusual group of pseudohalides of Li and Na contain OCP or SCP anions.

8.12 SOLUBILITY TRENDS

The solubility of a substance depends on the Gibbs energy of solution, i.e. $\Delta_{sol}G^\ominus$ for:

$$MX(s) \rightleftharpoons M^+(aq) + X^-(aq) \tag{8.6}$$

in the case of Group 1 metal halides, MX. If entropy contributions are disregarded as a first approximation, the enthalpy change for (8.6), $\Delta_{sol}H^\ominus$ is given by:

$$\Delta_{sol}H^\ominus = L(MX) + \Delta_{hyd}H^\ominus(M^+) + \Delta_{hyd}H^\ominus(X^-) = L + \Sigma\Delta_{hyd}H^\ominus \tag{8.7}$$

Since L here is positive (an endothermic change) whereas the second term is exothermic, the sign of $\Delta_{sol}H^\ominus$ depends on a difference between these two large quantities. Unfortunately both L and $\Sigma\Delta_{hyd}H$ vary in a similar way on changing the size of the ions or their charges. In some cases, one of the terms in equation (8.6) predominates and determines the solubility trend. Table 8.7 is a matrix of solubility data for the group 1 halides, the solubilities are expressed as the number of grams of the solid which produce a saturated solution with 1 kg of water at 20°C unless otherwise stated.

Table 8.7 Solubilities of Group 1 halides

	F^-	Cl^-	Br^-	I^-
Li^+	2.7	830	1770	1650
Na^+	40	360	910	1790
K^+	950	347	670	1440
Rb^+	1310	910	1100	1520
Cs^+	3700	1860	1080	790

Table 8.8 The standard enthalpies of solution of the Group 1 halides

	F^-	Cl^-	Br^-	I^-
Li^+	4.5	-37.2	-49.1	-63.3
Na^+	0.3	3.9	-0.6	-7.6
K^+	-17.7	17.2	20.0	20.5
Rb^+	-26.3	16.7	21.9	26.1
Cs^+	-45.9	17.9	25.9	33.2

The calculated standard enthalpies of solution for the group 1 halides are given in Table 8.8. A comparison of the data given in Tables 8.7 and 8.8 shows that although there is a general correlation between the standard enthalpies of solution and the solubilities there are some serious discrepancies. For example, there is a reasonable correlation for the fluorides, but there are discrepancies in the orders of solubility data as compared with the orders of enthalpy data, particularly with the data concerning the K, Rb and Cs chlorides and bromides. The poor solubility of LiF is due to its high L value whereas the relatively low solubility of CsI is due to the smaller $\Delta_{hyd}H^{\circ}$ of the two large ions. The discrepancies between enthalpy calculations and the solubility data must imply that the entropy terms cannot be disregarded in a detailed attempt to explain the observed trends.

One conclusion which can be reached from the data in Table 8.7 is that the least soluble halide salt is that where the sizes of the cations and anions are similar. This seems to be a general conclusion which is used in synthetic inorganic chemistry that the precipitation or crystallization of a compound is more easily achieved if the cation and anion radii are similar.

8.13 NITRIDES AND AZIDES

Since the electronegativity of N is lower than that of O or F, the heats of formation of the nitrides are lower than either the corresponding oxides or fluorides. Li is unique among the Group 1 metals in burning in N_2 or air forming Li_3N. Similarly, the Group 2 metals form M_3N_2 solids. On the other hand, azides which contain the linear $(N{=}N{=}N)^-$ ions are more stable for the heavier less polarizing s block metals; a trend already observed with other polyanions. The trend in the stability of the nitrides parallels that of the hydrides.

8.14 STRUCTURES OF SOME SOLID COMPOUNDS

The NaCl structure is displayed by most Group 1 metal halides except for CsCl, CsBr and CsI. Although LiCl, LiBr and LiI have a radius ratio less than 0.414 and five other halides have a ratio greater than 0.732, they have the NaCl structure. The three exceptions adopt the CsCl structure with coordination numbers 8:8, as expected from their radius ratios. Except for those of Be, all the Group 2 chalcogenides, have the NaCl structure, with the exception of MgTe whose zinc blende structure (coordination number 4:4) is in accord with its radius ratio. Because of the uncertain radius of Be^{2+} and the expected departure of its chalcogenides from ionic character, the radius ratio is of little relevance to its compounds. Five chalcogenides have a radius ratio > 0.732 but still adopt coordination numbers 6:6. It has been suggested that the predominance of this structure would maximize the p_π-p_π orbital overlap along the *x*, *y* and *z* directions. Even the Group 1 metal hydrides have the NaCl structure in spite of the variable radius of the hydride ion in the solids.

Other structures worth noting are those of the M_2O, the structures of which are described as antifluorite, i.e. the cation and anion positions in fluorite are reversed. Cs_2O is an exception having a layer lattice. Many M_2E solids have the structure of the M_2O compounds.

The Group 2 halides adopt various structures, with the more ionic *fluorite* found in the fluorides of Ca, Sr and Ba whereas Mg F_2 has the rutile structure. The other Mg halides have a layer structure, indicative of departure from ionic character (small cation and large anions), Be halides are even less ionic than MgX_2, and have a chain structure except BeF_2 whose structure is similar to that of quartz.
(For further discussion see *J. Chem. Educ.*, 1985, **62**, 215)

8.15 THERMAL STABILITY OF OXOSALTS

The stability trend of the oxosalts is the reverse of the trend in hydrides and nitrides, i.e. the more polarizing M^{2+} ions, especially the lighter (smaller) ones, destablize the large oxoanions more than the larger M^+ ions. Semi-quantitatively, this can be dealt with by considering the Born-Haber thermochemical cycle, shown in Fig. 8.2, in which $L(MCO_3)$ and $L(MO)$ refer to the lattice enthalpies of the carbonate and the oxide respectively, $\Delta_f H^\ominus(MCO_3)$, $\Delta_f H^\ominus(MO)$ and $\Delta_f H^\ominus(CO_2)$ are the enthalpies of formation of the salt, the oxide and the gaseous product, i.e. CO_2 in the decomposition:

$$MCO_3(s) \rightarrow MO(s) + CO_2(g) \qquad (8.8)$$

and $\Delta_{decomp}H^\ominus$ is the enthalpy change for equation (8.8). The only unknown is $\Delta_{dissoc}H^\ominus(CO_3^{2-})$ which is the enthalpy of dissociation of $CO_3^{2-}(g)$ to $CO_2(g)$ and $O^{2-}(g)$. By using the first law of thermodynamics on the Born-Haber cycle it can be readily shown that:

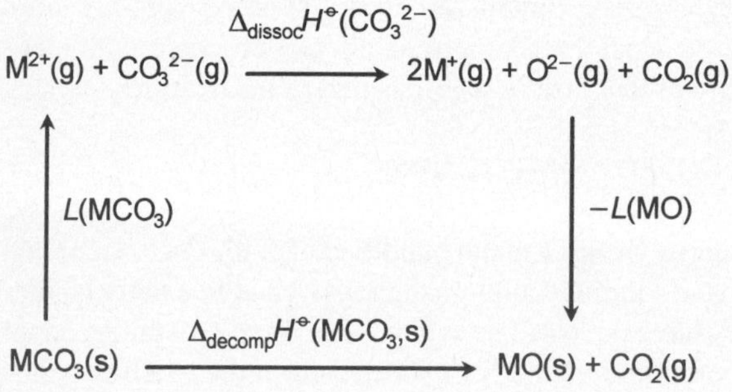

Fig. 8.2 A Born-Haber thermochemical cycle for the decomposition of Group 2 metal carbonates

$$\Delta_{dissoc}H^\ominus(CO_3^{2-}) = L(MO) - L(MCO_3) + \Delta_f H^\ominus(MO) - \Delta_f H^\ominus(MCO_3) + \Delta_f H^\ominus(CO_2) \qquad (8.9)$$

and

$$\Delta_{rdecomp}H^\ominus = L(MCO_3) + \Delta_{dissoc}H^\ominus(CO_3^{2-}) - L(MO) \qquad (8.10)$$

For the group 2 carbonates (Ca-Ba), $\Delta_{dissoc}H^{\circ}$ (CO_3^{2-}) has a value of about 670 kJ mol^{-1}. This implies that the two lattice energies in equation (8.10) are likely to be dominant in the determination of the enthalpy of dissociation of the carbonate into CO_2 and the metal oxide. The values of $\Delta_{dissoc}H^{\circ}$ (MCO_3) for the carbonates of Mg–Ba are 107, 178, 235 and 267 kJ mol^{-1} respectively indicating a general trend towards increasing thermal stability. Calculations for Be are not included because the compound $BeCO_3$ does not exist other than as a basic carbonate. The trend in thermal stability can be rationalized in terms of the general inverse relationship between L and the interionic distance $r^+ + r^-$, so that the value of $\Delta_{decomp}H^{\circ}$ for the decomposition of oxosalts can be regarded (consider the form of the Born-Landé equation, (7.10)) to be proportional to:

$$[1/(r_c + r_a)] - [1/(r_c + r_o)]$$

Since the radius of the oxide (r_o) is smaller than that of the anion, (r_a), the first term decreases less rapidly with increasing r_c, resulting in a more positive $\Delta_{decomp}H^{\circ}$ down a group.

Group 2 carbonates are less stable than group 1 carbonates, i.e. they decompose at lower temperatures. The stability of the carbonates increase down both groups. Similar considerations apply to the thermal stabilities of sulfates and nitrates.

8.16 ORGANOMETALLIC AND ELECTRON DEFICIENT COMPOUNDS

The s block metals have the lowest electronegativities among metals and their electronegativity coefficients decrease down both Groups. Hence, it is not surprising that these metals form the smallest number of organometallic compounds and their stability decreases down the two groups. Hence Be, Li and even Mg form relatively more organometallics than the rest of the group.

Methyls and alkyls

Methyl lithium, $Li_4(CH_3)_4$ has tetrahedral units of Li_4, with the C of CH_3 centred over the centre of each face. It is electron deficient and its bonding can be described in terms of 4-centre 2-electron bonds, with the possibility of additional Li...Li bonding. The polymeric $Be(CH_3)_2$ has CH_3 bridges between the Be atoms, with 3c2e bonds. Mg $(CH_3)_2$ is similar in structure and bonding. Other Mg organometallic compounds include RMgX, the alkyl and aryl Grignard reagents.

Generally, the heavier metals form more ionic and less stable organometallics. Sigma (σ) bonds can be formed between an s block metal and a carbon atom, e.g. in LiCCLi whereas electrostatic attraction is responsible for ionic $M^+(C_5H_5)^-$ or $Na^+(C_{10}H_8)^-$.

8.17 DIAGONAL SIMILARITIES

Many properties vary down a group in the opposite way to its variation across a period. If a movement is made diagonally from an s block metal to one in the period below it but one group to the right, the two elements may exhibit similarities. Thus Mg, which is diagonally related to Li, has a similar atomic radius to Li. The solubility of many salts of the two metals are similar, besides forming nitrides and organometallics. Both Al and Be have similar E°

values for the reduction of their aqueous ion to the metal (−1.66 V and −1.85 V respectively) and their oxides and hydroxides are amphoteric. Both form electron deficient methyls.

8.18 FRANCIUM AND RADIUM

The two heaviest s block metals are radioactive. Because the $t_{1/2}$ of the longest living Fr isotope (RAM = 212) is 21.8 minutes its chemistry was studied by tracer methods. It is expected to follow the same pattern found for the other Group 1 metals. On the other hand, the chemistry of Ra has been fully studied because of its use in cancer therapy. Its compounds and its reactivity are similar to Ba. Ra isotopes are formed in the radioactive decay series.

8.19 SOME APPLICATIONS OF s BLOCK ELEMENTS

The most important large scale industry based on the s block is the chlor-alkali process of electrolysis of brine using a mercury cathode. The sodium amalgam is treated with water to produce NaOH and dihydrogen. The chlorine is liberated at the anode. Another major process is the heating of limestone to produce CaO. Gypsum $CaSO_4.2H_2O$ is used in the building industry. Magnesium and to a lesser extent Be are useful alloying metals. Potassium compounds are used as fertilizers. Sodium bicarbonate is used in baking powder. Magnesium carbonate and sulfate are used for medical purposes for the treatment of stomach acidity and constipation respectively. Lithium carbonate is used to treat manic depression. Sodium and potassium ions are vital in many biological processes, Ca being essential for bones. In plants Mg is the central element in the chlorophyll molecule.

8.20 PROBLEMS

1. The following thermodynamic data (all in kJ mol^{-1} at 298K) are related to the decomposition of s block nitrates to their oxides.

Compound	$Ca(NO_3)_2$	$Ba(NO_3)_2$	CaO	BaO	NO_2	$LiNO_3$	Li_2O
$\Delta_f H^\circ$	−938.4	−992.1	−635.1	−553.5	33.9	−482	−596
$\Delta_f G^\circ$	−743.2	−796.7	−604	−525.1	51.3		

Calculate the temperature at which the first two nitrates would decompose and calculate ΔH° for the decomposition of $LiNO_3$. Discuss the stability of the nitrates of the s block metals.

2. (a) Calculate the solubility product of $Sr(OH)_2$ and its solubility in a solution of pH 13, given that (at 298K): ΔG° for: $Sr(OH)_2(s) \rightarrow Sr^{2+}(aq) + 2OH^-(aq) = 19.95$ kJ mol^{-1}
(b) Comment on the trend in the solubility of the Group 2 metal compounds, given the following solubility products:

Compound	Solubility product
$Ca(OH)_2$	1.3×10^{-6}
$CaSO_4$	2.4×10^{-5}
$SrSO_4$	2.8×10^{-7}

9

Chemistry of the p Block Elements

9.1 INTRODUCTION

Across each period, the electronegativity coefficient, χ, increases, but it decreases, though irregularly, down the groups of the p block, the elements of Groups 13–18 (see Chapter 3). High χ values characterize non-metals. Hence, χ is most relevant when discussing covalent bonding among non-metals. These trends in χ rationalize the occupation of the lower left hand section of the p block by metals, since metallic character is associated with low χ values.

Another general feature of the p block is the unique behaviour of the top element in each group. This is related, among other factors, to the limit on the capacity of the valence shell, which never exceeds 8 electrons. As is mentioned in Chapter 8, helium with its valence shell $2s^2$ is included with the other Group 18 (inert gases, noble gases) elements of the p-block. The usual vertical similarity, observed in the s and d blocks, is less marked in Group 13 (B – Tl) and especially so in Group 14 (C – Pb). However, increasing vertical similarity is shown by the later groups in the block.

The heaviest member of each group in the block has a tendency to be more stable in an oxidation state two units less than the maximum. This tendency is referred to as the 'inert pair effect'. On the other hand, members of period 4 (Ga – Kr) show some anomalous behaviour, e.g. higher χ values than expected. In spite of the wide range of the reactivity of the elements in this block, certain generalizations can be observed.

9.2 UNIQUE BEHAVIOUR OF THE TOP ELEMENT

Since the atomic radii decrease across a period, the p block atoms are smaller than their nearest s or d block atoms, thus F has the smallest radius. Associated with the small atoms, the 2p orbitals are very compact and influence the bonds formed. Interelectronic repulsions are thus more significant than in *n*p orbitals (where $n > 2$). Accordingly the N–N, O–O and F–F single bonds are comparatively weak, being weaker respectively than the P–P, S–S or Cl–Cl bonds. Another factor which can cause the latter set of bonds to be stronger than the former set is the use of d orbitals in the bonding. The instability of N_2H_4 and its derivatives, H_2O_2 and peroxo compounds and the high reactivity of F_2 can be thus explained.

A comparable trend is observed in the *electron attachment energies*, E_{ea}, of F, and the first electron attachment energies of O and N. The term *electron attachment energy* is defined as the enthalpy change when a gaseous atom, molecule or ion accepts an electron and is equivalent to the *electron affinity* of the atom, but with opposite sign. With the exception of these three top elements, E_{ea} becomes less negative with increasing size of the atom down a group, as expected from electrostatic considerations. The less negative values of E_{ea} of O and F than S and Cl respectively are explicable in terms of interelectronic repulsions in the 2p sub-shell. It has been estimated that the process: $N(g) + e^- \rightarrow N^-(g)$ is endothermic (7 kJ mol^{-1}) unlike the corresponding process for P which is exothermic (−71.7 kJ mol^{-1}).

Another tendency of the 2p orbitals is the efficient overlap of the p_π-p_π orbitals leading to π bonds between the same or different atoms of period 2, e.g. C=C, C≡N, C=O, C≡O, N≡N, N=N, N=O, and O=O. This tendency is not evident in the heavier elements, which accounts for the solid elements P_4 and S_8 (in which only σ bonds are used), unlike the gaseous N_2, O_2 and O_3 which have π bonding. Again, CO and CO_2 are gases unlike SiO_2 which is a solid with an infinite 3D structure. Gaseous N oxides contrast solids P_4O_6 and P_4O_{10}.

The small size of the atoms of N, O and especially F results in their high χ values. This is reflected in the formation of relatively strong hydrogen bonds, in X−H--Y, where X and/or Y may be N, O or F. As a result of this bonding, the hydrides NH_3, H_2O and HF have higher boiling points than the corresponding hydrides of the heavier atoms. By contrast, the boiling points of CH_4 and the other Group 14 hydrides increase from CH_4 to SnH_4 as the molecular size and magnitude of the van der Waals forces increase (larger molecules are more polarizable).

The valence shell capacity of the top elements limits the coordination number to a maximum of 4. However, in compounds of the lower elements, higher coordination numbers are attainable. Thus BH_4^- and BF_4^- contrast with $[AlF_6]^{3-}$, CF_4 contrasts $[SiF_6]^{2-}$ and NH_4^+ contrasts $[PCl_6]^-$. In $Be_4O(O_2CMe)_6$, O has a coordination number of 4 whereas 6-coordinated S is found in SF_6. Again, among the halogens coordination numbers of 5 and 7 are observed in ClF_5 and IF_7 whereas F can attain a coordination number of 2 in $[SbF_5]_4$ which contains Sb--F--Sb bridges. In the heavier members of each group, d orbitals are available for bonding and their participation may be envisaged in the attainment of the higher coordination numbers.

9.3 ALLOTROPY AND THE STRUCTURE OF THE SOLID ELEMENTS

Allotropes are different forms of an element which have a different number or arrangement of atoms in a molecule or crystal. Most of the p block elements, other than the metals, exhibit allotropy. The atoms of elements in Groups 14, 15, 16 and 17 can form four, three, two and one element-element bonds respectively. These are the coordination numbers of the atoms in the solid elements and can be generally achieved in more than one way, hence allotropic modifications are possible.

In Group 13, B is unique in having a number of modifications based on the regular B_{12} icosahedra, also found in some B compounds. The lower elements have metallic lattices, although in gallium each Ga atom in the solid has one close neighbour and 6 neighbours further away. The remarkably low melting point of Ga (30 °C) possibly represents the ease by which diatomic Ga_2 molecules are initially formed in the liquid phase rather than the complete loss of form that occurs when most metals melt. The boiling point of Ga (2397 °C) is entirely consistent with the usual decreasing values that apply to the metallic elements Al to Tl. Ga has the largest temperature range in which an element can exist in the liquid state.

Carbon, the top element in Group 14, has the well known allotropes: diamond, graphite and the newly discovered fullerenes. The hexagonal and pentagonal faces of the latter and the hexagons of the graphite layers have shorter C–C bonds than those in diamond in keeping with the delocalization of the π electrons throughout their structures. The four tetrahedrally arranged σ bonds in diamond produce a giant covalent lattice with the well-known exceptional hardness. The same structure is found in Si, Ge and the low temperature form of Sn. However, these elements are semi-conductors unlike diamond which is an insulator. The gradual change in electronic conductivity down the group leads to the metallic lattice of normal Sn and of Pb. The various crystalline forms of P have chains or rings of P_4 tetrahedra linked by P–P bonds showing a coordination number of 3, also found in the polymeric black P. Of the allotropes of As, Sb and Bi, the α forms have puckered sheets stacked in layers, with each atom having 3 nearest neighbours in the sheets.

Ozone, O_3, is a modification of oxygen. Whereas dioxygen supports life, O_3 in the lower atmosphere, as found in photochemically produced smogs in local industrialized zones, is poisonous, but the ozone layer in the stratosphere protects us from the more harmful ultraviolet solar radiation. It is formed from the photolysis of dioxgen by far-UV radiation, the oxygen atoms then reacting with dioxygen molecules to produce O_3. S, on the other hand, has a large number of allotropes, reflecting the variable strength and flexibility of the S–S σ bonds (their bond enthalpy could reach 430 kJ mol^{-1}). In the two common modifications of S_8, the crown-shaped rings are stacked differently. Of the numerous other cyclic forms, with 6–20 atoms per ring, the best known is the chair form of S_6. Infinite chains of S are obtained by quenching the free-flowing melt in water, the product being known as plastic sulfur. The polymeric state reverts slowly to the standard state of the element, rhombic sulfur which contains S_8 rings. The Se_8 molecules are arranged differently in α, β and γ Se, whereas grey Se has helical chains with a similar structure in the known form of Te. α-Po has a unique structure where the atoms are at the corners of a cube.

The solid halogens have molecular lattices where the covalently-bound X–X molecules are held in the crystal by relatively weak van der Waals forces. However, especially in I_2(s), there is evidence of weak intermolecular interactions, leading to weaker I–I bonds than in the vapour. Within the layer structure of the solid there seems to be some delocalization of the electrons which would normally contribute to the bonding of the diatomic molecules. This longer range interaction between the molecules weakens the intramolecular covalent bonds of the I_2 units in the solid.

9.4 HYDRIDES OF THE p BLOCK ELEMENTS

The formulae of the simple binary hydrides of the p block elements show a gradual decrease in the number of H atoms bound to an atom from four in CH_4 to three, two and one in NH_3, H_2O and HF respectively. These correspond to the number of E–H σ bonds. The shapes of these molecules and the bond angles are consistent with the VSEPR theory, the angles decrease from 109° in CH_4 to 107° in NH_3 and 104° in H_2O(g) (see the molecular orbital treatment in Chapter 6). However, for the corresponding heavier hydrides of Groups 15 and 16, the angles are nearer to 90°, suggesting that only p orbitals are involved in bonding in the heavier hydrides. The unique B_2H_6, which is electron deficient, has 4 H atoms coplanar with the two B's, with the other 2 H's bridging the 2 B's in a perpendicular plane (see Chapter 6).

Most of the hydrides are endothermic, i.e. have positive $\Delta_f H^{\circ}$ values except AlH_3(s), CH_4, NH_3, H_2E (E = O, S) and HX (X = F, Cl, Br). Many of the endothermic hydrides are very unstable, susceptible to attack by air and moisture.

Figs. 9.1 and 9.2 show plots of the E–H and E–E single bond enthalpy terms respectively, E being a p block element. The increase in the E–H bond enthalpy across a period is expected as the covalent radius of the atom E decreases and χ values of E increase, both leading to stronger bonds, due to better overlap of the orbitals. The very strong C–H bond is related to the maximum coordination of four and the utilization of all the available orbitals and valence electrons in bonding. However, in period 3 the Si–H and P–H bond enthalpy terms are nearly equal and in periods ≥4, there is no exception to the general increase across a period.

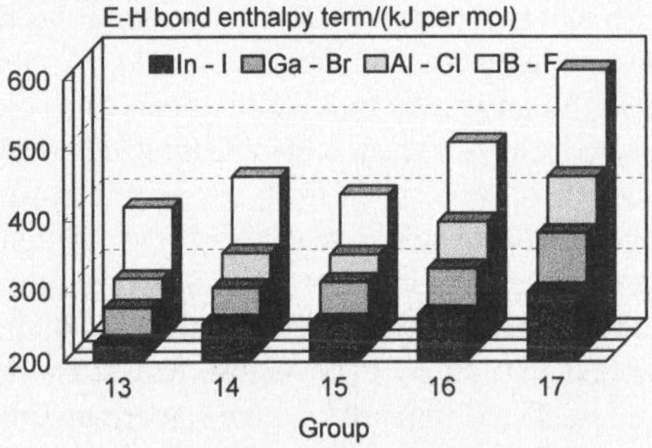

Fig. 9.1 The E–H bond enthalpy terms for the p block elements

In all the p block groups, the E–H bond enthalpy decreases down a group, as the covalent radius of E increases and as χ decreases. As a result, the thermal stability of the hydrides decreases down a group. The ready decomposition of the hydrides of As, Sb and Bi leading to the deposition of the element is a basis of the forensic detection of these poisonous elements. Parallel to the thermal instability, the reducing power of the hydrides increases down a group, even in the more stable hydrogen halides. Thus HI is a stronger reductant than HBr, reducing sulfuric acid to S and H_2S whereas HBr reduces the acid only to SO_2.

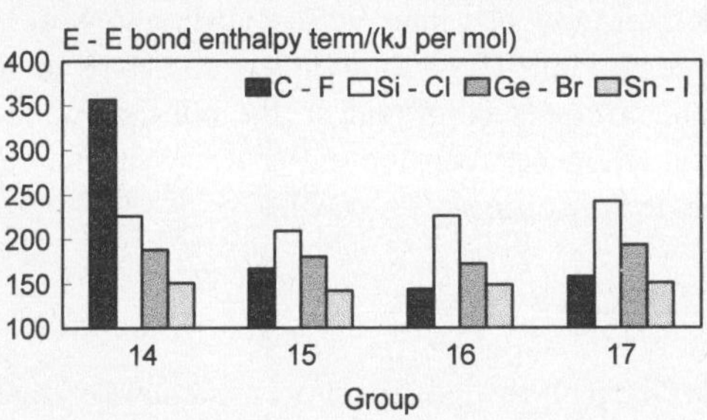

Fig. 9.2 The bond enthalpy terms for element–element bonds of the p block

The decrease in the basic strength and increase in acidic strength across a period from Group 15 to 17 is due to the increased χ of the atom E bound to H, allowing the release of H^+(aq) in solution. However, the basic strength decreases and acid strength increases down a group. Thus K_a (1.6×10^{-29}) and K_b (4×10^{-28}) for PH_3 are nearly equal, unlike NH_3. H_2S is a weak acid, but by contrast H_2O is weakly dissociated to H^+ and OH^-. But HF is a weak acid; the acid strength increases down the group. This trend, also found for Group 16 hydrides, is due mainly to the lower E–H bond enthalpy down the group.

Boron hydrides and related compounds

B_2H_6, like the other B hydrides is electron deficient in the sense of its having fewer valence electrons than are required to form conventional two-centre two-electron (2c2e) (single σ) bonds. The molecular orbital treatment of B_2H_6 is given in Chapter 6. Whereas the ethane molecule possesses 14 valence electrons—sufficient to form the seven single bonds required to bind the eight atoms together, the diborane molecule has only 12 valency electrons.

The higher boranes include the compounds B_4H_{10}, B_5H_9, B_5H_{11}, B_6H_{10}, B_6H_{12}, and $B_{10}H_{14}$. Examples of borate anions are $B_6H_6^{2-}$ and $B_{12}H_{12}^{2-}$ which have octahedral and icosahedral symmetries respectively. These are included because their structures are highly symmetric and have a bearing upon the structures and bonding of the less symmetric boranes. An understanding of the bonding in the higher boranes requires either a complex molecular orbital treatment or a simplified approach in which it is assumed that, in addition to the formation of terminal B–H and 3c2e bonds in BHB bridges as in diborane, there are two kinds of BBB bonds. These are B–B–B boron bridges and 'steering wheel' BBB triangular groups in which each of the B atoms presents an orbital to the centre of the triangle to form a 3c2e bond.

The M.O. approach to the octahedral *closo*-borate(6) anion typifies that which rationalizes all the borane and borate anion structures and their bonding. Each B atom is regarded as having two sp hybrid orbitals (this hybridization is feasible since the 2p–2s energy gap is small in boron) one of which is directed away from the cluster and the other of which is directed towards the centre of the cluster. The out-pointing hybrids are used in the formation of the B–H terminal bonds which are of the conventional 2c2e type. The inward pointing radial hybrid orbitals contribute to the binding of the cluster skeleton. The other orbitals of each B atom (p_x and p_z) are tangential to the skeletal atoms. The radial hybrids and the tangential atomic orbitals can be treated separately by symmetry theory and produce sets of molecular orbitals which transform as a_{1g}, t_{1u} and e_g (radial) and t_{1u}, t_{2g}, t_{2u} and t_{1g} (tangential) respectively and increase in energy in the orders given. The two sets have orbitals which transform as t_{1u} and these interact to give bonding and antibonding combinations. The M.O. diagram for the cluster orbitals and the terminal orbitals (the latter use n bonding orbitals containing $2n$ electrons) is shown in Fig. 9.3.

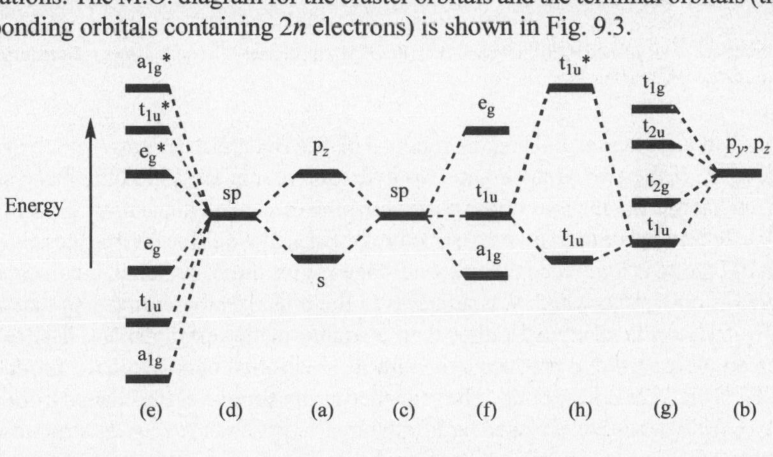

Fig. 9.3 The formation of terminal B–H and cage molecular orbitals in the borate(6) anion

The s and p_z orbitals of the boron atoms are shown at (a) and the p_x, p_y orbitals of the boron atoms are shown at (b). The s and p_z orbitals are allowed to form sp hybrid orbitals arranged radially so that six are inwardly pointing to the centre of the boron cage (c) and six are out-pointing towards the six hydrogen atoms (d). The out-pointing orbitals interact with the s orbitals of the hydrogen atoms to give sets of bonding and antibonding orbitals (e). The inwardly pointing boron sp hybrid orbitals form a set of radial group orbitals (f), one of which is an a_{1g} bonding orbital, three are non-bonding orbitals t_{1u} and two are antibonding orbitals e_g. The p_x and p_y orbitals of the boron atoms form tangential group orbitals (g), two sets of which are bonding and the other two sets are antibonding. Because the radial (f) and tangential (g) orbitals both contain orbitals which transform as t_{1u} these interact to give bonding and antibonding combinations (h). There are seven bonding orbitals which contribute to the bonding of the boron cage: a_{1g}, t_{1u} and t_{2g}. The other eleven orbitals concerning the cage are antibonding.

The 26 valence electrons of the *closo*-borate(6) dianion occupy the seven bonding orbitals of the cage-skeleton and the six orbitals reserved for B–H terminal bonding. It is a general rule that *closo*-clusters with *n* skeletal atoms require $2n + 2$ electrons for the cage structure of *n* atoms.

The actual structures of the higher boranes can be visualized in terms of the more symmetric borane anions, e.g. the structures of B_5H_9 and B_4H_{10} can be thought of as being derived from the structure of the borane anion $B_6H_6^{2-}$ by the removal one or two BH groups from their positions in the parent borane anion and with the necessary additions of $2H + 2H^+$ and 2H respectively as shown in Fig. 9.4.

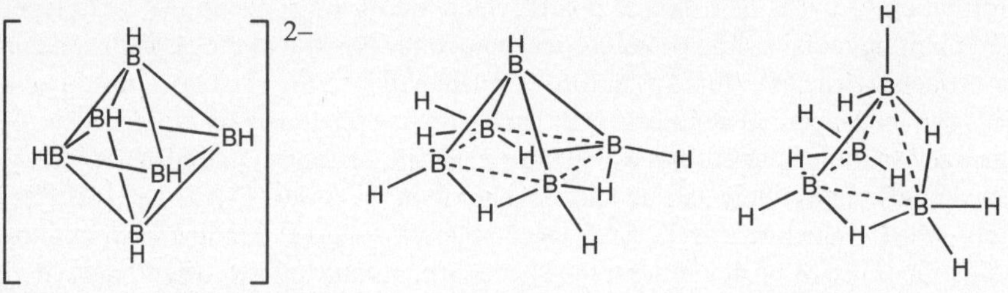

Fig. 9.4 Structures of the related $B_6H_6^{2-}$ ion, and B_5H_9 and B_4H_{10} molecules. The dotted lines indicate original B–B links in the borane ion structure

The $B_6H_6^{2-}$ anion is octahedral and the removal of one BH group leaves a square pyramidal B_5 framework. If the two protons and two hydrogen atoms are added the basic structure remains unchanged but the four boron atoms forming the square plane now all participate in B–H–B 3c2e bridge bonds to give the structure of B_5H_9. It would seem that there is a choice of which BH group is removed in the second stage to give the B_4 skeleton, but the one which is removed is one group which was adjacent to the one already removed so that a folded square B_4 skeleton is produced rather than a square planar arrangement. The two extra hydrogen atoms are used to produce two extra B–H terminal bonds with the production of an extra B–H–B 3c2e bridge bond. The sequence is an example of the change from a *closo*-structure (Greek: cage-like, a closed highly symmetric structure) to a *nido*-structure (Latin: nest-like) to an *arachno*-structure (Greek: web-like).

The structure of decaborane or borane(10), $B_{10}H_{14}$, can be thought of as being produced from the icosahedral $B_{12}H_{12}^{2-}$ ion by the removal of two vertices (2BH groups) and adding two protons and an extra two H atoms to the then three-connected (B–B) boron atoms. The structural relationship is shown in Fig. 9.5. In $B_{10}H_{14}$ all the B atoms are bonded to one terminal H and the two three-connected B atoms are also bonded to two of their B neighbours by B–H–B 3c2e bridges.

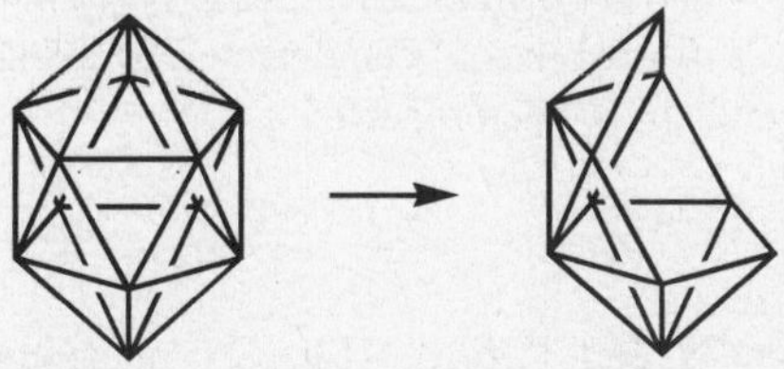

Fig. 9.5 A diagram showing the relationship between the boron skeletons of the icosahedral $B_{12}H_{12}^{2-}$ ion and of decaborane, $B_{10}H_{14}$. The actual structure of the molecule is not as regularly symmetrical as the one shown, the two 'end' BH groups are farther apart as the arachno structure opens up slightly. The bridging H atoms in the decaborane structure are shown in the lower diagram.

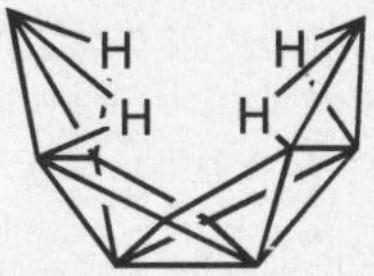

As a general rule, a *closo*-skeleton composed of n atoms requires a total of $2n + 2$ electrons in addition to the $2n$ electrons used to form terminal bonds outside the cluster. In the case of the anions $B_nH_n^{2-}$, there are $2n + 2$ electrons involved in cluster bonding (the two extra electrons arise from the charge on the anion). These electrons completely fill the $n + 1$ bonding m.o's. Their regular shape has n vertices. The m.o. diagram for $B_6H_6^{2-}$ shows that its 14 cluster electrons fill the 7 bonding m.o's. (the 6 B–H bonds are normal 2c2e bonds). The regular octahedral shape has 6 vertices which is the number of cluster B atoms which is one fewer than the bonding m.o.s. In the case of a molecule B_nH_{n+4}, Wade suggested that besides the $2n$ electrons from nBH groups, the 4 extra Hs (which may be extra terminal or bridging H atoms) contribute their 4 electrons to the cluster. Hence the total number of cluster electrons is $2n + 4$, which fill $n + 2$ bonding m.os. The cluster formed would have $n + 1$ vertices (one fewer than the number of m.o's). Since there are only n skeletal B atoms the polyhedron has 1 vertex unoccupied and the structure is described as *nido*. In a similar way a hydride B_nH_{n+6} will have $2n + 6$ cluster electrons which fill $n + 3$ m.o's. The polyhedron formed has $n + 2$ vertices but only n skeletal B atoms and is described as *arachno*, since 2 vertices are missing. As an example of these relationships the borane(4) molecule B_4H_{10} has four B–H terminal bonds requiring 8 electrons, leaving $22 - 8 = 14$ electrons for the skeletal bonding. These would be appropriate for an octahedral parent cluster requiring 6 + 1 bonding orbitals. There are two vertices missing from the octahedral parent giving the *arachno*-borane(4) structure.

The above generalizations are examples of what are known as Wade's rules, a detailed discussion of which are to be found elsewhere (see Mingos and Johnston in *Structure and Bonding*, **68**, 29, 1987). They have been put on a rational theoretical basis by Mingos so that they apply to any cluster molecules or ions and are more generally known as the Wade-Mingos rules (see Chapter 11 for examples of transition metal clusters).

Hydrides of Group 13 metals

A crystalline polymeric AlH_3 contains Al–H–Al 3c2e bonds. The compounds Ga_2H_6, InH_3 and TlH_3 are unstable gases.

Hydrides of Group 14 elements

Hydrocarbons are among the large number of organic compounds. Their stability is ascribed to the relatively high strengths of the C–H and C–C bonds. The tendency to form chains is referred to as catenation. This tendency is markedly reduced down the group, as the E–H and E–E bonds become longer and weaker (Figs. 9.1 and 9.2). In the hydrides E_nH_{2n+2}, n = up to 8 for Si, up to 5 for Ge and = 2 for Sn. Traces of PbH_4 have been detected by using radioactively labelled Pb. No simple E=E compounds have been reported.

Hydrides of Group 15 elements

Besides NH_3 and compounds mentioned earlier, N and P form $H_2E{-}EH_2$ unstable compounds, P_2H_4 burns in air and N_2H_4 is used as rocket fuel. Besides this hydride, N forms hydrazoic acid, HN_3, which contains the multiply bonded $-N{=}N^+{=}N^-$ group (π electron delocalization in m.o. language), and the unstable $H_2N{-}N{=}N{-}NH_2$. Because the P–P bond is stronger than the N–N bond, P can form cyclic hydrides: P_nH_x ($x = n + 1, n, n - 2$ or n - 4).

Hydrides of Group 16 elements

Although O forms only two hydrides: H_2O and H_2O_2, S forms H_2S and the endothermic H_2S_n (n = 2 - 8). These owe their formation to the strong S–S bonds. These hydrides decompose to $H_2S + nS$. The anions S_n^{2-} are found in compounds with s block metals. H_2E (E = Se, Te or Po) burn in air to give EO_2.

Halogen Hydrides

Hydrides E–H are stable when E = F, Cl or Br, the weak H–I bond leads to an equilibrium mixture of HI, H_2 and I_2(g). In aqueous solution the hydrides are strong acids (fully dissociated) except for HF which undergoes hydrogen bonding and dissociates weakly into H^+ and HF_2^-.

Problem 9.1

Calculate pK_a for HF(aq) using the following standard ΔG° values (in kJ mol^{-1} at 298K):

$$\Delta_{vap}G^\circ(\text{HF, aq}) = -23.9,\ D(\text{HF, g}) = 535.1,\ E_{ea}(\text{F}) = -347.5,$$
$$\Delta_{hyd}G^\circ(\text{H}^+, \text{g}) + \Delta_{hyd}G^\circ(\text{F}^-, \text{g}) = -1513.6,\ I(\text{H,g}) = 1312$$

Compare the calculated pK_a with those of the other HX(aq) given below. Calculate also K_b and pK_b for F^-(aq).

Acid	HCl	HBr	HI
pK_a	-7.0	-9.5	-10

Answer

$$\Delta G^\circ(\text{HF,aq.} \rightarrow \text{H}^+\text{,aq.} + \text{F}^-\text{,aq})$$
$$= 23.9 + 535.1 + 1320.2 - 347.5 - 1513.6 = 18.1 \text{ kJ mol}^{-1}$$
$$\Delta G^\circ = -RT\ln K_o = -2.303RT\log K_o$$
$$18100 = -8.314 \times 298 \times 2.303 \log K_o$$

$$\ldots K_a = 6.73 \times 10^{-4} \ \ldots pK_a = 3.17$$
$$K_b(F^-) = K_w/K_a = 10^{-14}/6.73 \times 10^{-4} = 1.49 \times 10^{-11}$$
$$pK_b = 10.83$$

The $\Delta G^{\ominus}$ values for HCl, HBr and HI are −39.7, −54.0 and −57.3 kJ mol^{-1} respectively. The acid strength increases down the group, with *D*(HX,g) (431, 366 and 299 kJ mol^{-1} for HCl, HBr and HI respectively) playing the major role. Qualitatively, thc weaker the H−X bond the more easily the dissociated acid is formed in solution.

Ammonium and phosphonium compounds

The NH_4^+ ion is intermediate in size between Rb^+ and K^+. NH_4^+ salts resemble the alkali metal salts in their solubility, but differ in the acidic reaction of the solutions of ammonium salts of strong acid. They are also unstable on heating giving NH_3 and the hydride of the non-metal e.g. HX. However, ammonium salts of oxoacids, decompose to give a product of the oxidation of NH_3 and the reduction of the anion, e.g. $NH_4NO_2 \rightarrow N_2O + 2H_2O$. Phosphonium salts are less stable than NH_4^+ salts, the stability increases in the series PH_4X (X = Cl < Br < I).

Intrinsic E−H energy terms

The bond enthalpy terms depicted in Figs. 9.1and 9.2 are based on experimental thermochemical data, hence the term, thermochemical bond enthalpies. The thermochemical cycles used in their estimation include $\Delta_a H^{\ominus}$, which refers to the production of the atoms of the element in their ground state. However in covalent bonds such as the E−H bonds, the electrons of the atom E could be thought of as being rearranged to occupy orbitals singly in an excited state, known as the *valence state*, ready to combine with H. To produce the gaseous atoms in their valence state requires an amount of *promotion energy*, *P*, which can be estimated from the atomic spectrum of the element concerned. The valence state is understood to be that in which, for example in the case of the carbon atom, has four unpaired electrons in tetrahedrally disposed orbitals which can overlap with the four hydrogen 1s orbitals to produce CH_4. Such valence states are imaginary, but the electronically excited state corresponding to four unpaired electrons in the carbon atom, ^5S with parallel spins, has not been observed. The lowest energy state corresponding to the $2s^1 2p^3$ configuration of carbon is the ^3D state which is 752 kJ mol^{-1} higher than the ground state, ^3P.

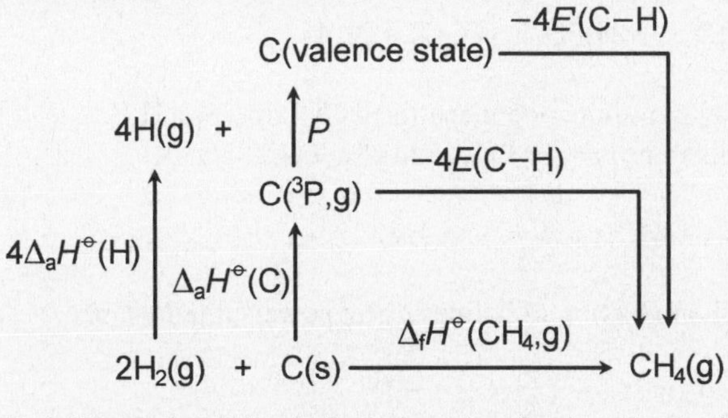

Fig. 9.6 Thermochemical cycles for the formation of methane showing the difference between thermochemical bond enthalpy and intrinsic bond enthalpy for the C−H bond

If the promotion energy is incorporated into the estimation of the C−H bond enthalpy as is shown in Fig. 9.6 the resulting value is referred to as intrinsic.

The normal bond enthalpy for the C–H bond is calculated as:

$$E(\text{C–H}) = \tfrac{1}{4}(4\Delta_a H^\circ(\text{H}) + \Delta_a H^\circ(\text{C}) + P - \Delta_f H^\circ(\text{CH}_4)$$
$$= \tfrac{1}{4}(4 \times 218 + 715 + 75) = 416 \text{ kJ mol}^{-1}$$

and the intrinsic bond enthalpy as:

$$E(\text{C–H}) = \tfrac{1}{4}(4\Delta_a H^\circ(\text{H}) + \Delta_a H^\circ(\text{C}) + P - \Delta_f H^\circ(\text{CH}_4)$$
$$= \tfrac{1}{4}(4 \times 218 + 715 + 752 + 75) = 604 \text{ kJ mol}^{-1}$$

The use of intrinsic bond enthalpies is extremely limited and debatable, since there are many cases where the concept is unnecessary, e.g. in compounds of Groups 15, 16 and 17 where only 3, 2 and 1 single bonds are required and where no electronic promotions are needed to supply the appropriate numbers of unpaired valence electrons.

9.5 PREPARATION OF HYDRIDES

1. The more reactive elements in Groups 16 and 17 combine directly with H_2, reactivity increasing across the period and decreasing down a group. Thus F_2 reacts explosively but the explosion of an H_2/O_2 mixture occurs for specified ratios while Cl_2 explodes when its mixture with H_2 is exposed to direct sunlight. On the other hand the combination of N_2 and H_2 takes place only under pressure and in presence of a catalyst at temperatures of about 450°C. This temperature is required to allow a fast rate of reaction, although it does not favour the formation of NH_3. A free radical chain mechanism explains the explosive reactions.
2. Acidification of a binary metal compound is a general method. Boranes and silanes were first prepared and studied by treating MgB_2 or Mg_2Si with an acid in a vacuum system. Acidic gaseous hydrides are displaced from their salts by non-oxidizing less volatile acids on warming or heating.

$$H_2SO_4 + MX \rightarrow MHSO_4 + HX \text{ (X = F, Cl, M = Group 1 metal)}$$

$$H_3PO_4 + MX \rightarrow MH_2PO_4 + HX \text{ (X = Br, I)}$$

Depending on the metal sulphide, either dilute or concentrated HCl liberates H_2S.
3. Warming with alkali solution liberates NH_3 or PH_3 from EH_4X (E = N, P)

$$EH_4X + OH^- \rightarrow EH_3 + X^- + H_2O$$

4. A general useful method is by the reduction of halides using powerful reductants:

$$SiCl_4 + LiAlH_4 \rightarrow LiAlCl_4 + SiH_4$$

$$4BF_3 + 3LiAlH_4 \rightarrow 2B_2H_6 + 3LiAlF_4$$

9.6 SOME REACTIONS OF HYDRIDES

1. The basic or acidic behaviour in aqueous solutions of hydrides has been mentioned above. Other p block hydrides undergo hydrolysis in water. Boranes and silanes are sensitive to water, giving H_3BO_3 and SiO_2 respectively with the liberation of H_2. Similar reactions take place with alcohols.
2. Pyrolysis or photolysis of hydrides often breaks the E–H bonds, the conditions depend on the strength of the bonds:

$$SiH_4(g) \rightarrow Si + 2H_2 \text{ at } 500°C$$

$$EH_3(g) \rightarrow E + {}^3/_2H_2 \quad (E = As, Sb, Bi)$$

$$H_2O(g) \xrightarrow{h\nu} H_2 + ½O_2$$

Electrolysis of HCl solution liberates H_2 at the cathode and chlorine at the anode.
3. The reducing power of a hydride depends on the E–H bond enthalpy. Thus NH_3 reduces CuO on heating whereas PH_3 or AsH_3 reduce some metal ions in aqueous solution to phosphide or arsenide or a mixture of either with the solid metal.
4. Reaction with O_2: boranes or silanes are sensitive to air and are rapidly, and in some cases, explosively oxidized. When impure, PH_3 burns in air to give P_4O_{10}, whereas the oxidation of NH_3, which requires heat and a catalyst to produce NO, is a stage in the manufacture of HNO_3.
5. Formation of adducts: The electron deficient B_2H_6 forms LBH_3 adducts, whose stability depends on the donor power of the group L. On the other hand, EH_3 especially NH_3 can act as donors e.g. in transition metal ammines. Group 14 hydrides are weak acceptors.

9.7 HALIDES OF THE p BLOCK ELEMENTS

Figs. 9.7 and 9.8 show how the E–F and E–Cl bond enthalpy terms (where E is an atom of the p block) vary down Groups 13-17 (excluding the p block metals which form more ionic halides). The energy terms depend on the overlap between the orbitals forming the σ bond, the electronegativity difference $\chi(X)-\chi(E)$ as well as interelectronic repulsion and back bonding.

It can be seen from Figs. 9.7 and 9.8 that the bonds E–F are stronger than the corresponding E–Cl bonds and the trend continues in the E–Br and E–I bonds.

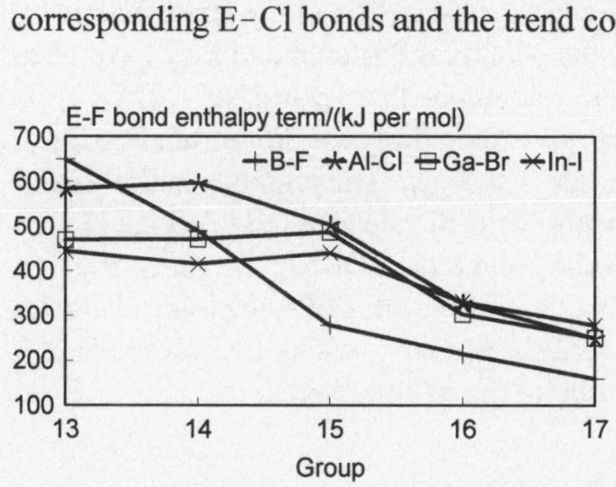

Fig. 9.7 E–F bond enthalpies for the p block elements

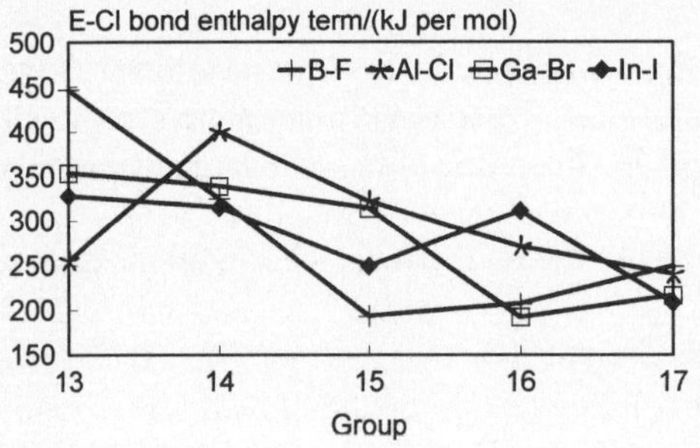

Fig. 9.8 E–Cl bond enthalpies for the p block elements

The strength of the E–F bonds is a consequence of their length (F being very small in size and it has the highest χ value). Considering the fluorides of period 2, there is a decrease in E–F energy across the period, partly due to the decrease in the difference: χ(F)–χ(E). However the very strong B–F bond and the other B–X bonds are due to π overlap (π orbitals of X and the vacant π orbital of B). This is also clear from the B–X bond length which is less than the sum of the covalent radii of B and X. On the other hand, the weak N–F and O–F bonds, like the F–F, N–N and O–O bonds, are ascribed to interelectronic repulsions. Along period 3, the bond enthalpy term also decreases across the period, presumably due to the decrease in χ(F)–χ(E) but the bonds are stronger than the corresponding bonds in the earlier period. The trends in the E–Cl bonds are roughly similar to the E–F bonds. The main difference is between the bonds to N, O and the halogen. These bonds increase in strength from E = N < O < F as expected from the increase in electronegativity.

Boron halides

The triangular planar BX_3 molecules have a valence shell of 6 electrons and can readily act as acceptors, forming adducts: LBX_3. The halides B_2X_4 contain a B–B bond, the molecules are planar, e.g. B_2F_4 and solid B_2Cl_4. A series of subhalides are known which are electron deficient e.g. B_nCl_n (n = 4, 7–12) and B_nBr_n (n = 7–10).

Halides of Group 14

The stability of the CX_4, of T_d symmetry, decreases from the very stable CF_4 to CI_4 which smells of I_2. As the RMM increases, so do the melting or boiling points, a general trend in BX_3 and in other covalent halides. An important group of mixed halides, CF_xCl_{4-x} (x = 1–3,) though used as refrigerants, allegedly participate in the destruction of the stratospheric O_3 layer. Their manufacture is now prohibited by the Montreal Protocol and they have been replaced by a range of hydrofluorocarbons. Although the hydrolysis of CCl_4 is thermodynamically feasible, it is kinetically stable towards hydrolysis. Silicon tends to form, besides SiX_4 and $SiX_nX'_{4-n}$, a series of higher halides, Si_nX_{2n+2}. The vigorous hydrolysis of $SiCl_4$ in contrast to CCl_4, is related to the expansion of the Si valence shell to accept H_2O in an intermediate step. Partial hydrolysis gives cyclic polymers $(SiOCl_2)_n$. Ge shows the first tendency to form divalent compounds, e.g. GeX_2, which are solids of some ionic character and which are reductants reverting to the more stable +4 GeX_4, which hydrolyse rapidly. Complex anions $[GeX_6]^{2-}$ (X = F or Cl) are similar to the Si complexes.

Group 15 halides

Among the N halides, NF_3 is the most stable, stability decreases with increase in RMM. NBr_3 is explosive and NI_3 has not been prepared. The unreactivity of the trigonal pyramidal NF_3 is related to its low polarity, the lone pair moment is in the opposite direction to the N–F bond moments. This situation is the reverse of the polar NH_3 molecule. The hydrolysis of most covalent halides split the E–X bonds with the formation of HX acid. However, NCl_3 is hydrolysed to NH_3 and HOCl, with the intermediate formation of H-bonded species, e.g. Cl_3N--HOH. The halides of the heavier elements are more stable and represent the +5 and +3 oxidation states. All EX_3 compounds are known, with the melting or boiling point increasing with RMM. Most have molecular lattices, with the triangular pyramidal molecules as the unit.

The tendency towards some ionic character with BiX_3 is expected from its metallic behaviour. Intermediate between the ionic and molecular lattices are the layer lattice of AsI_3. Although all EF_5 are known, the tendency to form pentahalides decreases down the group and from Cl to I. This is expected on the basis of the weaker E–X bonds especially when $\chi(E) = \chi(X)$. The fluxional behaviour of PF_5 (this is the rapid interchange between apical and equatorial positions by the F atoms) is shown by ^{19}F NMR spectroscopy. At room temperature, there is a single absorption by ^{19}F nuclei split by coupling interaction with the ^{31}P nucleus, i.e. there are two equivalent absorptions consistent with a single fluorine environment. At low temperatures there are two different ^{19}F absorptions in a 2:3 intensity ratio, but still showing ^{31}P coupling. At these temperatures the molecular structure is not fluxional and the two different F environments persist. $PCl_5(g)$ is trigonal bipyramidal but the solid is ionic $[PCl_4]^+[PCl_6]^-$. Because of the relative strength of the P–P bond, P_2X_4 (X = F, Cl, I) have been studied. Less stable are the corresponding compounds of the heavier elements, e.g. E_2I_4 (E = As, Sb), as expected from the weakness of the E–E bonds. Adducts and complexes are known. PF_3 is a powerful π acceptor ligand. $[ECl_4]^+$ and $[EF_6]^-$ (T_d and O_h symmetry respectively) are found either together or with different counter ions. The lower elements also form tetrahedral $[EX_4]^-$, square pyramidal $[SbX_5]^{2-}$ and octahedral $[EX_6]^{3-}$.

Group 16 halides

Compounds of O and the halogens, considered as halogen oxides, are dealt with together with other oxides of the p block, most of which are acidic. Among the other chalcogens (i.e. Group 16 elements), S tends to form unique compounds due to the strength of the S–S and S–F bonds. SSF_2 (C_s point group) contains an S=S bond with terminal S^{II} and a central S^{IV} with a lone pair. S–S single bonds occur in S_2F_4 with a S–S–F bond as well as a lone pair. A dimer of SF_2 is formulated as F_3SSF. In S_2F_{10} the long S–S bond suggests F to F repulsions. One F can be replaced by Cl in ClF_4SSF_5. SF_6, like SeF_6 and TeF_6, is octahedral and is an example of the stabilisation of the highest oxidation state by F. The observation that SF_6 is not reactive, i.e. kinetically inert, is due to the fact that it is coordinatively saturated. Although no ECl_6 compounds are known (the S–Cl bond is weaker than S–F bond), one Cl or even one Br can replace one F to give mixed hexahalides. The toxic S_2F_{10} disproportionates on heating to SF_4 and SF_6. Although the formal oxidation number of S in S_2F_{10} is +5, S is hexacoordinated to five F atoms and to the other S. The trend in the p block compounds is a change in oxidation number by two units. In the +4 state, the distorted tetrahedral EF_4 are formed by the three E elements but E_4X_{16} (where E = Se, Te and X = Cl, Br or I) have been prepared. The angular EX_2 molecules are not very stable and none of the iodides has been isolated. The E_2X_2 compounds are usually more stable than the EX_2 but even here not all the possible species have been prepared. S chains are found in S_nCl_2 and S_nBr_2 (n = 2–8), representing another example of catenation. The subhalides of Te, e.g. Te_2X, on the other hand, are based on helical chains of Te.

The halogens and interhalogens

Excluding the radioactive At, the most reactive of the four halogens is F_2 due to its small dissociation energy and the strong E–F σ bonds formed with non-metals and the high lattice enthalpy of MF_n solids formed with metals. Reactivity decreases from Cl_2 to I_2 as the physical state changes from gas to liquid to solid and E–X bond enthalpy decreases as the bond length increases. The high oxidizing power of F_2 ($E^{\ominus}$ for $\frac{1}{2}F_2(g) + e^- \rightarrow F^-(aq)$) is associated with low dissociation energy of F_2 and the highly negative $\Delta_{hyd}H^{\ominus}$ of the small F^- ion. The decrease in $E^{\ominus}$ from Cl_2 to I_2 (see Sec. 7.2 for a discussion of the reduction potential trend) explains the order of the replacement reaction: $\frac{1}{2}X_2(g,l) + X'^-(aq) \rightarrow \frac{1}{2}X'_2(g,l) + X^-(aq)$, where X′ is the halogen with the less positive $E^{\ominus}$.

The formation of interhalogen compounds, XX'_n (n = 1, 3, 5, 7) is not unexpected for the highly reactive halogens, the odd number n shows the tendency to pair electrons in σ bonds. As the size of X increases, the number n increases, e.g. BrF_5 and IF_7 (trigonal bipyramidal and pentagonal bipyramidal respectively). Halogen fluorides may act as ionizing solvents, the liquids are associated via F bridges. Self ionization, e.g. $BrF_3 \rightleftharpoons [BrF_2]^+ + [BrF_4]^-$ shows that $[BrF_2]^+[SbF_6]^-$ acts as an acid in this solvent whereas $AgBrF_4$ acts as a base. On mixing the two, the solvent BrF_3 is formed (similar to neutralization in aqueous solution). The XX'_3 are T-shaped (two equatorial lone pairs in VSEPR theory) whereas the dimer I_2Cl_6 is the corresponding stable compound.

The anions like $[BrF_4]^-$, termed polyhalides, are mentioned in Chapter 8. In addition to the halogens, molecules like C_2N_2, NCSSCN are termed pseudohalogens and CN^-, ECN^- (E = S or O), CNO^- or N_3^- are called pseudohalides.

Noble gas halides

Due to its lower ionization energy, Xe is the only noble gas which forms relatively stable fluorides. Depending on temperature and pressure, one of the fluorides XeF_n (n = 2, 4 or 6) is formed. VSEPR arguments correctly predict the linear shape of XeF_2 and the square planar XeF_4. However, XeF_6 is distorted octahedral and is fluxional. E(Xe–F) = 130 kJ mol^{-1} is only a little lower than E(F–F). The assumption that promotion of Xe electrons to its 5d orbitals (to form the excited valence states for the three fluorides) is necessary for bonding is energetically disadvantageous. An alternative explanation of bonding is the use of 3c4e bonds, where two electrons occupy a bonding and a non bonding molecular orbital. Because of the lower $\chi(Cl) < \chi(F)$, only unstable $XeCl_2$ and $XeCl_4$ have been identified by infrared spectroscopy. Kr, which has a higher ionization energy forms the less stable KrF_2. In addition to the neutral molecules, there are anionic species, e.g. XeF_5^- (planar pentagonal), $[XeF_8]^{2-}$ (square antiprism) and $[XeF_7]^-$, the latter two form ionic compounds with the lower alkali metals. Cationic species are also formed.

Problem 9.2

The following standard ΔH°'s in kJ mol^{-1} at 298 K are:
$\Delta_f H^{\circ}(NF_3,g) = -114$, $\Delta_f H^{\circ}(NCl_3,g) = 258$, $\Delta_f H^{\circ}(PCl_3,g) = -287$, $\Delta_a H^{\circ}(N_2,g) = 473$, $\Delta_a H^{\circ}(P,s) = 315$, $D(Cl_2,g) = 242$, $D(F_2,g) = 158$
Calculate:
(a) average N–F thermochemical bond enthalpy in NF_3;
(b) average N–Cl thermochemical bond enthalpy in NCl_3;
(c) average P–Cl thermochemical bond enthalpy in PCl_3;
Discuss the data provided and calculated related to the stability of the trihalides of N and P. The following electronegativities $\chi F = 4.1$, $\chi Cl = 2.8$, $\chi P = 2.1$ are useful for the discussion. Suggest a reason why NCl_3 hydrolyses by a different mechanism from the hydrolysis of PCl_3.

Answer

(a) $E(N-F) = 473 + 237 + 114)/3 = 275$ kJ mol^{-1}
(b) $E(N-Cl) = (473 + 363 - 258)/3 = 193$ kJ mol^{-1}
(c) $E(P-Cl) = (315 + 363 + 287)/3 = 322$ kJ mol^{-1}

The stability of NX_3 decreases from NF_3 to NI_3 as the bonds become longer and weaker. The negative $\Delta_f H^{\circ}(NF_3,g)$ unlike NCl_3 is due to the lower $D(F_2)$ compared to $D(Cl_2)$ as well as $E(N-F) > E(N-Cl)$. The greater stability of PX_3 compared to NX_3 is due to the lower $\Delta_a H^{\circ}(P,s)$ than $\Delta_a H^{\circ}(N_2,g)$. The availability of vacant orbitals on P permits the attack of water dipoles. NCl_3 hydrolyses to NH_3 and HOCl.

Problem 9.3

Calculate ΔH° for: $Cl_2,g + 2I^-,aq. = I_2,s + 2Cl^-,aq.$ using the following ΔH°'s (in kJ mol^{-1}) at 298 K:
$\Delta_a H^{\circ}(I) = 138$, $D(Cl_2) = 242$, $E_{ea}(I) = -320$, $E_{ea}(Cl) = -364$, $\Delta_{hyd} H^{\circ}(I^-) = 308$; $\Delta_{hyd} H^{\circ}(Cl^-) = 378$.
Discuss the replacement of one halogen by another in aqueous halide solutions.

Answer

$\Delta H^{\circ} = 242 - 728 + 616 + 2 \times 320 - 276 - 2 \times 364 = -234$ kJ mol^{-1}
With the exception of F_2 (which reacts with water, liberating O_2), the higher the halogen in the group will replace a lower one from its salt solutions because $\Delta_{hyd} H^{\circ}$ becomes less negative down the group (as the size of X^- increases) and because E becomes less exothermic down the group (both compensate the higher D for the higher X_2 molecules).

9.8 SOME PREPARATIONS OF HALIDES

1. Many elements combine directly with a halogen. F_2 gives the highest fluoride:

$$E + 3F_2 \rightarrow EF_6 \ (E = S, Se, Te)$$

Chlorine and bromine give lower halides, e.g:

$$E + 2X_2 \rightarrow EX_4 \ (E = Se, Te; X = Cl, Br)$$

Iodine is the least reactive:

$$E_4 + 6X_2 \rightarrow 4EX_3 \ (E = P, As, Sb, Bi; X = Cl, Br, I)$$

2. A halide may be prepared from a lower one by halogenation, e.g:

$$PX_3 + X_2 \rightarrow PX_5 \ (X = Cl, Br)$$

$$ClF_3 + F_2 \xrightarrow{h\nu} ClF_5$$

or from a higher halide by reduction, e.g:

$$2TeF_6 + Te \rightarrow 3TeF_4$$

$$GeBr_4 + Zn \rightarrow GeBr_2 + ZnBr_2$$

3. HX(g or aq) is often used to prepare halides, generally in a low oxidation state. Either the element, its oxide, or hydroxide can be used:

$$Sn + 2HCl \rightarrow SnCl_2 + H_2$$

$$GeO_2 + 4HCl(aq) \rightarrow GeCl_4 + 2H_2O$$

$$Ge(OH)_2 + 2HI \rightarrow GeI_2 + 2H_2O$$

HF may be produced in a reaction mixture, e.g:

$$2CaF_2 + 2H_2SO_4 + SiO_2 \rightarrow SiF_4 + 2CaSO_4.2H_2O$$

4. Halogen exchange or other exchange can be also used, e.g:

$$SnCl_4 + 4HF \rightarrow SnF_4 + 4HCl$$

$$SF_4 + SeO_2 \rightarrow SeF_4 + SO_2 \ \text{(more volatile)}$$

$$NaCl + 3F_2 \rightarrow ClF_5 + NaF$$

The inequality in lattice energies, $L(NaF) > L(NaCl)$, is important in the last reaction.

9.9 SOME REACTIONS OF HALIDES

In addition to some of the above reactions, which characterize halides, the following are noteworthy.

1. The E–X bond is subject to attack by water and hydroxylic solvents which break the bond with H forming HX. In addition an oxoacid of E or an oxide is formed:

$$BX_3 + 3H_2O \rightarrow H_3BO_3 + 3HX$$

$$SiCl_4 + 2H_2O \rightarrow SiO_2 + 4HCl$$

2. Formation of adducts. BX_3 molecules can readily accept an electron pair from a donor, e.g. NH_3. The order of the strength of the acceptor ability of B halides is: $BF_3 < BCl_3 < BBr_3$, which is opposite to what is expected from the electronegativity of the halogen. This paradox can be resolved by the above-mentioned π-bonding between X and B. On the other hand, the trend in SiX_4 is the reverse of that in BX_3. In Group 15, PX_3 (especially PF_3) can act as a σ donor and π acceptor in metal complexes. The heavier elements in their pentahalides act as acceptors, e.g.

$$SbF_5 + 2HF \rightarrow [SbF_6]^- + [H_2F]^+$$

3. Formation of cationic or anionic species. The structure of PCl_5(s) and the self-ionization of the halogen fluorides, mentioned above are examples where both cation and anion exist in the same system. The formation of cations of the halogens, especially iodine are known with various structures and different charges, e.g.

$$3I_2 + FSO_2OOSO_2F \rightarrow 2I_3^+ + 2SO_3F^-$$

The overlap of the I 5p orbitals and Cl 2p orbitals gives a σ bonding and two non-bonding π orbitals in ICl, but two σ bonding and one non-bonding π orbitals in ICl_2^+. Hence the bond orders are 1 and 2 respectively.

4. Halogenation. The halogen fluorides are strong fluorinating agents, e.g. BrF_3 reacts with some metal oxides giving the metal fluoride in addition to Br_2 and O_2.

9.10 OXIDES OF THE p block ELEMENTS

The maximum oxidation number in each group is achieved in oxides. This increases by one unit across a period. Thus the oxides E_2O_n ($n = 3$, 5 or 7) are formed by elements in Groups 13, 15 or 17 respectively, and oxides EO_n ($n = 2$, 3 or 4) are formed by elements in Groups 14, 16 or 18 respectively. Many of the lighter oxides are gases, especially for C, N and F. The gaseous nature of CO_2, CO, NO and NO_2 arises from the π bonds between E and O. By contrast, the lower elements from solid oxides: SiO_2, P_4O_6, P_4O_{10}. In the last compound, and in SO_2 and SO_3, the strong short bonds to O are ascribed to d_π-p_π overlap. Oxidation numbers two units lower are generally formed, especially down a group.

Bond enthalpy terms

Data are not as extensive as those for halides. Table 9.1 gives the available data for E–O, E=O and E≡O bonds. The decrease from B to C is related to the decrease in $\chi(O)-\chi(E)$ but the weak E–O bonds, where E=N, O or F, show the effect of interelectronic repulsions in the compact 2p subshell already noted. The increase in the bond enthalpy with bond order for the C to O and N to O bonds is parallel to the decrease in the E to O bond distance as the bond order increases. The E–O bonds of the elements of Period 3 show the same trend but they are all higher than those of Period 2 as was found for the halides. The strong B–O and Si–O bonds explain the large number and structures of borates and silicates. The E=O energies show a similar trend as the corresponding E–O energies. However Si=O is lower than C=O whereas the O=O is intermediate between S=O in SO_2 and in SO_3.

Table 9.1 Bond enthalpy terms (kJ mol^{-1}) for E–O and E=O bonds of the p block elements

	13	14		15		16		17
	E–O	E–O	E=O	E–O	E=O	E–O	E=O	E–O
Period 2	536	358	799	201	607	144	494	214
Period 3		452	638	360	544		469	218
Period 4		385		301	389			201
Period 5								201

With the exception of a few neutral oxides (CO, NO, N_2O and oxygen fluorides), the majority of the oxides are predominantly acidic, generally soluble in water to give oxoacids. They tend to combine with basic oxides to form oxosalts. Depending on the number of replaceable H atoms, the oxoacids may be monobasic, e.g. HNO_3, dibasic, e.g. H_2SO_4 or even tribasic, e.g. H_3PO_4.

Crystalline boron oxide, B_2O_3, consists of a network of trigonal planar BO_3 units joined through their O atoms.

Besides CO (poisonous gas) and CO_2 (which plays an important role in the atmosphere and in the carbon cycle of animals and plants), C forms the endothermic suboxide O=C=C=C=O. The common form of SiO_2 is α-quartz which has an infinite 3D structure based on the sharing of corners of SiO_4 tetrahedra. The phase diagram of silica contains other crystalline forms stable over specified temperature ranges. A metastable SiO has been reported but it is not well characterised, neither is GeO. On the other hand, one form of GeO_2 has a structure similar to β-quartz whereas another has the rutile structure. The oxides of Sn and Pb are dealt with together with the other p block metals.

Oxides of nitrogen

Nitrogen forms oxides in the oxidation states +1 to +5 inclusive. In gaseous N_2O_5, an O bridges $2NO_2$ groups but the ionic solid contains $[NO_2]^+[NO_3]^-$. In neither form does the valence shell of N exceed the octet. Although $\Delta_f H^\ominus$ of the solid is negative, its $\Delta_f G^\ominus$ is positive in the gas phase. The well known $NO_2 \rightleftharpoons N_2O_4$ equilibrium has been studied fully, the colourless N_2O_4 is a planar molecule with a long N—N distance (164 pm). The angular paramagnetic NO_2 has an angle of 134°. The N_2O_3 molecule has a long N—N bond (186 pm) joining the NO and NO_2 groups. It is unstable, disproportionating to NO and NO_2. Gaseous NO is an odd electron molecule but there is evidence of an associated N_2O_2. NO is readily oxidized to NO_2 and their toxic mixture, referred to as NO_x, is found in polluted atmospheres. However, it has been recently discovered that NO plays an important biological role in our bodies and it also forms a number of important metal complexes. The oxidation number of +1 is represented by N_2O which is a linear NNO molecule (the 'laughing gas' used in anaesthesia). N_2O_3 and N_2O_5 are the anhydrides of the acids HNO_2 and HNO_3 respectively whereas N_2O_4 (or NO_2) give both acids when they react with water. The variety of the oxides of N arises from the N-O and N-N bonds, including p_π-p_π interactions.

Oxides of Group 15

The oxides E_4O_6 (E = P, As, Sb) contain tetrahedral arrangement of the E atoms with 6 bridging O atoms, the E-O represent single bonds but Bi_2O_3 has a polymeric structure. The P_4O_{10} molecule has a similar structure to E_4O_6 with 4 extra O atoms in shorter P=O bonds. However, As_2O_5 contains cross-linked $[AsO_6]$ and $[AsO_4]$ units. Sb_2O_5 is poorly characterized and so is Bi^V oxide. P has additional oxides where the P_4O_6 structure includes one, two or three P=O terminal bonds.

Group 16 oxides

The well known gaseous SO_2 and SO_3 (V-shaped and triangular planar respectively) are acid anhydrides. Solid SO_3 has trimeric and polymeric forms which contain O bridges. SeO_3 is tetrameric whereas EO_2(E = Se or Te) and TeO_3 are polymeric solids. S is unique in forming S_nO which have a S=O terminal bond plus the ring of S_8 (crown shape) or S_6 (chair shape). The unstable, SO, S_2O or S_2O_2 are known also as ligands. SeO_2 is less stable than either SO_2 or TeO_2. Higher S oxides are derived from SO_3 by peroxo-bridges, giving SO_{3+x} ($x < 1$). SeO_3 is thermally unstable unlike EO_3 (E = S or Te). This is another example of what is termed a 'mid-row anomaly'.

Group 17 oxides

The OF_2 molecule is V-shaped like Cl_2O. The other oxide of F, i.e. O_2F_2, is very unstable and has an O-O bond which is very weak. Chlorine forms the largest number of oxides among the halogens. In ClO_2, the short O-Cl bonds indicate π bonding and the oxide is gaseous like Cl_2O. The other oxides are liquids: Cl_2O_6 (as a solid it is $[ClO_2]^+[ClO_4]^-$) and the unstable Cl_2O_7. The two oxides of Br are Br_2O (polymer at low temperature) and Br_2O_4. The most stable oxide of I is I_2O_5 which is a polymeric solid and similarly is I_2O_4 which is less stable. Both this and the unstable I_4O_9 disproportionate to I_2O_5 and I_2.

Xe oxides

The very weak Xe-O bond indicates that the 2 oxides: XeO_3 and XeO_4 are less stable than the fluorides of xenon.

Problem 9.4

Using the following standard ΔH° (in kJ mol^{-1} at 298 K):

Element	S	O	F	Cl
$\Delta_a H^\circ$	279	248	79	121

Compound	SO_2 (g)	SO_2Cl_2(g)	SF_4(g)	SF_6(g)
$\Delta_f H^\circ$	-297	-382	-763	-1220.5

Calculate:
(a) the S to O bond enthalpy term in SO_2(g);
(b) the S to Cl bond enthalpy term in SO_2Cl_2(g) assuming the S to O bond is the same as in SO_2;
(c) the S to F bond enthalpy term in SF_6(g) and in SF_4(g);
(d) the S to F intrinsic bond enthalpy in SF_4(g) taking the promotion energy of S to the valence state with four unpaired electrons as 454;
(e) the $\Delta_f H^\circ$ of the unstable SCl_4 and the hypothetical SCl_6(g) assuming the S to Cl bond enthalpy is the same as in SO_2Cl_2(g). Compare the latter two $\Delta_f H^\circ$ values with those of the corresponding fluorides.
Comment on the data given and the calculated energies. Discuss the statement that F and O stabilize the highest oxidation state of non metals.

Answer

(a) E(S to O) = (279 + 496 + 297)/2 = 536 kJ mol^{-1}
(b) E(S to Cl) = [(382 + 279 + 496 + 242) − 2 × 536]/2=163.5 kJ mol^{-1}
(c) E(S to F) = (279 + 1220.5 + (6 × 790)/6 = 329 kJ mol^{-1}
(d) E(S to F) = (279 + 763 + 316)/4 = 339.5 kJ mol^{-1}
Intrinsic E = (279 + 763 + 316 + 454)/4 = 453 kJ mol^{-1}
(e) $\Delta_f H^\circ(SCl_6)$ = 279 + 726 − (6 × 163.5) = 24 kJ mol^{-1}
$\Delta_f H^\circ$ (SCl_4) = 279 + 484 − 654 = 109 kJ mol^{-1}

The two endothermic values show instability compared the negative $\Delta_f H^\circ$ values for the fluorides E(S to Cl) < E(S to F) < in SF_4 or SF_6

Low $\Delta_a H^\circ$ for F compared to Cl and the stronger E to F bonds compared to E to Cl bonds (which are longer) also χ(Cl) < χ(F) enable F to form higher compounds. Oxygen can form double bonds and hence stabilize the higher compounds.

9.11 ELLINGHAM DIAGRAMS

Many elements occur in nature as oxides or as sulfides. The latter can be transformed to oxides by roasting in air at a relatively high temperature. The most economical method of extracting the metals from the oxides is by using a cheap reductant such as coke or CO, when high temperatures are required. This section contains material applicable to both p and d elements despite the chapter title.

A useful pictorial representation is to plot the changes in ΔG° with temperature for the oxidation of metals and elements like C, Si, and H_2 for the reaction involving 1 mole of O_2 at 1 atmosphere. Ellingham found that the plots are very nearly linear. This arises from the relationship:

$$\Delta G^{\ominus} = \Delta H^{\ominus} - T\,\Delta S^{\ominus} \tag{9.1}$$

since $\Delta H^{\ominus}$ and $\Delta S^{\ominus}$ vary little with temperature so that graphs of $\Delta G^{\ominus}$ against T are linear with a slope equal to $-\Delta S^{\ominus}$.

Recalling that entropy is a measure of the randomness of a system, entropies of gases are much higher than those of liquids and these have higher entropies than solids. Hence, melting is accompanied by increase in entropy and boiling a liquid is accompanied by a large increase in entropy. The entropy change for a chemical reaction is given by:

$$\Delta S = \Sigma S(\text{products}) - \Sigma S(\text{reactants}) \tag{9.2}$$

A reaction accompanied by an increase or a decrease in the number of gaseous molecules is associated with an increase or a decrease in entropy, respectively. Thus, for the reactions:

$$2C(s) + O_2(g) \rightarrow 2CO(g) \tag{9.3}$$

$$2CO(g) + O_2(g) \rightarrow 2CO_2(g) \tag{9.4}$$

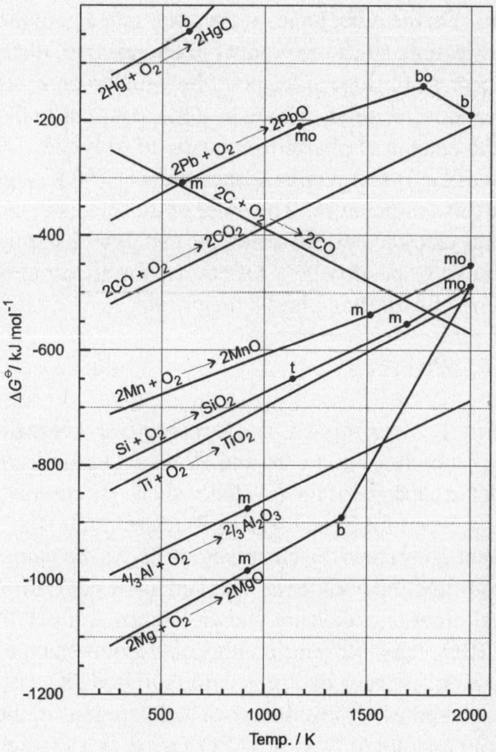

Fig. 9.9 An Ellingham diagram for the oxides discussed in the text

$\Delta S^{\ominus}$ is positive and negative respectively.

Fig. 9.9 shows a simplified Ellingham diagram where $\Delta G^{\ominus}$ is plotted against T for various oxidations. A lower metal (or element) will reduce the oxide of element above it since $\Delta G^{\ominus}$ for the reduction is negative. Its magnitude is equal to the difference in $\Delta G^{\ominus}$ of the plots of the two elements at the chosen temperature. When the oxidation of an element gives a solid oxide, as is usually the case, the line slopes upwards since 1 mol O_2 is lost by the reaction. The line for equation (9.4) slopes similarly but the line for equation (9.3) slopes downwards. It can be seen from the diagram that Mg, Al, Ti or Si can reduce MnO to Mn even at room temperature, although the reactions are not observed unless heat is applied to accelerate the reaction kinetically.

A usual method for preparing small quantities of metals (the thermit process) is to heat the mixture of oxide and Al powder with a flux, the reaction being

$$M_2O_3 + 2Al \rightarrow Al_2O_3 + 2M \quad (9.5)$$

Although C does not reduce MgO at low temperatures, at 1920K where the lines of the oxidation of Mg and C meet, the equilibrium constant, K_p for:

$$MgO + C \rightarrow Mg + CO \quad (9.6)$$

is equal to 1 since $\Delta G^\circ = 0$. At higher temperatures, the reduction takes place provided the products are quickly cooled to quench the reverse reaction. Because the line for the oxidation of C slopes downwards it crosses other lines of oxidation of metals at even lower temperatures, although technically the reduction of some metal oxides, such as TiO_2, are not feasible (a TiC could be formed). At temperatures ≤ 2000 K, CO is not a suitable reductant for the metal oxides below it. However, in the blast furnace coke is the reductant of iron oxides at lower temperatures, but CO is the efficient reductant at higher temperatures (the line for the oxidation of Fe is not shown).

It is worth noting that the line for the oxidation of Mg at points **m** and **b** shows an increase in slope when Mg melts at **m** and especially when it boils at **b**, both processes are accompanied by an increase in S° of the reactant. The same applies at the points m in the oxidation line for Si, Mn and Pb, where the metals melt and at the points **b** for the lines of the oxidation of Pb and Hg, where they boil. On the other hand, at the point mo, e.g. in the line for Pb, the line slopes in the opposite direction, since the product of the reaction, PbO, melts. There is even a more pronounced decrease in slope at the point **bo** which represents the boiling point of PbO produced in the reaction. A small change in slope at point **t** in the line for the oxidation of Si, representing the change of phase from quartz to tridymite. A special feature of the line for the oxidation of Hg is that it crosses the O line at 750 K, and HgO is thermodynamically unstable above this temperature. The order of the metals in the diagram is roughly parallel to their order in the electrochemical series, the metals with highly negative E° below the ones with less negative or positive E°. Of course, the situation is different in aqueous solution where enthalpy of hydration plays a major role.

9.12 OXOACIDS OF THE p BLOCK ELEMENTS

The typical oxoacids of Period 2, H_nEO_3 ($n = 1, 2$ or 3 for N, C and B respectively) contain the triangular planar EO_3 group, with p_π-p_π bonding in H_2CO_3 and HNO_3. The limiting coordination number of three is dictated by the capacity of the E valence shell. By contrast, the typical acids of Period 3, H_nEO_4 ($n = 1, 2, 3$ or 4 for E = Cl, S, P or Si respectively) have the tetrahedral EO_4 group. This arrangement dominates the chemistry of P. As the atoms become larger down the groups, a higher coordination number is observed in the octahedral $Te(OH)_6$ and $[IO(OH)_5]$. However, several other oxoacids are known ranging from HOF (which decomposes at -50°C) to the solid HIO_3. Generally the number of O atoms changes by 1 as the oxidation number of the central atom changes by 2, e.g. in the series $HClO_x$ (x = 1, 2, 3 or 4 for oxidation numbers +1, +3, +5 and +7, respectively). The hydrogens in the formulae can be attached to the central atom and not to O as in H_3PO_x ($x = 2$ or 3), which have two and one P–H bonds respectively.

The strength of the oxoacids

Generally the strength of oxoacids increases across a period, e.g.

$$H_3AsO_4 < H_2SeO_4 < HBrO_4 \text{ and } Te(OH)_6 < IO(OH)_5$$

Across the series, the formal oxidation number of the central atom increases, favouring proton release. Pauling observed that the strength of acids $O_pE(OH)_q$ increases as $(p-q)$ increases. An empirical relation: $pK_a = 8 - 5p$ gives pK_a values of 8, 3, −2 and −7 for $p = 0$, 1, 2 or 3, the latter two values indicate a strong acid. In the series $HClO_x$, $pK_a = 7.5, 2, -1.2$ and −10 for $x =$ 1, 2, 3 or 4. This can be rationalized by considering: O=E−OH as an example. If the E−OH bond is considered formally non-polar, the O=E bond represents formally the loss of two electrons from E, hence the net loss in the molecule is one electron from E, thus the formal positive charge favours the loss of the proton. However, in cases like H_3PO_3 or H_3PO_2, one or two H atoms are directly bound to P and should not be counted in q in the generalized formula above. In addition, in cases like H_2CO_3 the solution contains a small per cent of the acid and the majority is physically dissolved as CO_2. H_3BO_3 is a very weak acid and is best considered as OH^- acceptor, i.e. it dissociates to $[B(OH)_4]^- + H_3O^+$. In presence of compounds with −C(OH)C(OH)− (*gem*-diol) groupings, the acid strength increases markedly, because of the formation of complexes, enabling its titration against a strong alkali.

Phosphorus oxoacids

P forms the largest number of oxoacids, which contain tetrahedral arrangement of at least one P=O and one P−OH. Two peroxoacids are known, H_3PO_5 which has a POOH group and $H_4P_2O_8$ which contains an O−O bonding the two $P(OH)_2O$ fragments. The two lower acids are solids, H_3PO_3 (phosphonic) with two ionizable H atoms and H_3PO_2 (phosphinic) has one such H. In $H_4P_2O_5$, an O joins a $PO(OH)_2$ and a PH(OH)O group. These P−H bonds confer a reducing property which is greater for two P−H groups than one P−H bond. Catenation can take place either by a P−O−P group as above and in $HPO(OH)OP(OH)_2O$, or in its isomer which has a P−P bond. Condensation of orthophosphoric acid gives di-, tri and poly-acids, $H_4P_2O_7$, $H_5P_3O_{10}$ and $H_{n+2}P_nO_{3n+1}$ (n = 4 - 14). Cyclic metaphosphates are derived from $(HPO_3)_n$ (n = 3 or 4) which contain rings whereas the polymeric phosphates are derived from $(HPO_3)_n$, which contains O-linked chains.

Oxoacids of sulfur

These are less numerous than those of P. In many S oxoacids, S is surrounded tetrahedrally by O and/or S atoms. The two peroxoacids, like those of P, contain an S−OOH in the mono H_2SO_5 or an S−O−O−S in the diperoxo $H_2S_2O_8$. Condensation of H_2SO_4 gives $H_2S_2O_7$ whereas replacing an O by a S leads to $H_2S_2O_3$ (with a distinguishable terminal S). A series of polythionic acids, $H_2S_{n+2}O_6$ contain a chain of S atoms joining the HSO_3 groups. The acid with $n = 6$ is known only as its salts. The lower S acids, also known only as salts, include: H_2SO_3 (trigonal pyramidal), $H_2S_2O_5$ or $(OH)O_2SSO(OH)$ and $H_2S_2O_4$ or (OH)OSSO(OH). In the latter three acids one or two S atoms have a lone pair.

9.13 STRUCTURES OF BORATES AND SILICATES

B and Si, having similar electronegativities, exhibit diagonal similarity, notably in the range of structures of oxosalts. However, the borates have triangular planar BO_3 groups joined by sharing O atoms to form rings or chains, e.g. $B_3O_9^{3-}$ rings and CaB_2O_4 which has infinite chains. In some cases BO_3 and tetrahedral BO_4 units share O atoms. The structures of the silicates are based on SiO_4^{4-} tetrahedra which are rarely found as discrete units in orthosilicates or when two units share an O in pyrosilicates. In cyclic silicates, SiO_4^{4-} tetrahedra share O atoms with two neighbours to form 3, 4, 6 or 8-membered rings, e.g. in beryl or $Be_3Al_2(Si_6O_{18})$:

$$6\ SiO_4^{4-} - 6O^{2-} \rightarrow Si_6O_{18}^{12-}$$

The structure of the silicate ion $Si_6O_{18}^{12-}$ in the mineral beryl is shown in Fig. 9.10.

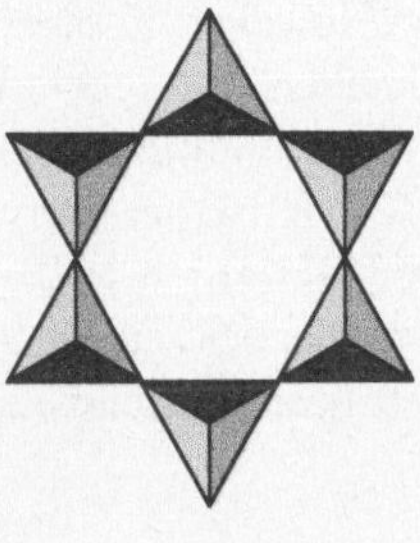

Fig. 9.10 The structure of the $Si_6O_{18}^{12-}$ ion

When tetrahedra share two corners with neighbouring tetrahedra to form an infinite chain, the structures are classified as pyroxenes:

$$nSiO_4^{4-} - nO^{2-} \rightarrow (SiO_3^{2-})_n$$

If two pyroxene chains cross-link on alternate tetrahedral units to form a double chain the amphibole (e.g. asbestos) structures are produced. These have the general formula:

$$4(SiO_3^{2-})_n - nO^{2-} \rightarrow (Si_4O_{11}^{6-})_n$$

Alternatively tetrahedra may share three corners forming infinite sheets in two dimensions (Fig. 9.10 extended in two dimensions) as in micas:

$$n[2SiO_4^{4-} - 3O^{2-}] \rightarrow (Si_2O_5^{2-})_n$$

A 3-dimensional network is formed when the tetrahedra share four corners, e.g. in quartz. AlO_4^{5-} tetrahedra can also share corners forming a layer which may alternate with the SiO_4 layers. The ionic radii of Si and Al are similar and in the case of the latter, the radius ratio Al/O is close to the tetrahedral/octahedral limiting ratio.

9.14 LATIMER AND VOLT-EQUIVALENT DIAGRAMS

The volt-equivalent/oxidation number diagrams, sometimes called Ebsworth or *Frost* (Free energy oxidation state) diagrams (see Section 7.4) are possibly more useful two-dimensional form of the linear Latimer diagrams. The one-dimensional Latimer diagrams contain the same amount of information as the volt-equivalent diagrams with regard to the most stable oxidation state of an element—that which separates the positive and negative reduction potentials, and which states undergo disproportionation or comproportionation reactions.

These points are exemplified by the hypothetical Latimer diagram shown below.

$$\mathrm{M^{V}} \overset{E_{54}}{\text{——}} \mathrm{M^{IV}} \overset{E_{43}}{\text{——}} \mathrm{M^{III}} \overset{E_{30}}{\text{——}} \mathrm{M(0)}$$

If $E_{54} > E_{43}$ and E_{30} is negative, M^{III} is the most stable form of M in solution and there will be a comproportionation reaction if M^{V} is mixed with M^{III}:

$$\mathrm{M^{V} + M^{III} \rightarrow 2M^{IV}}$$

because $E_{54} - E_{43}$ is positive, leading to a negative value for the Gibbs energy change for the reaction. If $E_{54} < E_{43}$ M^{IV} has the potential to oxidize itself and reduce itself and it would disproportionate according to:

$$\mathrm{2M^{IV} \rightarrow M^{V} + M^{III}}$$

In the next subsections both Latimer and volt-equivalent diagrams are included.

Group 15 diagram

Latimer diagrams for the standard reduction potentials (volts) of nitrogen in its various oxidation states in 1 *M* H^+ solution and 1 *M* OH^- solution and the acid solution diagrams for P, As and Sb are given below.

$$\mathrm{NO_3^-} \overset{0.8}{\text{——}} \mathrm{N_2O_4} \overset{1.07}{\text{——}} \mathrm{HNO_2} \overset{1.0}{\text{——}} \mathrm{NO} \overset{1.59}{\text{——}} \mathrm{N_2O} \overset{1.77}{\text{——}} \mathrm{N_2} \overset{-1.87}{\text{——}} \mathrm{NH_3OH^+} \overset{1.42}{\text{——}} \mathrm{N_2H_5^+} \overset{1.275}{\text{——}} \mathrm{NH_4^+}$$

$$\mathrm{NO_3^-} \overset{-0.86}{\text{——}} \mathrm{N_2O_4} \overset{0.88}{\text{——}} \mathrm{NO_2^-} \overset{-0.46}{\text{——}} \mathrm{NO} \overset{0.76}{\text{——}} \mathrm{N_2O} \overset{0.94}{\text{——}} \mathrm{N_2} \overset{-3.04}{\text{——}} \mathrm{NH_2OH} \overset{0.73}{\text{——}} \mathrm{N_2H_4} \overset{0.1}{\text{——}} \mathrm{NH_4OH}$$

$$\mathrm{H_3PO_4} \overset{-0.94}{\text{——}} \mathrm{H_4P_2O_6} \overset{0.38}{\text{——}} \mathrm{H_3PO_3} \overset{-0.51}{\text{——}} \mathrm{H_3PO_2} \overset{-0.5}{\text{——}} \mathrm{P} \overset{-0.07}{\text{——}} \mathrm{PH_3}$$

$$\mathrm{H_3AsO_4} \overset{0.56}{\text{——}} \mathrm{HAsO_2} \overset{0.25}{\text{——}} \mathrm{As} \overset{-0.6}{\text{——}} \mathrm{AsH_3}$$

$$\mathrm{Sb_2O_5} \overset{0.58}{\text{——}} \mathrm{SbO^+} \overset{0.21}{\text{——}} \mathrm{Sb} \overset{-0.51}{\text{——}} \mathrm{SbH_3}$$

Volt-equivalent diagrams for the same oxidation states and the same conditions are shown in Fig. 9.11. Except for P, the elements occupy a local minimum, (i.e. minimum G°) indicating their stability towards oxidation or reduction. For N in acid solution, the reduction of the element to NH_4^+ (the lowest point) does not occur for kinetic reasons. The strong N≡N bond in N_2 explains its inertness. On the other hand, white P_4 is a weakly bonded form of P and is at a point above the line joining PH_3 and the oxoacids. Although the red and the black forms of the element are thermodynamically more stable, the yellow P_4 form has been chosen to represent the standard state. The element P disproportionates, in accordance with a decrease in Gibbs energy, especially in alkaline solution to PH_3 and H_3PO_4.

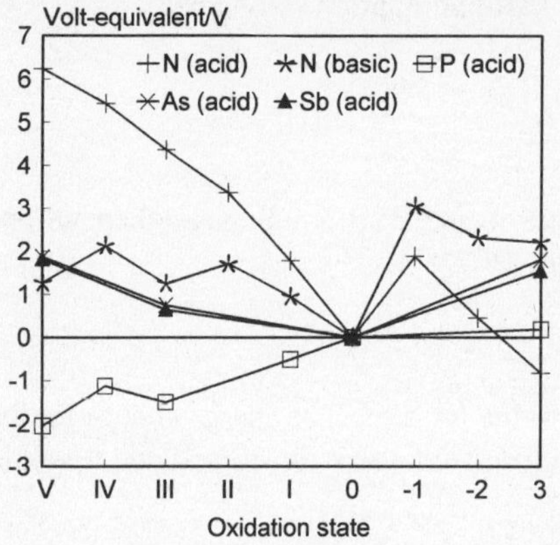

Fig. 9.11 A volt-equivalent versus oxidation state diagram for the Group 15 elements

Oxidation number − 1 for N (in NH_2OH) is unstable towards disproportionation in either acid or alkaline solution (leads to decrease in $G^{\ominus}$), giving $NO_2 + NH_4^+$ in acid and $N_2 + NH_3$ in alkali. On the other hand, $N_2H_5^+$ is on a relatively low point in acid solution and hence stable with respect to disproportionation to its neighbours. However, compared to N_2 and NH_4^+, it is on a high point. Like NH_3OH^+, $N_2H_5^+$ can act as a reductant or an oxidant, but it usually acts as a reductant being oxidized to N_2. Only with strong reductants, e.g. Ti^{3+}, could it act as an oxidant. The couple NH_3OH^+/N_2 is a stronger reductant than $N_2H_5^+/N_2$ as can be seen by comparing the slopes of the lines joining the couples. However, a strong reductant, e.g. V^{2+} can reduce NH_3OH^+ to NH_4^+.

Exercise

The $E^{\ominus}$ value for the reduction of HN_3 (hydrazoic acid, not shown in the diagrams) to dinitrogen is − 3.1 V—more negative than any other reductant in acid solution. Draw a volt-equivalent diagram for nitrogen which incorporates the HN_3 molecule (consider the oxidation state of the nitrogen atoms to be − ⅓). The compound can be only prepared by the oxidation of $N_2H_5^+$ by HNO_2, an example of *comproportionation* (HN_3 is below the line joining the two species).

$H_2N_2O_2$, another example of the +1 oxidation number, decomposes to N_2O and H_2O but it can only be prepared from its salts, obtained by reducing nitrites with Na/Hg. All the oxidation states of N between +1 and +4 inclusive lie above the line joining N_2 and HNO_3. Hence they tend to disproportionate generally to these two species. The disproportionation:

$$N_2O_4 + H_2O \rightarrow HNO_2 + HNO_3$$

is slow in acid solution but rapid in alkali where it is on a conspicuous maximum. HNO_2 also disproportionates to NO_3^- and NO on heating in acid solution but is stable in alkaline media (on a minimum). In solution there is an equilibrium:

$$NO + \tfrac{1}{2}N_2O_4 + H_2O \rightleftharpoons 2HNO_2 \rightleftharpoons N_2O_3 + H_2O$$

the first three N species lie on nearly a straight line. The disproportionation:

$$3NO \rightarrow N_2O + NO_2$$

though feasible, takes place only on heating at high pressure.

It can be seen that the diagram for P is unique since H_3PO_4 is the most stable state and has no oxidizing properties unlike the other acids in the group. The lower acids, on the other hand, are reducing, H_3PO_2 being more powerful than H_3PO_3, since the latter has one P–H bond compared to two bonds in the former. PH_3 as well as the other hydrides are reductants, the stability decreases down the group. The +4 state is not represented in As, nor in Bi which is not shown in Fig. 9.11 since neither Bi^V is well characterized nor is BiH_3 stable. The +4 acid $H_4P_2O_6$ is on a maximum and disproportionates. The oxidizing power of the +5 states decreases in the order: Bi > N > Sb > As. The behaviour of Sb is parallel to its neighbours in groups 16 and 17. The oxidizing power is markedly reduced in alkaline solution, as can be seen for N and this is a common feature for other elements. An example of the effect of pH is the oxidation of I^- to I_2 by H_3AsO_4, whereas in alkaline solution AsO_2^- reduces I_2 to I^-.

The oxidizing power of HNO_3 is demonstrated by the reactions:

$$3Cu + 2HNO_3(\text{dilute}) + 6H^+ \rightarrow 3Cu^{2+} + 2NO + 4H_2O$$

$$Cu + 2HNO_3(\text{concentrated}) + 2H^+ \rightarrow Cu^{2+} + 2NO_2 + 2H_2O$$

It can also oxidize elemental C or S when concentrated and hot. Other metals may reduce it to other products. HNO_2, on the other hand, can act as a reductant, e.g. with acidified MnO_4^- or as an oxidant:

$$2HNO_2 + 2I^- + 2H^+ \rightarrow 2NO + I_2 + 2H_2O$$

Group 16 diagrams

The Latimer diagrams for the acidic and alkaline conditions for the oxidation states and standard reduction potentials of oxygen are show below. The volt-equivalent diagram is shown in Fig. 9.12. The oxidizing power of dioxygen is clearly very high and the instability

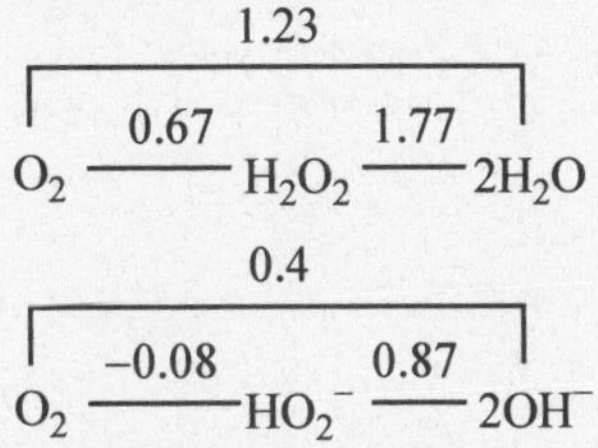

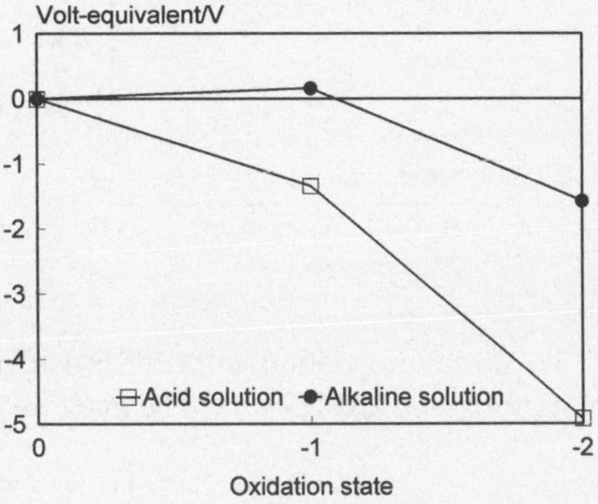

Fig. 9.12 The volt-equivalent diagram for oxygen

of hydrogen peroxide in both acid and alkaline solutions is indicated by the values of $G^\ominus$ for H_2O_2 and its conjugate base HO_2^- being higher than the straight line joining the zero and −2 oxidation states. Under both conditions hydrogen peroxide is unstable with respect to disproportionation and must be kept cool in solution to avoid the decomposition into dioxygen and water.

The Latimer diagrams for the more common oxidation states of S, Se and Te are shown below.

$$1\,M\,H^+ \qquad SO_4^{2-} \stackrel{-0.22}{\text{—}} S_2O_6^{2-} \stackrel{0.57}{\text{—}} H_2SO_3 \stackrel{-0.08}{\text{—}} HS_2O_4^- \stackrel{0.88}{\text{—}} S_2O_3^{2-} \stackrel{0.5}{\text{—}} S \stackrel{0.14}{\text{———}} H_2S$$

$$1\,M\,OH^- \qquad SO_4^{2-} \stackrel{-0.93}{\text{————}} SO_3^{2-} \stackrel{-0.58}{\text{————}} S_2O_3^{2-} \stackrel{-0.74}{\text{—}} S \stackrel{-0.51}{\text{——}} S^{2-}$$

$$SeO_4^{2-} \stackrel{1.15}{\text{————}} H_2SeO_3 \stackrel{0.74}{\text{————————}} Se \stackrel{-0.4}{\text{——}} H_2Se$$

$$H_6TeO_6 \stackrel{1.02}{\text{————}} TeO_2 \stackrel{0.53}{\text{————————}} Te \stackrel{-0.72}{\text{——}} H_2Te$$

Fig. 9.13 depicts a simplified *-nE*$^\ominus$/oxidation number diagram for the S group. In acid solution, S is on a concave point and does not disproportionate to its neighbours but H_2S is the lowest point. On the other hand, Se and Te are the most stable state in acid solution whereas the unstable H_2Se and H_2Te contrast H_2S and the very stable H_2O. The decrease in stability of the hydrides down the group is expected from the bond lengths and energies.

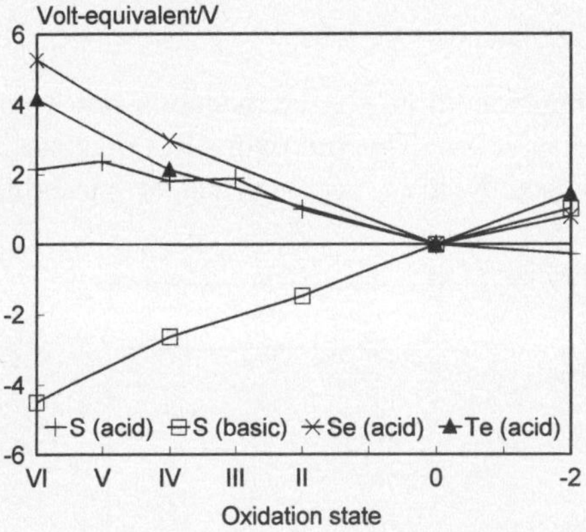

Fig. 9.13 A volt-equivalent diagram for sulfur (acid and alkaline solutions), selenium and tellurium (both for acid solutions)

The positive oxidation states of S in acid solution are characterized by two relatively high points at $H_2S_2O_6$ and $HS_2O_4^-$. The latter disproportionates readily:

$$S_2O_4^{2-} + H_2O \rightarrow S_2O_3^{2-} + HSO_3^-$$

whereas a solution of $H_2S_2O_6$ is moderately stable but it disproportionates on heating:

$$H_2S_2O_6 \rightarrow H_2SO_4 + SO_2$$

The ion, $S_2O_3^{2-}$ is on a slightly high point compared to S and H_2SO_3, hence:

$$S_2O_3^{2-} \rightarrow SO_3^{2-} + S$$

However, boiling SO_3^{2-} (aq) with S gives $S_2O_3^{2-}$. Thc latter is more stable in alkaline solution, being nearly collinear with S and SO_3^{2-} but the most stable state in this medium is SO_4^{2-}. In acid solution HSO_4^- is a moderate oxidant, the oxidizing power of the +6 state increases from $H_2SO_4 < Te(OH)_6 < H_2SeO_4$, the latter acid is another example of a 'middle row anomaly.' The position of the +4 state in acid solution follows the same order but they become more stable towards disproportionation down the group. The oxidizing power of hot concentrated H_2SO_4 cannot be appreciated from Fig. 9.13. In fact it can oxidize C, S, Cu, etc., itself being reduced to SO_2. On the other hand SO_2 and its aqueous solution generally act as reductants, reducing $Cr_2O_7^{2-}$ to Cr^{III}. The oxidation of H_2S by SO_2 to give elemental sulfur and water is an example of comproportionation.

Group 17 diagram

The Latimer diagrams for the oxidation states of the halogens in 1 *M* H^+ solutions are given below, together with the diagram for chlorine in standard alkaline solution. The volt equivalent/oxidation number diagram for Group 17 is shown in Fig. 9.14. The decreasing tendency of X_2(g, l, s) to pass into solution as X^-(aq) is related to $-\Delta_f H^\circ(X^-,g)$ and $-\Delta_{hyd}G^\circ$ which decrease in the series as shown in Table 9.2.

Table 9.2 Thermodynamic data for the halide ions

	F^-	Cl^-	Br^-	I^-
$-\Delta_f H^\circ(X^-,g)$/kJ mol^{-1}	255	234	219	195
$-\Delta_{hyd}G^\circ$ /kJ mol^{-1}	459	337	308	286

$$\tfrac{1}{2}F_2 \xrightarrow{2.87} F^-$$

$$ClO_4^- \xrightarrow{0.36} ClO_3^- \xrightarrow{1.21} HClO_2 \xrightarrow{1.65} HClO \xrightarrow{1.63} \tfrac{1}{2}Cl_2 \xrightarrow{1.36} Cl^-$$

$$ClO_4^- \xrightarrow{0.36} ClO_3^- \xrightarrow{0.33} ClO_2^- \xrightarrow{0.66} ClO^- \xrightarrow{0.4} \tfrac{1}{2}Cl_2 \xrightarrow{1.36} Cl^-$$

$$BrO_4^- \xrightarrow{1.74} BrO_3^- \xrightarrow{1.49} HBrO \xrightarrow{1.6} \tfrac{1}{2}Br_2 \xrightarrow{1.07} Br^-$$

$$H_5IO_6 \xrightarrow{1.6} IO_3^- \xrightarrow{1.34} HIO \xrightarrow{1.45} \tfrac{1}{2}I_2 \xrightarrow{0.54} I^-$$

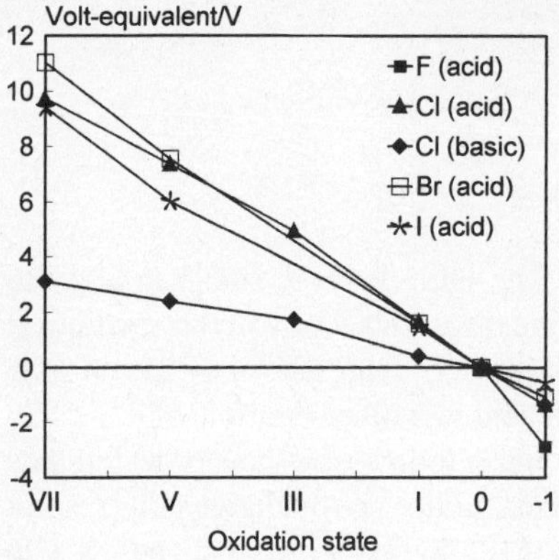

Fig. 9.14 A volt-equivalent diagram for the oxidation states of the halogens in acid solution and for those of chlorine in alkaline solution

It can be seen that the latter play a more important role in giving $E^\circ(\frac{1}{2}F_2 + e^- \rightarrow F^-)$ the most positive value, decreasing down the group. For F_2 and Cl_2, E° values indicate that either can liberate O_2 from water but Cl_2 does not (in absence of sunlight) for kinetic reasons. The higher overvoltage for O_2 evolution in this case, allows Cl_2 to be liberated at the anode in the electrolysis of brine. For the three heavier halogens, all the positive oxidation states are strongly oxidizing. The oxidizing power of the +7 state decrease in the series:

$$BrO_4^- > ClO_4^- > H_5IO_6$$

where the middle row anomaly is similar to the behaviour of the acids of Group 16.

The solution of Cl_2, Br_2 or I_2 in water depends on two equilibria:

$$X_2(g, l, s) \rightleftharpoons X_2(aq) \qquad K_1$$

$$X_2(aq) + H_2O \rightleftharpoons H^+ + X^- + HOX \qquad K_2$$

Table 9.3 Composition of aqueous solutions of the halogens and some equilibrium constants

	Cl_2	Br_2	I_2
K_1	0.062	0.21	0.0013
K_2	4.2×10^{-4}	7.2×10^{-9}	2.0×10^{-13}
Total solubility	0.091	0.21	0.0013
$[X_2,aq]$	0.061	0.21	0.0013
[HOX]	0.030	1.15×10^{-3}	6.4×10^{-6}
$K[X_2 + 2OH^- \rightleftharpoons X^- + OX^- + H_2O]$	7.5×10^{15}	2×10^{8}	30
$K[3XO^- \rightleftharpoons 2X^- + XO_3^-]$	1027	1015	1020

Table 9.3 gives the values of K_1, K_2 as well as the total solubility, $[X_2, aq]$ and [HOX] and the disproportionation constants of the X_2 molecules and the OX^- ions. The physical solubility of $Br_2(l)$ is the reason for its higher concentration. However, [HOX] decreases from X = Cl > Br > I. The base hydrolysis of all three halogens is favourable but becomes less so as the group is descended. The situation is complicated by the even more favourable disproportionation of HOX. Kinetics plays a major role here because BrO^- and

IO^- disproportionate rapidly, but ClO^- solutions are fairly stable unless the solution is heated. The +3 oxidation state is unique to Cl. Although the formation of ClO_2^- in the reaction:

$$2ClO^- \rightarrow Cl^- + ClO_2^-$$

is favourable, the transformation of OCl^- to ClO_3^- is more favourable. Hence chlorites, dioxochlorate(III), are prepared indirectly, e.g.

$$Na_2O_2 + 2ClO_2 \rightarrow 2NaClO_2 + O_2$$

The rate of oxidation by the oxoanions of the three halogens becomes faster as the oxidation number of the halogen decreases. The oxidizing power is reduced as the pH is raised (see the lower plot for Cl in alkaline solutions in Fig. 9.14). Acidified perchlorates and $HClO_4$ can react explosively with traces of organic matter. On the other hand periodates react rapidly and smoothly in oxidation reactions. Iodates are also useful oxidants widely used in analyses unlike bromates which are sometimes involved in oscillatory reactions in which a redox catalyst participates.

9.15 THE p BLOCK METALS

As χ increases across the period in the p block, metallic character decreases but it generally increases down the group, although the trend in χ down the group is irregular. All Group 13 elements, except B, are metals whereas in Group 14 only Sn and Pb are metals and in Group 15 only Bi can be described as a metal. In all these metals, the lower compounds are ionic but as the oxidation number increases and χ also increases, departure from ionic character becomes evident. Parallel to this, the lower oxides are basic, the basic character decreases with an increase in oxidation number.

These features are also exhibited by the d block metals. However, the highest oxidation number in the p block metals is 3, 4 and 5 in Groups 13, 14 and 15, with the lower number being two units lower. In this respect, these metals show the same trend as the other p block elements. The greater stability of this lower oxidation state in Tl, Pb and Bi is another example of the inert pair effect.

Reduction potential trends

$E^\ominus$ for $M^{3+} + 3e^- \rightarrow M(s)$ becomes less negative from Al down the group, $E^\ominus$ becomes +ve for Tl but $E^\ominus$ for $Tl^+ + e^- \rightarrow Tl(s)$ is negative, emphasizing the stability of the +1 oxidation state. The trend is the reverse of Group 3 where $E^\ominus$ becomes more negative down the group. In Group 14, $E^\ominus$ for $M^{2+} + 2e^- \rightarrow M(s)$ is a little more negative for Sn than Pb, but the +4 state is very unstable for Pb in acid or alkaline solution. Bi is different from the other p block metals in having a positive $E^\ominus$.

Oxides, hydroxides and oxoanions

Aluminium forms α-Al_2O_3 and γ-Al_2O_3, the former being a hard solid and the latter is used as an adsorbant and as the stationary phase in some chromatographic separations. The other

Group 13 oxides, similar to Al_2O_3, are amphoteric and the hydroxides tend to form oxoanions in alkaline media, e.g.

$$Al^{3+} + 4OH^- \rightleftharpoons Al(OH)_3 + OH^- \rightleftharpoons [Al(OH)_4]^-$$

This behaviour is found in the oxides and hydroxides of Sn, Pb and Bi. For the two former metals, the monoxides are more basic, especially PbO. In all cases, oxoanions are known, even for Bi. The oxoanion of Bi^V is more stable than Bi^V oxide, and can be prepared from Bi_2O_3:

$$Bi_2O_3 + 2Na_2O_2 \rightarrow Na_2O + 2NaBiO_3$$

The structures of the oxides of Bi, Sn and Pb exhibit distortions, related to the presence of a stereochemically active lone pair. The tendency to form a lower oxide increases from Ga to Tl, from Ge to Pb and from As to Bi. The mixed valence Pb_3O_4 contains Pb^{II} and Pb^{IV}. TlOH is soluble like Group 1 hydroxides.

Halides

Aluminium halides illustrate the effect of the size of the halogen and its polarizability on the structure and properties of the halide. Thus AlF_3 is an ionic solid of high melting point whereas $AlCl_3$ forms a layer lattice and the anhydrous chloride sublimes. The Al_2X_6(X = Cl, Br, I) are molecular solids soluble in organic solvents. Tl^I halides also illustrate this point, e.g. the difference between the calculated L and the experimental L increase from 71 to 76 to 84 kJ mol^{-1} in TlCl, TlBr and TlI, respectively. It is worth noting that the radius of Tl^+ is near to that of Rb^+ but the latter has a noble gas configuration unlike the former, which is accordingly more polarizing. The halides of Sn and Pb exhibit some interesting features. In MX_2(g), the angular molecule has a lone pair which is stereochemically active in the structure of the solids. This leads to distortions especially with the smaller halide ions. Whereas $SnCl_2$ is a useful reductant, $PbCl_2$ is a stable solid. On the other hand, PbX_4 halides are unstable, whereas SnX_4 have highly negative values of $\Delta_f H^\ominus$. The corresponding values for BiX_3 become less negative from BiF_3 as the size of the halide increases, an indication of greater departure from ionic character.

Oxosalts

The p block metals form oxosalts, e.g. carbonates, sulfates and nitrates, in the usual oxidation states. $Al_2(SO_4)_3$ forms a series of double salts, the alums, in which M_2SO_4, $Al_2(SO_4)_3$ and H_2O form well-characterized crystals, M = Na, K, Rb, Cs, NH_4, Ag and Tl, of general formula, $MAl(SO_4)_2.12H_2O$. $PbSO_4$ resembles $BaSO_4$ in being insoluble in water, the radii of Ba^{2+} and Pb^{2+} are similar. $Pb(NO_3)_2$ is one of the few water-soluble Pb compounds. The compound $Bi(NO_3)_3$ is the common Bi salt in the laboratory. All these salts hydrolyse in aqueous solutions, as expected from the amphoteric nature of the oxides and hydroxides of the metals.

9.16 XENON COMPOUNDS

As mentioned above, Xe reacts with F_2 under different conditions giving XeF_2, XeF_4 or XeF_6. The latter reacts with water:

$$XeF_6 + 3H_2O \rightarrow XeO_3 + 6HF$$

In alkaline solution, $HXeO_4^-$ is formed but it slowly converts to XeO_6^{4-} ions. Under different conditions, XeF_6 reacts with water to give $XeOF_4$. The action of H_2SO_4 on Na_4XeO_6 gives the explosive XeO_4. These reactions illustrate the ability of F and/or O to stabilise the higher oxidation even of the noble gas Xe.

Since Xe has one electron more than I, the similarity of the structure of Xe compounds and I-containing anions is not unexpected. Examples are: XeF_2, $[ICl_2]^-$, XeF_4, $[ICl_4]^-$, XeF_6, $[IF_6]^-$, XeO_3, IO_3^-, XeO_4, IO_4^-, and $[XeO_6]^{4-}$, $[IO_6]^{5-}$.

9.17 POLONIUM, ASTATINE AND RADON

The heaviest elements in Groups 16, 17 and 18, i.e. Po, At and Rn are only known as radioactive isotopes. The longest living At isotope has a $t_{1/2}$ of 8.3 hours. This limits the study of its chemistry. In addition, its high activity requires special precautions. Although there is evidence of formation of compounds similar to those of I, apparently no At^{VII} species have been identified, which is expected from the inert pair effect. Rn isotopes are formed in the natural radioactive series by the α decay of Ra. Again tracer methods are required to study its reactions, which are expected to be similar to those of Xe. High levels of Rn in buildings are thought to be responsible for some lung cancers.

On the other hand, Po has been more fully studied. The appearance of metallic properties in Po is shown by the solubility of its dioxide in HCl(aq) to give $PoCl_4$. It seems the highest oxide is PoO_2, another manifestation of the inert pair effect.

9.18 SOME APPLICATIONS OF p BLOCK ELEMENTS

Most large scale industrial chemistry processes are based on the elements of the p block. Nitric and sulfuric acid manufacturing are carried out on a very large scale. The Haber process for ammonia production is the first stage in the production of nitric acid. Diitrogen and dioxygen are produced from liquid air, dinitrogen being the liquid which is produced in greatest abundance. Silicon is used in the production of silicone and as the element in 'chips'. Silicates and aluminosilicates, whether found naturally or produced chemically, are used in the manufacture of ceramics. Among the halogens chlorine is the most important and is produced on a vast scale for water treatment, bleaching processes and in plastic manufacture (PVC). The metals of the block such as Al and Pb have uses in aviation (corrosion resistance and low density) and in electronics (low melting point and electrical conductance) industries respectively. The nitrogen and carbon cycles control life.

9.19 PROBLEMS

1. Use the following standard ΔH° values at 298 K (all in kJ mol^{-1}):
   $\Delta_a H^{\circ}(Si(s) \rightarrow Si(g)(3s^2 3p^2)) = 452$, $\Delta H^{\circ}(Si,g(3s^2 3p^2) \rightarrow Si(g)(3s^1 3p^3)) = 399$,

$\Delta H^{\circ}(SiH_4,g) = 34$, $D(H_2,g) = 436$,
to calculate:
(a) the average Si–H intrinsic bond enthalpy,
(b) the average Si–H thermochemical bond enthalpy,
(c) $\Delta_f H^{\circ}$(hypothetical SiH_2,g),
(d) ΔH° for: $2SiH_2(g) \rightarrow SiH_4(g) + Si(s)$,
Assume that Si–H thermochemical bond enthalpy in SiH_2(g) is the same as in SiH_4(g). Comment on your results.

2. (a) Use the following standard thermochemical ΔH° values (all in kJ mol^{-1})
$\Delta_f H^{\circ}(H_2O,g) = -242$, $\Delta_f H^{\circ}(H_2O_2,g) = -133$, $D(H_2,g) = 436$, $D(O_2,g) = 496$,
$\Delta H^{\circ}(O_3,g \rightarrow 3O,g) = 602$, $\Delta_a H^{\circ}(S,s) = 272$, $\Delta_f H^{\circ}(H_2S,g) = -21$, $\Delta_f H^{\circ}(H_2S_2,g) = -18$, to calculate:
(i) the average O–H bond enthalpy in H_2O(g) and the average S–H bond enthalpy in H_2S(g);
(ii) the oxygen–oxygen bond enthalpy in O_3(g) and in H_2O_2(g) assuming that the O-H bond energy in the latter is the same as that in H_2O(g);
(iii) the sulfur-sulfur bond enthalpy in H_2S_2(g), assuming that the S–H bond enthalpy is the same as that in H_2S(g);

3. The following ΔH° values at 298 K (all in kJ mol^{-1}) refer to B and Al fluorides.
$\Delta_a H^{\circ}(B) = 590$, $\Delta_a H^{\circ}(Al) = 314$, $D(F_2,g) = 158$, $E_{ea}(F) = -348$,
$\Delta_f H^{\circ}(BF_3,g) = -1111$, $\Delta_f H^{\circ}(AlF_3,s) = -695$, $I_{1\rightarrow 3}(B) = 6879$, $I_{1\rightarrow 3}(Al) = 5137$.
Calculate:
(a) E(B–F (average thermochemical bond enthalpy) and $\Delta_f H^{\circ}$(BF, g, hypothetical)
(b) $L(AlF_3,s)$ lattice enthalpy
(c) Comment on the observation that BF_3 is a covalent gas whereas AlF_3 is an ionic solid (the following electronegativities are useful).
$\chi B = 2$, $\chi F = 4.1$, $\chi Al = 1.5$, $\chi Cl = 2.8$, $\chi Br = 2.7$, $\chi I = 2.2$.
(d) Correlate the data with the observed structures of Al halides.

10

Coordination Complexes

10.1 INTRODUCTION

Alfred Werner (Nobel Prize for Chemistry, 1913) recognized in 1893 that compounds such as $CoCl_3.6NH_3$ should be more correctly formulated as coordination complexes, in this case as $[Co(NH_3)_6]Cl_3$ with the central metal ion retaining its normal or primary valency of three (with regard to the three chloride ions), but possessing a secondary valency of six (with regard to the six ammonia molecules). The ammonia ligand molecules are bonded to the central cobalt(III) ion by coordinate or dative bonds in which the electrons are provided by the ligands in thc production of a complex ion, $[Co(NH_3)_6]^{3+}$. Coordination complexes play a dominant role in modern inorganic chemistry with those with central transition metal ions representing a large area of chemistry. The material in this chapter is restricted to the theoretical treatment of the three main types of complex—those in which the *coordination number* of the metal ion (the number of ligand atoms attached to the metal) is six (octahedral, O_h) or four (square planar, D_{4h}, or tetrahedral, T_d). The point group symbols are used to denote the symmetry of the central metal ion and its immediate environment of donor ligand atoms, rather than necessarily expressing the true symmetry of the complex as a whole.

In all complexes the metal to ligand bonding relies mainly upon pairs of electrons which are donated from the ligand to the metal. In some instances reverse donation from metal to ligand is important. The conventional view of the bonding describes it as coordinate bonding in which the electrons of the metal are not allowed any importance. This is a great oversimplification of the apparent situation where a particular oxidation state of a metal accepts a number of pairs of ligand electrons to allow complex formation.

The subjects dealt with in this chapter are:

(i) bonding in coordination complexes,
(ii) their electronic spectra,
(iii) their magnetic behaviour,
(iv) their thermodynamic stability, and
(v) the kinetics of (a) ligand exchange and
(b) oxidation-reduction reactions.

The formation of coordination complexes is not restricted to the d and f block metals. The s block metal ions (hard acids) also form complexes, particularly with cyclic ligands with oxygen atom donors (hard bases).

10.2 MOLECULAR ORBITAL TREATMENT OF THE METAL-LIGAND BOND

This section is restricted to a discussion of the three main geometries of complexes (octahedral, square planar, and tetrahedral) containing one transition metal centre. Similar treatments can be applied to any of the other geometries adopted in other complexes.

Emphasis is placed upon the relative energies of the d orbitals since they lose their free-atomic five-fold degeneracy in the presence of an environment of ligands of non-spherical symmetry. This has consequences for the electronic spectra and magnetic properties of the complexes so formed.

Complexes with octahedral symmetry

A complex such as $[FeF_6]^{3-}$ possesses true octahedral symmetry (O_h) in that the ligands are monatomic. Complexes with molecular ligands, such as $[Co(NH_3)_6]^{3+}$, are not strictly octahedral. In such a case the random orientations of the hydrogen atoms of the six ammonia ligands as the molecules undergo rotational and vibrational motion prevents the complex from having true octahedral symmetry at any particular time. Such motion by the ligands does not seem to be important. Either the rotational and vibrational motions of the non-ligated atoms produce a statistical balancing effect or the effects upon the d orbitals are produced by the 'local' symmetry of the donor atoms. The replacement of the six monodentate ammonia ligands of the $[Co(NH_3)_6]^{3+}$ complex by three bidentate (because the molecule possesses two donor nitrogen atoms) diaminoethane ($NH_2CH_2CH_2NH_2$ = en) ligands does not alter the 3d orbital energies significantly even though the $[Co(en)_3]^{3+}$ complex has only D_3 symmetry. It would appear that the local symmetry (approximating to O_h) of the six ligating nitrogen atoms is predominant in affecting the 3d orbitals of the cobalt(III) ion at the centre of the complex.

Molecular orbital treatment of ML_6 (O_h) complexes

The molecular orbital treatment of an ML_6 complex follows the usual pattern of classifying the orbitals of the central metal atom, and the ligand group orbitals, with respect to the O_h point group, and allowing those orbitals with the same symmetry to form bonding and anti-bonding combinations.

The orbitals of the central metal atom are, for a member of the first transition series, the 4s, 3d and 4p atomic orbitals. They transform, with respect to the O_h point group, as follows:

4s(M):	a_{1g}
$3d_{xy,xz,yz}$ (M):	t_{2g}
$3d_{z^2}$, $3d_{x^2-y^2}$ (M):	e_g
$4p_{x,y,z}$ (M):	t_{1u}

The six s orbitals from the six ligands may be placed in diametrically opposed pairs along the Cartesian axes. In Fig. 10.1 the positive ψ portions of those orbitals are shown.

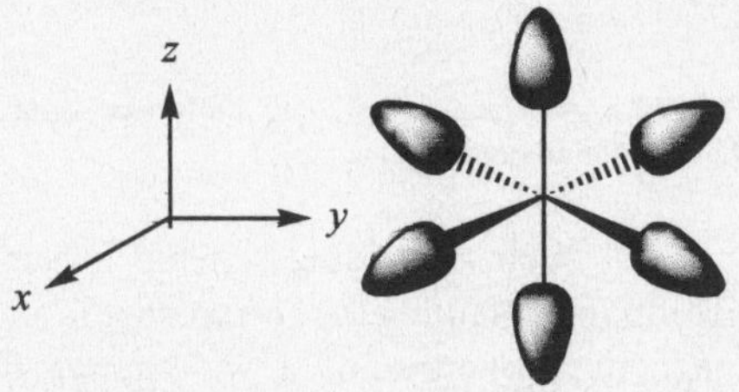

Fig. 10.1 Ligand orbitals in an octahedral complex

Their character with respect to the symmetry operations of the O_h point group may be written down in the usual fashion in terms of the number of orbitals unaffected by each operation:

	E	C_3	C_2	C_4	C_2	i	S_4	S_6	σ_h	σ_d
$6 \times \sigma(L)$	6	0	0	2	2	0	0	0	4	2

Some indication of the disposition of the various elements of symmetry is helpful in understanding the derivation of the above character. The C_3 axes are best visualized as passing through the centre of the triangular faces of the octahedron and the central atom of the complex. Rotation around such an axis by 120° moves all six orbitals. The C_2 axes are those which bisect any two of the Cartesian axes and lie in one or other of the planes, *xy*, *xz* or *yz*. Rotation around such an axis by 180° affects all six σ orbitals of the ligands. The C_4 axes are the three which are coincident with the Cartesian axes and each contain the central metal atom and two diametrically opposed ligand atoms. Rotation around such an axis by 90° leaves the two ligand orbitals, centred on that axis, unchanged. The next C_2 axes are those which are coincident with the C_4 axes—rotation around them by 180° leaves two ligand orbitals unchanged.

The operation of inversion causes all six ligand orbitals to move. The S_4 axes are co-axial with the C_2 axes which bisect the pairs of Cartesian axes. The S_4 operation moves all six ligand orbitals. The S_6 axes are co-axial with the C_3 axes. The S_6 operation moves all six ligand orbitals.

Reflexion of the six ligand orbitals through any of the three horizontal planes of symmetry (represented by the *xy*, *xz* and *yz* planes) leaves four of the ligand orbitals (the ones in the particular plane chosen) unaffected by the operation. The dihedral planes (there are six) bisect the pairs of horizontal planes (each of which contains a C_2 axis) and reflexion in any one of them causes four ligand orbitals to move, leaving two unaffected.

The character of the six σ ligand orbitals is reducible to the sum:

$$6 \times \sigma(L) = a_{1g} + e_g + t_{1u}$$

The a_{1g} and t_{1u} combinations have the correct symmetries for interaction with the 4s and 4p orbitals, respectively, of the metal atom. Likewise the e_g orbitals of the ligands and the metal atom may combine to give bonding and anti-bonding orbitals. The remaining t_{2g} orbitals of the metal atom are non-bonding in a σ-only complex. A σ-only m.o. diagram is shown in Fig. 10.2.

It is conventional in this area of m.o. theory to dispense with the Mulliken numbering of levels and to use asterisks (*) to indicate anti-bonding character.

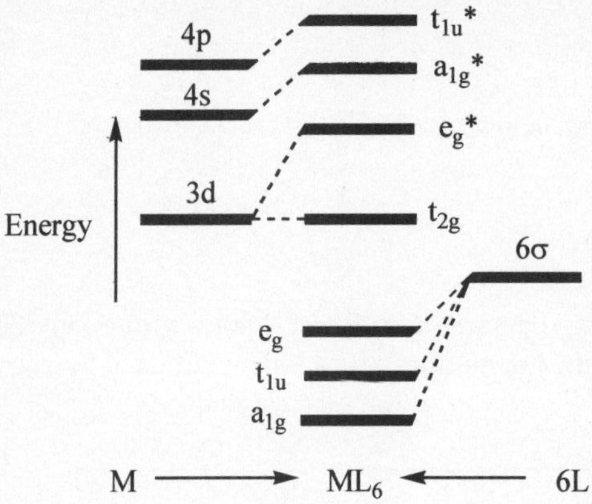

Fig. 10.2 A sigma-orbital only molecular orbital diagram for an octahedral complex

Such a bonding scheme is very helpful in explaining the spectroscopic and magnetic properties of a wide range of complexes. The main determinant of those properties is the energy difference between the anti-bonding e_g* and the non-bonding t_{2g} orbitals. Such a gap in energy allows for *d-d electronic transitions* which happen to lie mainly in the visible region of the electromagnetic spectrum and account largely for the highly colourful chemistry observed for most transition metal complexes.

The energy gap referred to above is the equivalent of the $\Delta_{octahedral}$, or Δ_{oct} or Δ_o (or sometimes 10Dq) of crystal field theory (which treats ligands as though they are point charges, see page 221).

The magnetic properties of any compound depend upon the number of unpaired electrons it possesses. Consider the hexaaquo(III) ion, $[Fe(H_2O)_6]^{3+}$, in which there are five electrons supplied by the Fe^{III} ion and twelve electrons by the six water ligands making a total of seventeen electrons to be placed in the molecular orbitals of Fig. 10.2.

If the *aufbau* principle is obeyed this would lead to the electronic configuration:

$$[Fe(H_2O)_6]^{3+}:\ a_{1g}^2 t_{1u}^6 e_g^4 t_{2g}^5$$

In this configuration the five electrons of highest energy enter the triply degenerate t_{2g} orbitals with one of the electrons being unpaired. The pairing of electrons in the t_{2g} orbitals would only occur if the energy gap between the anti-bonding e_g* level and the t_{2g} orbitals were to be sufficient to enforce pairing in the lower level. For the example in question this does not seem to be the case because the complex has a magnetic moment which indicates that there are *five* unpaired electrons, rather than the single unpaired electron which would result from a t_{2g}^5 configuration. The manner in which the five electrons may be distributed so that they may occupy orbitals singly is to make use of the anti-bonding e_g* orbitals to give the configuration: $t_{2g}^3(e_g^*)^2$. In compliance with Hund's rules the five electrons have parallel spins which is consistent with the observed magnetic moment. The energy gap between the e_g* and t_{2g} orbitals in the complex, $[Fe(H_2O)_6]^{3+}$, must be smaller than the interelectronic repulsion energy which would be experienced in the t_{2g}^5 configuration plus the loss of exchange energy from $10K$ to $4K$ when pairing occurs. This conclusion arises from the following calculation.

The energy of the t_{2g}^5 configuration is given by:

$$E(t_{2g}^5) = -\ 5E(t_{2g})\ + 2P - 4K \qquad (10.1)$$

where P is the increase in internuclear repulsion energy or pairing energy when two otherwise unpaired electrons pair up in one orbital and K is the unit of exchange energy between two electrons of the same spin. In the t_{2g}^5 configuration there are two electrons with one spin and three electrons with the opposite spin. The pair of electrons are stabilized by one unit of exchange energy and the triplet of electrons are stabilized by $3K$ since there are three pair-wise interactions in that case. In general the number of pair-wise interactions between n electrons of the same spin is given by the value of $^nC_2 = n!/(2!(n-2)!)$, i.e. the number of combinations of two electrons, the particles being indistinguishable from each other.

The energy of the $t_{2g}^3(e_g*)^2$ configuration is given by:

$$E(t_{2g}^3(e_g*)^2) = -3E(t_{2g}) + 2(-E(t_{2g}) + \Delta) - 10K = -5E(t_{2g}) + 2\Delta - 10K \quad (10.2)$$

The difference in energy between the two possible configurations is thus:

$$\begin{aligned} E(t_{2g}^3(e_g*)^2) - E(t_{2g}^5) &= -5E(t_{2g}) + 2\Delta - 10K - (-5E(t_{2g}) + 2P - 4K) \\ &= 2\Delta - 2P - 6K \end{aligned} \quad (10.3)$$

If the value of Δ is less than $P + 3K$ the $t_{2g}^3(e_g*)^2$ configuration will be more stable than the t_{2g}^5 one and a high spin state results. On the other hand, if the value of Δ is larger than $P + 3K$ pairing will result to give t_{2g}^5 as the more stable configuration, a low spin state. In the latter case the large Δ value counteracts the loss of energy due to the pairing of four electrons and the loss of $6K$ of exchange energy stabilization.

The d^4 case is another example of the possibility of spin pairing occurring if the value of Δ is sufficiently large. The energy of the high spin $t_{2g}^3(e_g*)^1$ configuration is given by:

$$E(t_{2g}^3(e_g*)^1) = -3E(t_{2g}) + (-E(t_{2g}) + \Delta) - 6K = -4E(t_{2g}) + \Delta - 6K \quad (10.4)$$

That of the low spin t_{2g}^4 is given by:

$$E(t_{2g}^4) = -4E(t_{2g}) + P - 3K \quad (10.5)$$

A plot of the two equations as functions of Δ shows that the high spin is lower in energy than the high spin case if $\Delta = 0$, but as Δ increases in value a cross-over occurs making it the less stable configuration.

Exercise. Sketch the plots of equations (10.4) and (10.5) and show that the above statement is correct. Similar graphs should be derived for d^{5-7} cases.

Both P and K are difficult to estimate and so the decision as to whether pairing in the d^{4-7} cases occurs is generally made by considering the experimentally derived values of Δ and magnetic moment of the complex under study.

The complexes of iron(III) and cobalt(III) with six ligands which are either fluoride ion, F^-, ammonia or cyanide ion, CN^-, serve to extend this discussion. A summary of the magnetic properties of the six possible homoleptic (i.e. where the metal ion is complexed by only one particular ligand) ML_6 complexes is shown in Table 10.1.

The substitution of F^- ion for NH_3 in the Fe^{III} complex has no effect upon the electron arrangement, but the substitution of CN^- causes electron pairing to be preferable to population of the e_g* level.

Table 10.1 The numbers of unpaired electrons for some metal ion-ligand combinations

Ligands	**Fe^{III}, d^5**	**Co^{III}, d^6**
$6F^-$	5	4
$6NH_3$	5	0
$6CN^-$	1	0

The pattern is somewhat different in the case of the cobalt(III) complexes, where both the ammonia and cyanide complexes have a large enough e_g*$-t_{2g}$ energy gap to enforce electron pairing. It is clear that the energy gap depends upon the nature of the metal ion (it also depends upon the oxidation state of the metal, see later) as well as upon the nature of the ligand. Observations of the visible spectra of complexes allow estimates of the energy gap to be made and the results are shown in Table 10.2.

Table 10.2 e_g* $-$ t_{2g} energy differences for some metal ion-ligand combinations/ kJ mol^{-1}

Ligands	**Fe^{III}, d^5**	**Co^{III}, d^6**
$6F^-$	150	196
$6NH_3$	209	272
$6CN^-$	285	370

By comparing the data in Tables 10.1 and 10.2 it is clear that electron pairing takes place (in these systems) when the e_g*$-t_{2g}$ gap is greater than ~ 270 kJ mol^{-1}.

In general ligands have the same relative effect upon the e_g*$-t_{2g}$ splitting irrespective of the metal ion. Some common ligands in the order of their increasing splitting effect are: $I^- < Br^- < Cl^- \sim SCN^-$(ligand atom, S) $< F^- < OH^- < C_2O_4^{2-} \sim H_2O <$ edta $< NCS^-$ (ligand atom, N) $\sim H^- < NH_3 \sim$ py $<$ en $< NO_2^-$ (ligand atom, N) $<$ bipy $<$ phen $< CO \sim C_2H_4 \sim CN^-$ (phen is the abbreviation for 1,10-phenanthroline). There are negative ligands at either end of this series and neutral ligands in between. The more basic (in terms of the Lewis base electron pair donation sense of the term) ligands are centrally placed in the series. The series itself is called the *spectrochemical series*. This is because the values upon which it is based are derived from measurements of absorption spectra, from which the e_g*$-t_{2g}$ energy gaps may be determined.

The variations in the e_g*$-t_{2g}$ energy gap may be understood in terms of the relative basicities (in terms of electron pair donation tendency) of the ligands and the type of π-type bonding which is involved. The more basic the ligand the stronger will be the bonding and the larger the e_g*$-t_{2g}$ gap should be. The effects of π-bonding explain why the cyanide ion, and the neutral CO molecule, with low basicity are at the top end of the spectrochemical series.

There are two cases of π-bonding to be considered. These are:
(i) π donation of electrons from the ligand to the metal atom, and
(ii) π donation of electrons from the metal atom to the ligand.

Ligand-to-metal atom π donation of electrons

The diagram in Fig. 10.3 shows the possible overlap between a metal atom d_{xy} orbital (one of the t_{2g} set) and a ligand p_x orbital.

Fig. 10.3 The pi-overlap of a metal d_{xz} orbital with a ligand p_z orbital in the *xz* plane

It is clear from the signs of ψ (positive shaded, negative white) that such an overlap would lead to m.o. formation. Ligand p orbitals are usually doubly occupied so that for any π–interaction to be energetically advantageous it is necessary for the t_{2g} orbitals of the metal ion to be either vacant or only singly occupied. Such interaction can only be advantageous for the t_{2g}^{0-3} configurations of metals - where any electrons in excess of three must occupy the e_g* orbitals (i.e. high spin d^{4-5} configurations).

The energies of the ligand orbitals which engage in ligand-metal π-bonding are normally lower than those of the t_{2g} orbitals of the metal. This means that when π-interaction occurs the lower (bonding) orbitals have a majority contribution from the ligand orbitals. The higher (anti-bonding) combinations have energies greater than those of the original t_{2g} orbitals (which should be labelled strictly as t_{2g}*) and, in consequence, the e_g*–t_{2g}* energy gap is reduced. The effect is shown on the left-hand side of Fig. 10.4.

Metal atom-ligand π donation of electrons

The cyanide ion and the carbon monoxide molecule (see Section 5.5) are isoelectronic and possess occupied π_u bonding orbitals and vacant anti-bonding π_g* orbitals. The latter orbitals may interact with the metal t_{2g} orbitals to give metal-ligand bonding and anti-bonding π m.o.'s.

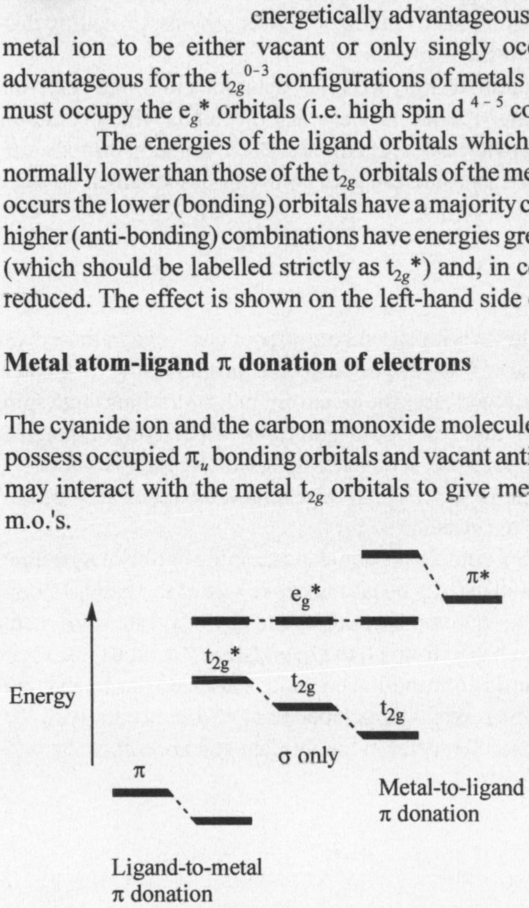

Fig. 10.4 A molecular orbital diagram showing the effects of π interaction on the energy of the t_{2g} orbitals of an octahedral complex

The anti-bonding ligand π_g* orbitals are normally of higher energies than the metal t_{2g} orbitals, so it is the latter which contribute more to the metal-ligand π-bonding orbitals. The ligand π_g* orbitals have a major contribution to the metal-ligand anti-bonding m.o.'s. This has the effect of increasing the e_g*–t_{2g}* energy gap as is shown on the right-hand side of Fig. 10.4.

This ligand-to-metal donation of electrons is sometimes known as π **back donation**, as if to offset the normal or forward ligand-metal donation. Another description of the same phenomenon is π-acid behaviour of the ligand, one definition of an acid being that it is a compound which has electron-pair accepting properties (see section 7.5). The back donation effect explains, in electronic terms, the production of relatively strong bonds by such ligands as CN^- and CO. Their normal tendency to be weakly basic is overcome because any loss of electron pairs by ligand-metal donation is balanced, to some extent, by the metal-ligand back donation. One effect feeds the other and is known as *synergism*. The *synergic effect* of one type of bonding reinforcing the other contributes to very strong bonding and a very large Δ_o value.

Complexes with back bonding are produced when the metal t_{2g} orbitals are filled so that the electrons therein are stabilized by the formation of the π orbitals, providing that suitable receptor orbitals are available on the ligands.

The ethene molecule, C_2H_4, bonds strongly to some metal ions in a side-ways-on mode. Because of the terminal C–H bonds it does not have any σ orbitals which could be used in ligand-metal donation. It does, however, possess filled π_u C–C bonding orbitals, the electrons of which may be used to form a ligand-metal sigma bond (this is explained on page 212).

The Jahn-Teller effect

The Jahn-Teller effect applies to complexes with an odd number of electrons in the e_g* or t_{2g} orbitals. It is more important in the e_g* cases (because the orbitals have a greater interaction with the ligands) and, as such, it concerns the electronic configurations: high spin d^4, low spin d^7, and d^9, with $(e_g*)^1$, $(e_g*)^1$ and $(e_g*)^3$ configurations respectively. It may be stated in the form: 'If, in a non-linear molecule, a degenerate set of orbitals is unevenly occupied, and a distortion is possible which removes their degeneracy, then such a distortion will occur and lead to a stabilization of the system.'

In the case of regularly octahedral complexes which possess the doubly degenerate e_g* anti-bonding level the distortion which usually occurs is where two of the bonds (along the *z* direction) become elongated, and the four bonds in the square plane (*xy*) are shortened. Such a distortion causes the symmetry to change from O_h to D_{4h} and the e_g* orbitals lose their two-fold degeneracy. The orbital with the d_{z^2} contribution becomes stabilized (the ligands are further away from the metal ion along the *z* axis) and is labelled a_{1g}*. The orbital with the $d_{x^2-y^2}$ is destabilized (the ligands are nearer the metal in the *xy* plane) and becomes the b_{1g}* orbital. Fig. 10.5 demonstrates these effects.

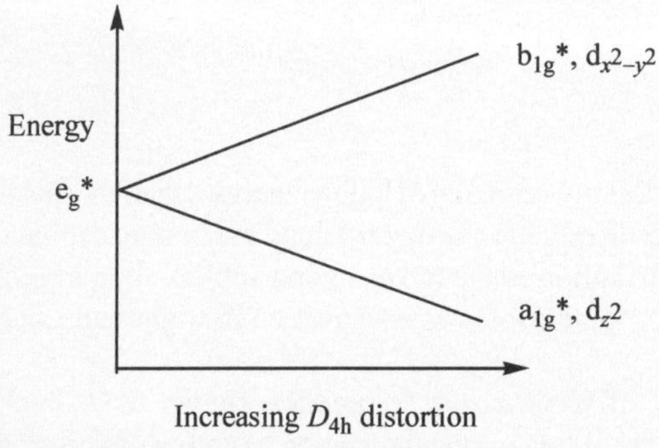

Fig. 10.5 The Jahn-Teller effect on the energies of the e_g* orbitals as D_{4h} distortion increases (to give four shortened bonds and two longer ones)

The distortion to give the more stable a_{1g}* orbital leads to the 'distorted' complex being more stable than the regular octahedral form when it is possible for there to be more electrons in it than occupy the destabilized b_{1g}* orbital.

The reverse effect produced by a distortion of the two short bond, four long bond configuration of a complex does occur but only in the solid state where crystal packing effects are an added consideration.

The case of the even filling of the e_g* orbitals of an O_h complex may lead to stabilization by distortion to D_{4h} symmetry only if the distortion is sufficient to force electron pairing in the lower a_{1g} orbital. If the difference in energy between the b_{1g}* and a $_{1g}$* orbitals is sufficiently high then an $(e_g{}^*)^2$ (O_h) configuration would be more stable as $(a_{1g}{}^*)$ (D_{4h}) with consequent changes in geometry and in magnetic properties. Such an effect is only observed in square planar complexes which may be thought of as distorted octahedral complexes with two infinitely long bonds along the z axis.

Molecular orbital treatment of square planar, D_{4h}, complexes

The complex is set up with its C_4 axis coincident with the z axis. Inspection of the D_{4h} character table allows the classification of the orbitals of the central metal ion as follows:

4s(M):	a_{1g}
$3d_{xy}$(M):	b_{2g}
$3d_{z^2}$(M):	a_{1g}
$3d_{x^2-y^2}$ (M):	b_{1g}
$3d_{xz,yz}$ (M):	e_g
$4p_z$ (M):	a_{2u}
$4p_{x,y}$ (M):	e_u

so that mixing of the 4s and $3d_{z^2}$ orbitals is possible. The reduction of the symmetry from O_h to D_{4h} causes there to be more irreducible representations to deal with, and with a consequent lowering of degeneracy, which leads to a more complicated m.o. diagram.

The four σ ligand orbitals lie along the metal-ligand directions (coincident with the x and y axes) and are classified as:

$$4 \times L(\sigma) = a_{1g} + b_{1g} + e_u$$

The four ligand p_z (or π orbitals with a z component) are classified as:

$$4 \times L(\pi, z) = a_{2u} + b_{2u} + e_g$$

The other four p or π ligand orbitals (those in the xy plane at right angles to the metal-ligand directions) have a character which reduces to the sum:

$$4 \times L(\pi, xy) = a_{2g} + b_{2g} + e_u$$

The m.o. diagram, for the *sigma* orbitals only, is shown in Fig. 10.6. As in the case of octahedral complexes there is the possibility of π bonding which would possibly alter the order of the energy levels.

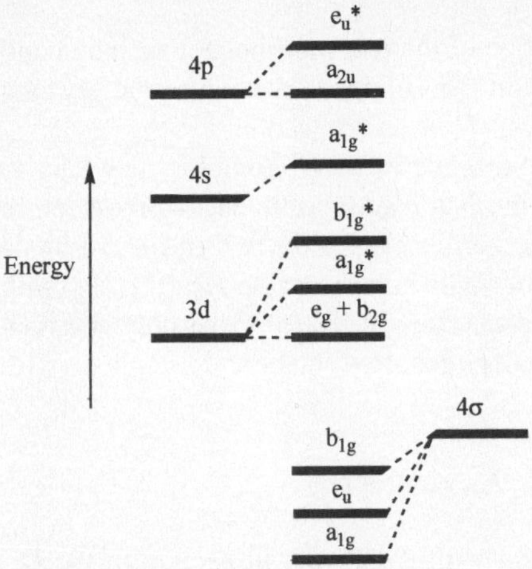

Fig. 10.6 A molecular orbital diagram for a square planar complex (sigma orbitals only)

The bonding of alkenes (the simplest being ethene, C_2H_4) to metals is particularly important in square planar complexes. One notable example is that of the trichloro(η^2-ethene)platinate(II) ion, $[(C_2H_4)PtCl_3]^-$. The ethene molecule is bonded so that its carbon-carbon axis is perpendicular to the plane containing the platinum(II) and chloride ions. The geometry is shown in Fig. 10.7. The two carbon atoms of the ethene molecule are bonded to, or are within bonding distance of, the platinum centre. It is for this reason that the nomenclature of the complex makes use of the Greek letter eta, η, (from ηαπτειν, = haptein, to fasten) with a superscript (2 in this case) to indicate the number of donor atoms bonded to the central metal ion.

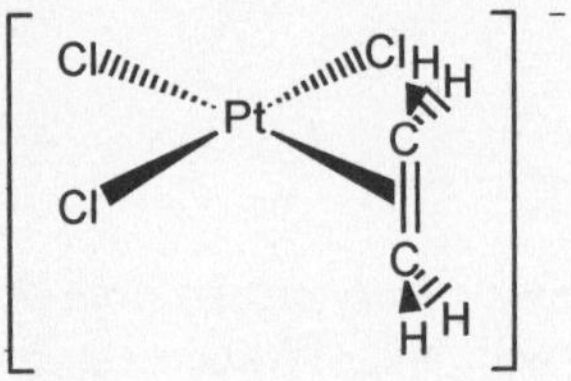

Fig. 10.7 The structure of the trichloro-η^2-(ethene)platinum(II) ion

The ethene molecule has a filled π bonding orbital which is in the plane which is perpendicular to the molecular plane and which contains the two carbon atoms. The vacant anti-bonding π* orbital is in the same plane and both are shown in Fig. 10.8. It is possible for electrons to be donated to a σ platinum orbital ($d_{x^2-y^2}$ is used in Fig. 10.8) forming an ethene-platinum bond. The ethene-to-platinum electron-pair donation reduces the effectiveness of the π-bonding between the carbon atoms of the ethene ligand. The $5d_{xz}$ orbital of the platinum has the correct symmetry to form bonding and anti-bonding (Pt-ethene) combinations with the anti-bonding π* orbital of the ethene molecule to promote π back bonding (donation from platinum to ethene) and produce the synergism required to ensure efficient ethene to platinum donation in the σ bonding.

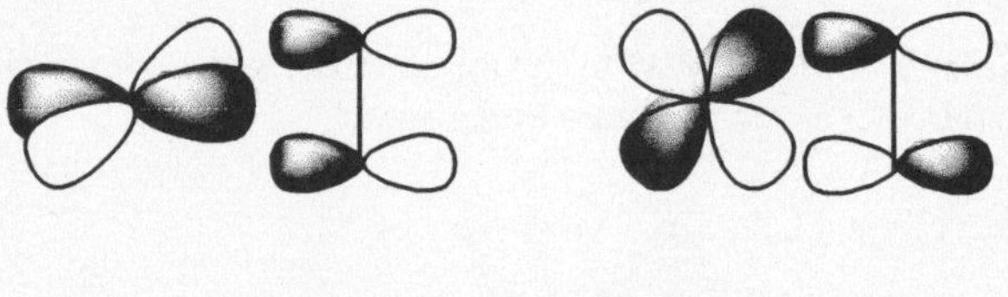

$Pt(d_{x^2-y^2}) \leftarrow C_2H_4(\pi)$ $Pt(d_{xz}) \rightarrow C_2H_4(\pi^*)$

Fig. 10.8 Forward (ethene to Pt) donation and back (Pt to ethene) donation in the bonding between platinum and ethene

The promotion of an electron from π - π* in a free ethene molecule produces a change of symmetry from D_{2h} to D_{2d} (as would be produced if one CH_2 group underwent a 90° rotation with respect to the other). The distortion of the ethene molecule when it acts as a ligand is not consistent with such an electronic transition, the four hydrogen atoms being co-planar due to internuclear repulsion.

Alkenes (and alkynes) do not form complexes with typical acceptor metals (those with vacant π orbitals) but do form many complexes with metals possessing a full d^{10} complement of electrons. In such cases the ligand to metal donation can only be to the s and p orbitals of the metal. Back donation of a π–type is responsible for the formation of the strong linkages which are observed.

Molecular orbital treatment of tetrahedral, T_d, complexes

In a tetrahedral environment the orbitals of a central metal atom transform as:

4s(M):	a_1
$3d_{xy,xz,yz}$(M):	t_2
$3d_{z^2,x^2-y^2}$(M):	e
$4p_{x,y,z}$ (M):	t_2

If the four σ ligand orbitals are arranged along the formal metal-ligand directions their character reduces to the sum:

$$4 \times L(\sigma) = a_1 + t_2$$

The other eight ligand (π) orbitals transform as the sum:

$$8 \times L(\pi) = e + t_1 + t_2$$

The m.o. diagram may then be constructed as in Fig. 10.9. There is a low energy a_1 m.o. and a higher energy t_2 set which is a mixture of σ and π contributions from the ligand orbitals. The ligand t_1 set is non-bonding since there are no metal orbitals of that symmetry. Because of the general disposition of the orbitals of the metal and ligands none of the bonding interactions are as efficient as those observed in octahedral and square planar complexes.

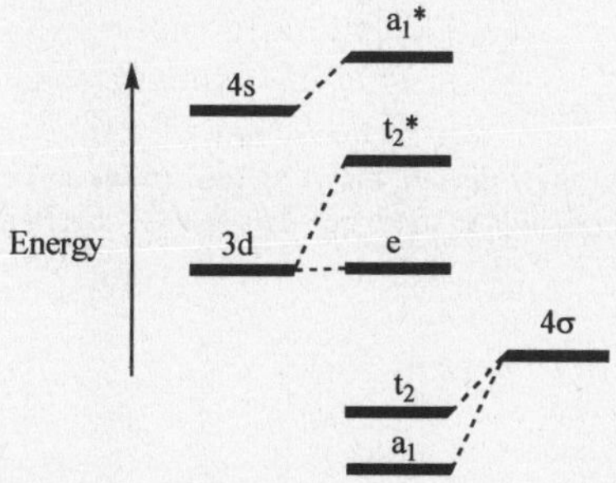

Fig. 10.9 A molecular orbital diagram for a tetrahedral complex

10.3 THE ANGULAR OVERLAP APPROXIMATION

The m.o. diagrams derived for O_h, D_{4h} (square planar) and T_d symmetries of the above sections suffer from being qualitative and even by present day capabilities it is not possible to derive them on a fully quantitative basis. The angular overlap approximation serves to produce a semi-quantitative version of such diagrams and may be used to great effect in deciding which of several symmetries gives the lowest energy for any particular metal atom-ligand system.

The basis of the angular overlap approximation is to consider the maximum σ axial overlap between a metal orbital and a ligand orbital to stabilize the ligand-rich bonding orbital by one unit of energy, denoted by ϵ_σ, and to *destabilize* the metal-rich antibonding orbital by the same amount. The various situations for the overlap of a ligand orbital and the d_{z^2} orbital of a metal are shown in Fig. 10.10.

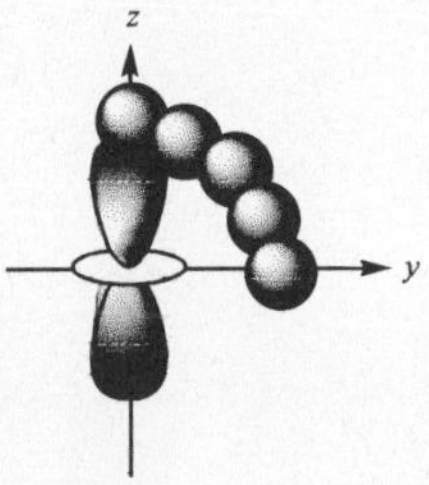

Fig. 10.10 A diagram showing the variation in overlap of a metal d_{z^2} orbital and a ligand sigma orbital as the ligand location moves from along the *z* axis to along the *y* axis

As the angle θ between the metal-ligand direction and the *z* axis increases the overlap between the two orbitals becomes less efficient such that at a 90° angle the stabilization/destabilization energy is only one quarter of that when the angle is zero.
This result is derived from the equation:

$$E\,(d_{z^2}) = {}^1/_4(3\cos^2\theta - 1)^2\,\epsilon_\sigma \qquad (10.6)$$

by making θ = 0° for the full end-on σ overlap of a ligand orbital with the d_{z^2} orbital along the *z* axis, for which $E\,(d_{z^2}) = \epsilon_\sigma$, and θ = 90° for the overlap of the ligand orbital and the d_{z^2} orbital in the *xy* plane, in which case $E\,(d_{z^2}) = \frac{1}{4}\epsilon_\sigma$. Such position-dependent energies can be calculated for the ligand positions for any symmetry and any stoichiometry by using equation (10.6) and the following equations for the energies of destabilization of the other d orbitals by interaction with ligand orbitals in any position, their positions being defined by the angles θ and ϕ, these angles being illustrated in Fig. 10.11.

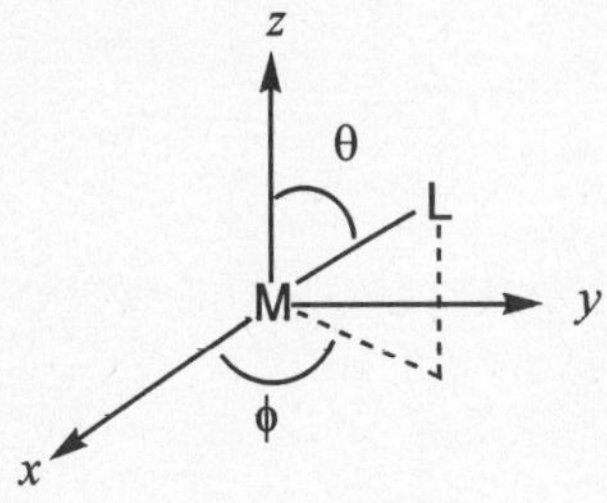

Fig. 10.11 The definitions of the angles θ and ϕ as used in the polar coordinates of the ligand positions used in the angular overlap approximation

The angle θ is that between the positive *z* axis and the line joining the metal and donor atom centres. The angle ϕ is that between the positive x axis and the projection of the metal–donor atom line on the xy plane. The angles are indicated by the diagrams shown in Fig. 10.12, together with numbered positions for the ligands that can approach the metal atom.

$$E(d_{x^2-y^2}) = {}^3/_4[\sin^2\theta(\cos^2\phi - \sin^2\theta)]^2\epsilon_\sigma \quad (10.7)$$

$$E(d_{xy}) = 3[\sin^2\theta\cos\phi\sin\phi)]^2\epsilon_\sigma \quad (10.8)$$

$$E(d_{xz}) = 3(\sin\theta\cos\theta\cos\phi)]^2\epsilon_\sigma \quad (10.9)$$

$$E(d_{yz}) = 3(\sin\theta\cos\theta\sin\phi)^2\epsilon_\sigma \quad (10.10)$$

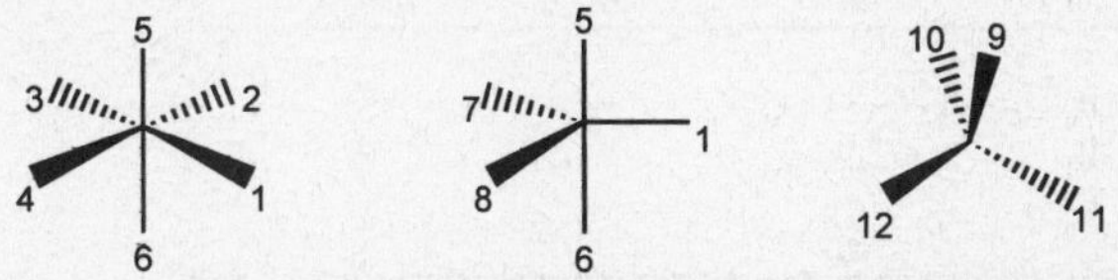

Fig. 10.12 Ligand positions for octahedral, trigonally bipyramidal and tetrahedral complexes

The ligand positions referred to in Fig. 10.12 may be used to calculate the orbital energies for complexes with the more common symmetries. Positions 1–4 are for ligands placed along the x and y axes and are used for square planar, D_{4h}, complexes. Positions 5 and 6 along the z axis are useful for linear complexes and with the additional use of positions 1–4 allow octahedral complexes to be dealt with. Other useful positions are defined by the angles given in Table 10.3.

Table 10.3 Values of the angles θ and ϕ for ligand positions used in D_{3h} and T_d complexes

Ligand position	Value of θ°	Value of ϕ°
D_{3h}		
7	90	120
8	90	240
T_d		
9	54.733	45
10	54.733	225
11	125.267	135
12	125.267	315

For calculations of the orbital energies of trigonally planar arrangements of ligands would use positions 1, 7 and 8, positions 5 and 6 being used additionally for trigonally bipyramidal complexes. For tetrahedral complexes positions 9, 10, 11 and 12 are used.

Similar calculations are available for π interactions, but are not included in this treatment as the appropriate energy term, ϵ_π, is substantially smaller than ϵ_σ.

Table 10.4 contains the results of the angular overlap calculations for the defined positions.

Table 10.4 Angular scaling factors for the d orbitals in terms of the energy unit, ϵ_σ

Ligand position	Metal d orbital destabilization energy / ϵ_σ				
	d_{z^2}	$d_{x^2-y^2}$	d_{xy}	d_{xz}	d_{yz}
Octahedral					
1	¼	¾	0	0	0
2	¼	¾	0	0	0
3	¼	¾	0	0	0
4	¼	¾	0	0	0
5	1	0	0	0	0
6	1	0	0	0	0
Trigonally planar					
7	¼	3/16	0	0	9/16
8	¼	3/16	0	0	9/16
Tetrahedral					
9	0	0	⅓	⅓	⅓
10	0	0	⅓	⅓	⅓
11	0	0	⅓	⅓	⅓
12	0	0	⅓	⅓	⅓

Using the energy terms in Table 10.4 leads to the σ-only m.o. diagrams for O_h, D_{4h} and T_d symmetries as shown in Fig. 10.13.

The π interactions are of a minor nature compared to the σ and it is the latter which are responsible for determining the structure preferred by any metal-ligand system. The stabilization resulting from the formation of any metal-ligand interaction may be calculated for any d electron configuration. In the case of an octahedral complex it may be considered that four electrons occupy the e_g bonding orbitals and are stabilized to the extent of $4 \times 3\epsilon_\sigma = 12\epsilon_\sigma$ by occupying the six bonding orbitals. This stabilization is offset by the destabilizations suffered by the electrons occupying the anti-bonding d orbitals.

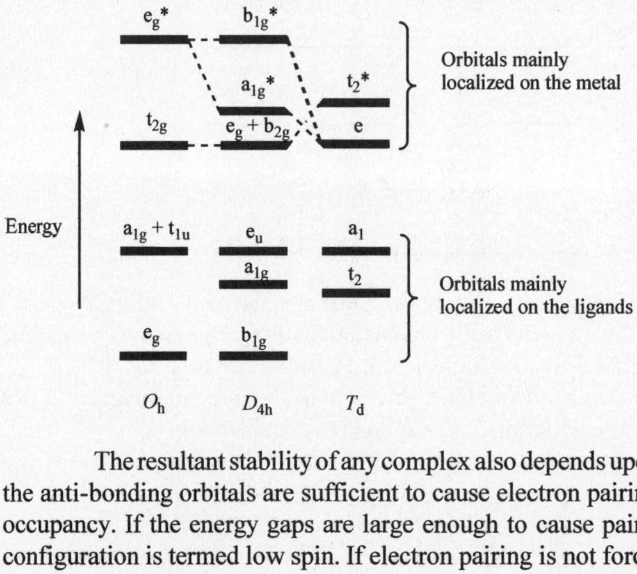

Fig. 10.13 Sigma-only molecular orbital diagrams for octahedral, square planar and tetrahedral complexes

The resultant stability of any complex also depends upon whether the gaps between the anti-bonding orbitals are sufficient to cause electron pairing to be preferable to single occupancy. If the energy gaps are large enough to cause pairing the resulting electronic configuration is termed low spin. If electron pairing is not forced upon the system, i.e. the energy gaps are too small to cause pairing, the electronic configuration is termed high spin.

In the D_{4h} square planar case four electrons are accommodated in the a_{1g} and b_{1g} bonding orbitals, leading to an overall stabilization of $8\epsilon_\sigma$. In the tetrahedral geometry four electrons occupy the e bonding orbitals with the consequent stabilization of $8\epsilon_\sigma$. The stabilization energies are represented as negative quantities in the following treatment, the destabilizations of the antibonding electrons being positive amounts.

Table 10.5 Stabilization energies (MOSE) for octahedral, square planar D_{4h}, and tetrahedral high spin (H.S.) and low spin (L.S.) complexes in units of ϵ_σ

d	O_h		D_{4h}		T_d
	High spin	Low spin	High spin	Low spin	High spin
0	-12	-12	-8	-8	-8
1	-12	-12	-8	-8	-8
2	-12	-12	-8	-8	-8
3	-12	-12	-8	-8	$-6^2/_3$
4	-9	-12	-7	-8	$-5^1/_3$
5	-6	-12	-4	-8	-4
6	-6	-12	-4	-8	-4
7	-6	-9	-4	-7	-4
8	-6	-6	-4	-6	$-2^2/_3$
9	-3	-3	-3	-3	$-1^1/_3$
10	0	0	0	0	0

It is useful to compare the calculated *molecular orbital stabilization energies* (MOSE) for the low and high spin configurations of octahedral and square planar complexes, together with the high spin configurations of tetrahedral complexes here are practically no low spin T_d complexes since the t_2^* – e energy gap is relatively low in all cases. Calculations of the stabilization energies for the above-mentioned cases are shown in Table 10.5. The results given in the table indicate the reason for the very great predominance of the octahedron in the chemistry of complexes. The following generalizations may be made concerning the likely complex to be formed by a metal-ligand system depending upon the number of d electrons originally possessed by the metal.

High spin cases. (Small gaps between sets of d orbitals)

(i) For d^{0-8} configurations the complex formed is likely to be of the ML_6, O_h, type. Complexes of metal ions with d^4 configurations are subject to Jahn-Teller distortion to D_{4h} symmetry.

(ii) Complexes with d^9 configurations can be either ML_6, O_h, or ML_4, D_{4h}, types. A d^9 configuration is subject to the Jahn-Teller effect which favours the production of a D_{4h} distorted ML_6 complex or, if the distortion is of the two infinitely long axial bonds type, a D_{4h} ML_4 complex.

(iii) Complexes with d^{10} configurations can have one of the three symmetries, O_h, D_{4h} (square planar) or T_d.

Low spin cases. (Large gaps between sets of d orbitals)

(i) Complexes with configurations d^{0-7} should be octahedral. A d^7 complex would be subject to a Jahn-Teller distortion to D_{4h}.
(ii) Complexes with d^8 or d^9 configurations could be either octahedral or square planar. The d^9 configuration is subject to Jahn-Teller distortion and is most likely to produce a D_{4h}, ML_4 complex.
(iii) Complexes with d^{10} configurations can be either octahedral, square planar or tetrahedral. In the tetrahedral cases it is possible for extra stability to be derived from the interaction of the 4p orbitals and the $3d_{xy,xz,yz}$ orbitals since both groups transform as t_2. Such p orbital interaction favours tetrahedral symmetry in complexes with a d^{10} configuration in which the anti-bonding d orbitals are completely filled. The 4p participation increases the stability of the t_2 bonding orbitals

Other factors influencing stoichiometries and shapes.

The above generalizations are subject to considerations of three other factors which may influence the preferred symmetry of a complex.

The Pauling electroneutrality principle

The electroneutrality principle has as its basis the operation of Coulomb's law. The positive metal centre has an attraction for the ligand system (which is composed of negative ions or is essentially the negative end of a dipolar system) and, depending upon the strength of the attraction, will result in the stabilization of the system until the attractive forces are balanced by the repulsive forces which operate at short internuclear distances. If the metal charge is low the position of electrostatic equilibrium will be reached by the participation of fewer negative ligands than would be the case for a highly charged metal. This ignores the other factors.

Lattice energy and/or hydration energy

If a complex has an overall charge its stability is enhanced by the interaction with its counter ion (in the case of a solid) or with the solvent (in solution). Highly charged (positive or negative) complexes are favoured by this factor (in spite of the electroneutrality principle). Metals in their zero oxidation states forming complexes with neutral ligands are exempt from this consideration.

Interligand repulsion and/or steric hindrance

Interligand repulsion may well influence the formula of a complex but is difficult to separate from steric hindrance (in which ligands would be closer together than indicated by their normal Van der Waals radii). There is some evidence for the ligands Cl^- and O^{2-} being too large in the Van der Waals sense to form ML_6 complexes where M is a first-row transition element. Iron(III) forms $[FeCl_4]^-$ with the larger chloride ion, but forms $[FeF_6]^{3-}$ with the smaller fluoride ion. In non-aqueous solution and in the crystalline state the $[FeCl_4]^-$ ion is tetrahedral, but in aqueous solution two water molecules are included in the axial positions. The $[CoCl_4]^{2-}$ complex ion is tetrahedral in the solid state and in aqueous solution. There are no cases of complexes with greater than four O^{2-} ligands. This may be due to steric hindrance or to the large repulsion forces in operation or to both effects.

Generalizations

Bearing in mind the above calculations and factors there are only three general cases which apply to the majority of existing complexes.

Weak metal-ligand bonding

If the metal–ligand interaction is weak the energy gaps between the d-type orbitals are relatively small. This allows all he anti-bonding orbitals to be occupied if necessary. It has the consequence that the complexes in this category are those with high spin—they have the maximum number of unpaired electrons with parallel spins. The angular overlap calculations indicate that the most likely complex to be formed by metals with d^{0-8} configurations is the octahedral ML_6 type, although metals with d^4 configurations will be Jahn-Teller distorted. Square planar complexes are a possibility for d^9 configurations while both square planar and tetrahedral complexes may be formed by metals with full d^{10} configurations.

If the complex formed is ML_6 the number of electrons in the valence shell is given by 12 (each ligand contributing two σ-electrons) plus the number of d electrons originally possessed by the metal ion. The range is therefore from 12 (d^0) to 22 (d^{10}) electrons in the valence shell. The occupation of the e_g* anti-bonding orbitals would be detrimental to stability, particularly in the d^{10} case.

In the case of d^9 metals, there being no difference between the O_h and D_{4h} stabilization energies, inter–ligand effects may favour the production of ML_4 (D_{4h}) complexes. In that case the number of valence electrons would be 8 + 9 = 17.

In the case of a d^{10} metal the most likely complex to be formed is a tetrahedral ML_4. Such a complex would have favourable interligand repulsion energy and p orbital participation would stabilize the T_d symmetry relative to either of the alternatives. Complexes of this type would possess 18 valence shell electrons.

Examples of weakly bonded complexes are given in Table 10.6. The formal oxidation states of the metals are indicated, together with the number of valence shell electrons (VSE).

Strong metal-ligand σ bonding

In a strongly bonded complex the energy differences between the various sets of anti-bonding orbitals are large enough to force electron pairing to occur and low spin complexes result. The higher anti-bonding orbitals (e_g in O_h) possess too high an energy to be occupied in a stable complex. The twelve electrons from the six ligands are joined by up to six electrons from the metal so that the maximum number of electrons in the valence shell is eighteen. The angular overlap calculations indicate that metal centres with d^{0-6} configurations should form octahedral complexes (12–18 VSE). Metals with d^7 configurations should also form octahedral complexes but with a single e_g electron, they would be subject to the Jahn-Teller effect, and are easily oxidised to the more stable d^6 configuration. If interligand effects are important enough to restrict the coordination number to four the angular overlap calculations indicate that for d^8 and d^9 configurations D_{4h} complexes are the most likely alternative structures. The four ligands supply eight electrons so that the total number of valence shell electrons would be either 16 or 17 in such cases. Some examples are given in Table 10.7.

Table 10.6 Maximum spin, weakly bonded complexes

d	Complex	Point group	Valence electrons
0	$[TiF_6]^{2-}$	O_h	12
1	$[VCl_6]^{2-}$	O_h	13
2	$[V(C_2O_4)_3]^{3-}$	O_h	14
3	$[Cr(H_2O)_6]^{3+}$	O_h	15
4	$[Mn(H_2O)_6]^{3+}$	O_h	16
5	$[Fe(C_2O_4)_3]^{3-}$	O_h	17
6	$[Fe(H_2O)_6]^{2+}$	O_h	18
7	$[Co(H_2O)_6]^{2+}$	O_h	19
8	$[Ni(H_2O)_6]^{2+}$	O_h	20
9	$[Cu(NH_3)_4]^{2+}$	D_{4h}	21
10	$[ZnCl_4]^{2-}$	T_d	22

Table 10.7 Some minimum spin, strongly bonded complexes

d	Complex	Point group	Valence electrons
0	$[ZrF_6]^{2-}$	O_h	12
1	$[WCl_6]^{-}$	O_h	13
3	$[TcF_6]^{2-}$	O_h	15
4	$[OsCl_6]^{2-}$	O_h	16
5	$[PtF_6]^{-}$	O_h	17
6	$[PtF_6]^{2-}$	O_h	18
8	$[PtCl_4]^{2-}$	D_{4h}	16

Notice that the reduction of the coordination number to four in the case of $[PtCl_4]^{2-}$ may be rationalized from considerations of electroneutrality, in addition to the angular overlap expectations.

Strong metal-ligand σ bonding with metal to ligand π donation

If the ligand has vacant π* (anti-bonding) orbitals which may accept electrons from the metal centre the ligand→ metal σ donation is encouraged by metal→ ligand π back donation or ligand π-acid behaviour. Such synergism enables weakly basic ligands (CN^-, CO, NO, alkenes, alkynes, arenes and numerous classes of heteroatomic aromatic molecules, and various substituted phosphines, PR_3, where R represents F, alkyl or aryl groups) to form very strongly bonded complexes, the π interaction contributing to the large energy gaps between the various sets of anti-bonding d-type orbitals. For this kind of bonding to be fully efficient the d orbitals which may engage in p-type interaction should be fully occupied. This means that metal centres should possess d^6 configurations to maximize the back donation effect in the production of octahedral complexes with 18 valence shell electrons. If the metal contributes more than six electrons the coordination number of the complex is reduced to that which is consistent with the number of valence shell electrons being maintained at eighteen.

A d^8 metal centre configuration would produce a five-coordinate, ML_5, complex which would have D_{3h} symmetry. A d^{10} metal centre would produce a tetrahedral complex because in T_d symmetry all five d orbitals may participate in π bonding. Inter-ligand effects would be consistent with T_d symmetry in that case.

This 'eighteen electron rule' does apply to the d^7 and d^9 cases. A d^7 metal would form an ML_5 complex with 17 valence shell electrons which would be paramagnetic. Although charged paramagnetic complexes are very common, neutral ones are rare. The 17-electron ML_5 unit would gain stability by using its odd electron to form a metal-metal bond, the resulting dimer, M_2L_{10}, being diamagnetic. Each metal centre would then be sharing a valence shell electron complement of 18 electrons. A d^9 metal centre might be expected to produce a monomeric ML_4 unit with 17 valence shell electrons which would dimerize to give an M_2L_8 diamagnetic product containing a metal-metal bond.

Such dimerization does not occur with charged complexes because the electrostatic repulsion between the monomers outweighs any potential stability which would be produced by metal-metal bond formation.

There are exceptions to the eighteen electron rule. One is $V(CO)_6$ with five electrons from the vanadium atom and twelve from the six ligands making a total of 17 valence shell electrons. The compound is paramagnetic but does not form a dimer because of the steric problems which would ensue. It very easily accepts an electron to give the ion, $V(CO)_6^-$, which does conform to the rule. The vast number of organometallic compounds with central transition metal centres generally conform to the eighteen electron rule.

Ligand Field Stabilization Energy (LFSE)

The consideration of ligands as point charges surrounding a metal ion in octahedral or tetrahedral symmetries forms the basis of crystal field theory. In the octahedral case the six negative charges affect electrons in the five 3d, 4d or 5d orbitals so that the d_{z^2} and $d_{x^2-y^2}$ (e_g) electrons are repelled to a greater extent than those occupying the $d_{xy,xz,yz}$ (t_{2g}) orbitals. If the d orbital energy in a totally spherical field (with respect to a zero corresponding to ionization) is denoted by $-E_s$, the total energy of a d^5 configuration, with one electron occupying each of the five orbitals, is $-5E_s$. In an O_h environment, the e_g orbitals would be destabilized to an energy, $-E_s+x$, so that the e_g^2 configuration has an energy, $2(-E_s+x)$. The three t_{2g} orbitals become stabilized to an energy, $-E_s-y$, the energy of the t_{2g}^3 configuration is then, $-3(E_s+y)$. The orbital energies in spherical and octahedral fields are shown in Fig. 10.14.

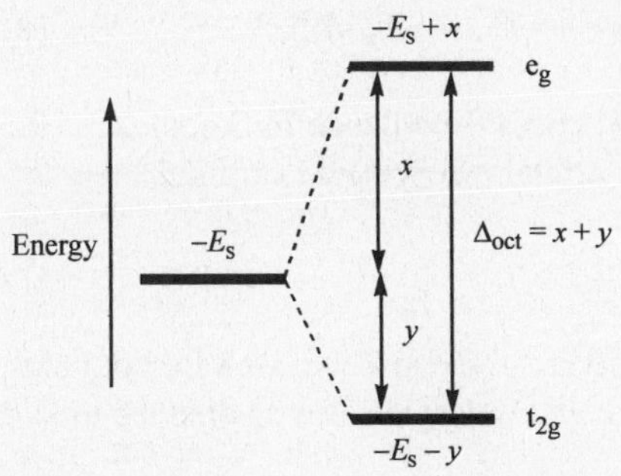

Fig. 10.14 The splitting of the d orbital energies in an octahedral environment

The total energy must be the same as that of the spherical field as represented by the equation:

$$-5E_s = 2(-E_s + x) - 3(E_s + y) \quad (10.11)$$

This equation simplifies to:

$$3y = 2x \quad (10.12)$$

If the crystal field splitting energy, Δ_o, is defined as the difference in energy between the e_g and t_{2g} orbitals:

$$\Delta_o = x + y \quad (10.13)$$

Putting $x = \Delta_o - y$ into equation (10.13) gives:

$$3y = 2(\Delta_o - y) \text{ or } y = {}^2/_5\Delta_o \quad (10.14)$$

and

$$x = {}^3/_5\Delta_o \quad (10.15)$$

Real complexes do not have ligands which are point charges; they have ligands which are either molecules or negative ions. Crystal field theory is a very simple approximation which yields useful results, but cannot take π bonding into account. Molecular orbital theory and the angular overlap approximation are more applicable to real complexes, but yield very similar results to those of crystal field theory when applied generally.

The actual energy of an ion with a configuration $t_{2g}{}^m e_g{}^n$ is given by the equation:

$$E_{\text{actual}} = n \times 3\epsilon_\sigma \quad (10.16)$$

The *ligand field stabilization energy* (LFSE) is the difference between the actual energy and the average energy that the electrons would have in a spherical field. If all five orbitals were singly occupied the average energy would be given by:

$$E_{\text{av}} = (2 \times 3\epsilon_\sigma + 3 \times 0)/5 = {}^6/_5\,\epsilon_\sigma \quad (10.17)$$

The value of LFSE in general is then given by the equation:

$$E_{\text{actual}} - E_{\text{av}} = n \times 3\epsilon_\sigma - (n + m)\,({}^6/_5\,\epsilon_\sigma)$$

$$= {}^9/_5\, n\,\epsilon_\sigma - {}^6/_5\, m\,\epsilon_\sigma = ({}^3/_5\, n - {}^2/_5\, m)\Delta_o \quad (10.18)$$

The final stage of the derivation of equation (10.18) arises from the definition of Δ_o as the difference in energy between the e_g* (in molecular orbital theory) and t_{2g} orbital energies:

$$\Delta_o = 3\epsilon_\sigma \quad (10.19)$$

The LFSE values for various d orbital configurations are given in Table 10.8 for both high spin and low spin cases.A plot of the LFSE for high and low spin cases against the total d configuration is shown in Fig. 10.15.

Table 10.8 LFSE values for d electron configurations in high spin (HS) and low spin (LS) cases; the units of energy are $\Delta_o/5$

d	$m(t_{2g})$	$n(e_g)$	LFSE
0	0	0	0
1	1	0	-2
2	2	0	-4
3	3	0	-6
4 (HS)	3	1	-3
4 (LS)	4	0	-8
5 (HS)	3	2	0
5 (LS)	5	0	-10
6 (HS)	4	2	-2
6 (LS)	6	0	-12
7 (HS)	5	2	-4
7 (LS)	6	1	-9
8	6	2	-6
9	6	3	-3
10	6	4	0

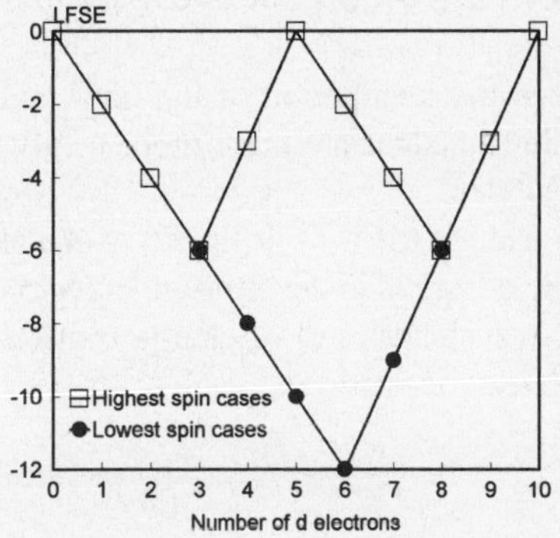

Fig. 10.15 Plots of the ligand field stabilization energies for 0–10 d electrons

The consequences of the ligand field stabilization of various d configurations upon the ionic radii of the M^{2+} ions of the first transition series are shown in Fig. 10.16. The radii of the M^{2+} ions are subject to a general decrease from Ca^{2+} to Zn^{2+} because the effectiveness of the increasing nuclear charge is not offset by the additional electrons, d electrons being less effective at nuclear shielding than s or p electrons. Superimposed upon the general decrease in radius are the ligand field effects which are very similar to those on the LFSE of Fig. 10.15. Other examples of ligand field effects are shown in Fig. 10.16 for the lattice energies of the M^{II} chlorides and enthalpies of hydration of M^{2+} ions respectively of the series Ca to Zn.

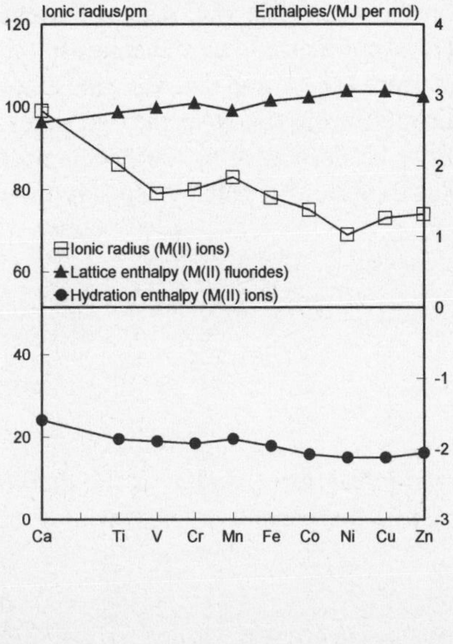

Fig. 10.16 Plots of the ionic radii and standard enthalpies of hydration for the M^{II} ions of the first transition series and the lattice enthalpies for their fluorides

The ligand field effects upon ionic radii can be understood in terms of the different shielding characteristics of the e_g and t_{2g} electrons. In the series of M^{2+} ions from Ca^{2+} to V^{2+} electrons are progressively added to the t_{2g} orbitals which lie along the bisectors of the coordinate axes. These electrons are not very effective at shielding the ligands from the effect of the nuclear charge and so the radii become progressively smaller. For low spin cases, this trend continues in the ions Cr^{2+} (t_{2g}^4), Mn^{2+} (t_{2g}^5) and Fe^{2+} (t_{2g}^6). In the high spin cases of Cr^{2+} and Mn^{2+}, the additional electrons are added to the e_g orbitals which, because they are mainly directed along the coordinate axes, are better at shielding the ligands from the nuclear charge than are electrons in the t_{2g} orbitals. The radius trend is reversed in the two ions and continues in the general downwards trend in the high spin Fe^{2+} ($t_{2g}^4e_g^2$) case where the added electron occupies the t_{2g} level. The high spin Co^{2+} and Ni^{2+} ions show a decrease in radius as electrons are added to the t_{2g} orbitals. The low spin Co^{2+} and the remaining ions of the series (Ni^{2+}, Cu^{2+} and Zn^{2+}) partake in an increasing radius trend as the additional electrons are added to the e_g orbitals. The ligand field effects on the radii of the M^{II} ions explains the trends observed in the hydration enthalpies and the lattice energies of the MCl_2 compounds.

10.4 ELECTRONIC SPECTRA OF COMPLEXES

There are two types of electronic transitions, characteristic of transition metal complexes, which may be observed in the visible and ultra-violet regions of the spectrum; d–d transitions and charge transfers.

1. d–d Transitions which are associated with changes in the occupation of the non- and anti-bonding (essentially d character) molecular orbitals localized mainly on the metal centre. They have low intensities because they are orbitally forbidden.
2. Charge transfer transitions are either from metal to ligand (MLCT) or ligand to metal (LMCT). They are usually of high intensity and are usually observed in the ultra-violet region of the spectrum of a complex. In some cases, the long-wavelength tail of charge transfer transitions can mask the higher wavelength d–d low intensity transitions.

d–d Transitions

There are two selection rules which apply to transitions between electronic states of atoms that are relevant to d–d transitions of complexes. These are that allowed changes in the values of the quantum number l (for the electron transferred) are ± 1, and that there must not be a change in the multiplicity (i.e. the number of unpaired electrons) between the two states.

The basis of the Δl and multiplicity rules may be dealt with as the problem of determining the intensity of an electronic transition. The transition probability depends upon the square of the integral over all space:

$$\int_0^\infty \Psi_f M \Psi_i \, d\tau$$

where Ψ_f and Ψ_i are the wavefunctions of the final and initial states, and M is the dipole moment operator. The Ψ functions can be expressed as the product of Φ (the orbital wavefunction) and σ (the spin wavefunction): Φ × σ. The above integral can then be expressed as:

$$\int_0^\infty \Phi_f M \Phi_i d\tau . \int_0^\infty \sigma_f \sigma_i d\sigma$$

where the subscripts have their previous meanings and dσ is is an element of the spin coordinates. The first integral is concerned with *orbital* changes, the second being concerned with changes of *spin*. The two integrals can be dealt with separately to give the selection rules appropriate to orbital and spin changes.

The integral concerned with orbital changes can be seen to be either zero or finite by the application of symmetry theory. If the discussion is limited to the one electron which is transferred in the electronic transition the only symmetry property which is relevant is the inversion centre. Atomic orbitals are either symmetric to the inversion operation (s and d, for example) or anti-symmetric (p and f, for example). The dipole moment operators (in the *x*, *y* and *z* directions) are always anti-symmetric to inversion (like arrows). In order that the integral should be non-zero it should have *g*-type symmetry. Since the dipole moment operators are always *u*-type it must be that for a non-zero value to result the symmetry of the product $\Phi_f . \Phi_i$ must be *u*-type. The only way that can be achieved is by Φ_f and Φ_i having different symmetries. This means that the transitions, p → s, d → p, etc., are allowed and that s → s, p → p, etc. are forbidden. This is consistent with the *Laporte rule* which restricts electronic transitions in atoms to changes in l of ±1.

The spin integral is easily seen to be either zero or finite. There are only two values of the spin quantum number (±½) and these may be represented by α and β. The spin integral for an electronic transition in which there is conservation of spin is $\int \alpha^2 d\sigma$, which has a non-zero value (think of the area under a plot of the equation: $y = x^2$ which is always positive). The spin integral for a transition in which there is a spin change is $\int \alpha\beta d\sigma$ which has a zero value (similar to the area under a plot of $y = x$ which is zero when all values of x are included, positive and negative). This is the basis of the spin rule which implies that no spin changes take place in a spin-allowed transition.

d–d Transitions are essentially 'atomic' in nature and the rules of atomic spectroscopy apply to them. Since they involve no change in the value of the l quantum number the transitions are orbitally forbidden and are observed to have absorption coefficient (ϵ) values of the order of 1 m^2 mol^{-1}—a factor of about 10^3 smaller than the value expected for a fully allowed transition.

Digression concerning the Beer–Lambert law of radiation absorption

The Beer-Lambert law governs the manner in which monochromatic (i.e. of a particular frequency) radiation is absorbed by a species of concentration (molarity, *M*), *c*, with an optical path (cell thickness) of l usually measured in centimetres rather than the S.I. unit of metres. If the intensity of the radiation incident upon the cell containing the absorbing solution is I_0 and the intensity of radiation emerging from (transmitted by) the cell is I_t, the law can be written in the form:

$$I_0 = I_t . 10^{-\epsilon cl} \tag{10.20}$$

where the constant ϵ is termed the decadic absorption (or extinction) coefficient and has a value dependent upon the nature of the absorbing species and the operation of the appropriate selection rules. Equation (10.20) is converted to one which is of practical use by rearrangement and the taking of logarithms of both sides to give:

$$A = \log\left(\frac{I_0}{I_t}\right) = \epsilon cl \tag{10.21}$$

the quantity, A, being termed the *absorbance* of the particular absorbing species. Equation (10.21) is useful analytically in that the absorbance of an absorbing species is directly proportional to its concentration. It is used to estimate the value for ϵ for any given substance at its frequency of maximum absorption, a plot of A against frequency being the absorption spectrum of the system.

Since the term ϵcl is dimensionless the units of ϵ are dependent upon whether or not the path length is measured in centimetres or metres. If, as is usual, the path length is in centimetres the units of ϵ are L mol^{-1} cm^{-1}. The use of S.I. units, much to be encouraged although in this case the many ϵ values in the literature are in the old units, leads to the ϵ values being a factor of 10 smaller, the S.I. units being m^2 mol^{-1}.

Their small intensities are thought to be due to the influence of vibrational changes accompanying the electronic changes. This allows the incorporation of two vibrational wavefunctions in the intensity integral which could arrange for its value to be finite. Such transitions are described as vibronic.

The d-d spectrum of a particular complex is dependent upon the number of d electrons in the valence shell of the metal centre. The d^1 and d^9 cases are straightforward, the other configurations being more complicated. The d^1, d^9 and d^2 cases are treated in some detail in the following sections. Other cases are dealt with only in a general way with sufficient description
for the main features of their spectra to be appreciated.

The spectra of d^1 ions

In an octahedral d^1 ion, the electron in its ground state occupies the t_{2g} level. In its excited state the electron occupies the e_g* level. In orbital terms the transition may be written as:

$$t_{2g}^{\ 1} \rightarrow (e_g^*)^1$$

The d^1 configuration in an atom gives rise to the electronic state, ^2D, and this, in an octahedral environment, splits (because the orbital degeneracy breaks down) into the two states, $^2T_{2g}$ (the ground state) and ^2E_g (the excited state). In terms of changes in electronic state the transition is written as:

$$^2T_{2g} \rightarrow {}^2E_g$$

Such a transition is *orbitally forbidden* since the triple product representing the intensity integral is $g \times u \times g$, which is u in character, (the dipole moment operators transforming as T_{1u}) and cannot contain the fully symmetric representation (A_{1g}).

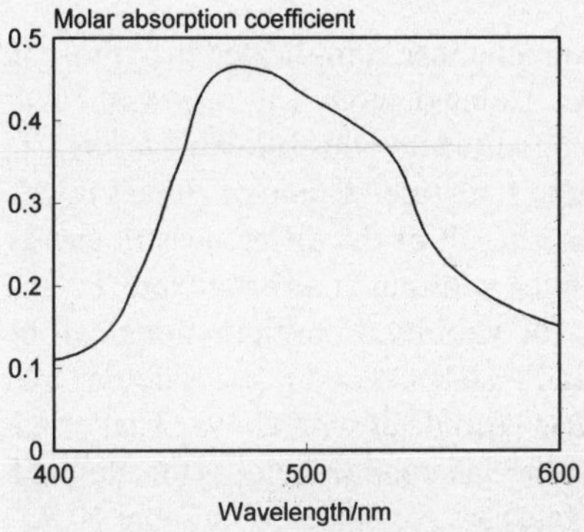

Fig. 10.17 The absorption spectrum of the $[Ti(H_2O)_6]^+$ ion in the visible region

Fig. 10.17 shows the visible absorption spectrum of the $[Ti(H_2O)_6]^+$ ion, with its single absorption band with a maximum absorbance at 476 nm and a shoulder at 526 nm. The absorption band may be regarded as the sum of two overlapping bands. The two bands may be interpreted in terms of the Jahn-Teller effect. The e_g level consists of the two degenerate orbitals, d_{z^2} and $d_{x^2-y^2}$, so that the excited state, ^2E_g, in orbital terms could be either $d_{z^2}{}^1$ or $d_{x^2-y^2}{}^1$. The two possible configurations are degenerate in O_h symmetry and so the Jahn-Teller effect applies. If a distortion of the excited state occurs such that it assumes D_{4h} symmetry with two long axial bonds and four short bonds in the horizontal plane the d_{z^2} orbital is stabilized and the $d_{x^2-y^2}$ orbital is destabilized with respect to their energies in O_h symmetry. The bonding along the z axis will be weaker so the d_{z^2}, anti-bonding orbital will have a lower energy. The reverse is true for bonding in the xy plane so that the $d_{x^2-y^2}$, anti-bonding energy is raised.

The difference in energy between the two states, ^2E_g and $^2T_{2g}$, is, in the d^1 case, equal to the difference between the energies of the e_g* and t_{2g} orbitals here being no interelectronic effects in the one electron case). That difference is known as Δ_o and is called the *ligand field splitting energy*. It is the splitting produced when the ligand environment interacts with the metal orbitals to produce the molecular orbitals of the complex. Equations (10.15) and (10.16) imply that, as Δ_o increases, the energy of the t_{2g} orbitals decreases (x increases) and that of the e_g orbitals increases (y increases). These relationships are shown in Fig. 10.18 which is the **Orgel diagram** for the d^1 case.

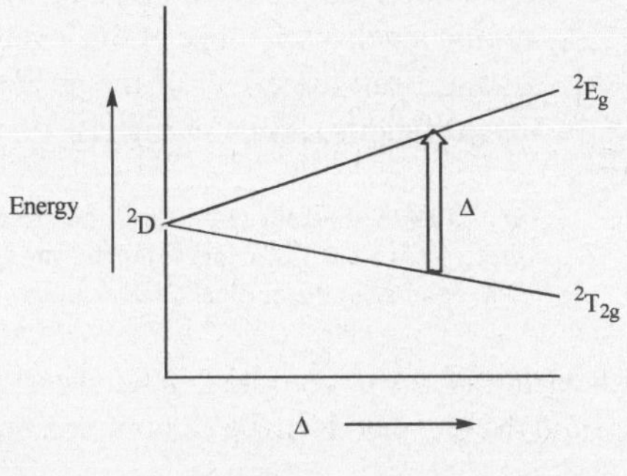

Fig. 10.18 An Orgel diagram for a d^1 complex

The line defining the stabilization of the t_{2g} orbitals has a slope of $-(^2/_5)\Delta_o$ and that defining the destabilization of the e_g* orbitals has a slope of $+(^3/_5)\Delta_o$, the difference between the two lines at any value of Δ_o being Δ_o. The energy of the transition observed in d^1 complexes is a direct measure of Δ_o (the actual value would be the average of the energies of the two absorptions which are caused by the operation of the Jahn-Teller effect).

The spectra of d² ions

Cases involving more than one electron are always much more complicated than those in which there is only one electron to consider. An isolated atom (or ion) with a d^2 configuration gives rise to 45 microstates which may be sorted into the following terms: ^3F, ^1D, ^3P, ^1G and ^1S, the ^3F state representing the ground state (application of Hund's rules).

In an O_h environment atomic states split as a result of the differences in energy between the orbitals. The previous section deals with the splitting of a D state into T_{2g} and E_g states. The main features of spectral diagrams for various d configurations can be appreciated by considering only the splittings of D and F states. S and P states do not split in an O_h environment and the splitting of a D state has been dealt with above. That of a F state is to produce A_{2g}, T_{1g} and T_{2g} states which carry the same multiplicity as the atomic F state. The states produced by atomic S, P, D and F states are summarized in Table 10.9.

Table 10.9 Effects of an octahedral environment on some atomic states

Atomic state	States in an O_h environment
S	A_{1g}
P	T_{1g}
D	$E_g + T_{2g}$
F	$A_{2g} + T_{1g} + T_{2g}$

When the metal atom is placed in an O_h environment by bonding with six ligands the anti-bonding d orbitals have different energies, the t_{2g} set being lower than the e_g set. This allows for there being three configurations of different energies: t_{2g}^2, $t_{2g}^1(e_g^*)^1$ and $(e_g^*)^2$. The correlation diagram connecting the atomic levels with those of the levels arising from the effect of the strong ligand field is shown in Fig. 10.19.

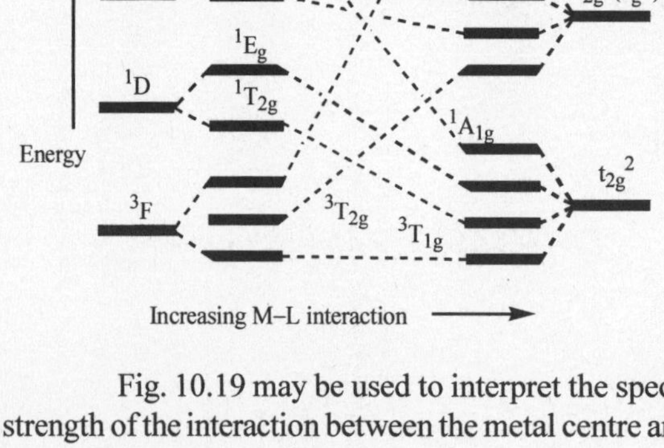

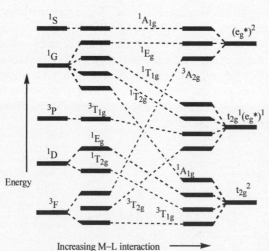

Fig. 10.19 A diagram correlating the states arising from weak (left-hand side) and strong (right-hand side) metal–ligand interactions

Fig. 10.19 may be used to interpret the spectrum of any d^2 complex. Whatever the strength of the interaction between the metal centre and the six ligands in the O_h complex the ground state must be $^3T_{1g}$. Any orbitally allowed transitions can only be to higher energy triplet states. Transitions involving a change in multiplicity are *spin forbidden* and would not be expected to be observed. That restricts the allowed transitions to three:

(a) $^3T_{1g} \rightarrow {}^3T_{2g}$

(b) $^3T_{1g} \rightarrow {}^3T_{1g}$

(c) $^3T_{1g} \rightarrow {}^3A_{2g}$

The hexaaquavanadium(III) ion is an example of a d^2 complex. In aqueous solution transition (a) is observed at a wavelength of 588 nm and transition (b) at 400 nm, but transition (c) is overlain by an intense (allowed) *charge transfer transition*. Transition (c) may be observed in crystalline V_2O_3 (where no charge transfer is possible) at 263 nm.

Orgel and Tanabe-Sugano diagrams

Diagrams such as the one in Fig. 10.19 are available for all d configurations but they are not normally used for the interpretation of spectra. There are two modified forms of the diagrams which are to be found in the literature. These are (i) Orgel diagrams and (ii) Tanabe-Sugano diagrams. An Orgel diagram for the d^2 case is shown in Fig. 10.20. The singlet states are omitted since they are only relevant to spin forbidden transitions. The relationship between Figs 10.19 and 10.20 is clear, the former being a more complicated version of the latter, only the states with the same multiplicity as the ground state being included.

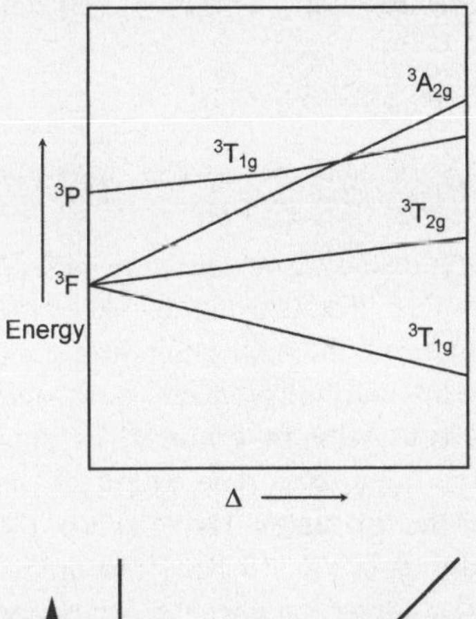

Fig. 10.20 An Orgel diagram for an octahedral d^2 complex

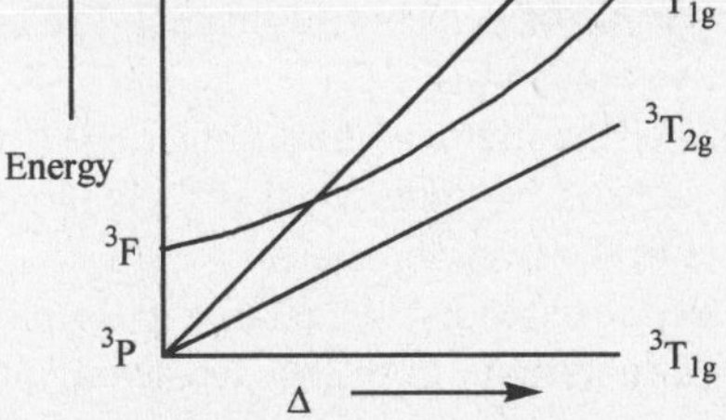

Fig. 10.21 A Tanabe-Sugano diagram for d^2 complexes; the $^3T_{1g}$ state line forms the x axis

The appropriate Tanabe-Sugano diagram is shown in Fig. 10.21. The basis of the diagram is that the ground state correlation line for the $^3T_{1g}$ state coincides with the x axis of the diagram, with the other correlations being relative to the ground state. The quantities which are plotted against each other in a Tanabe-Sugano diagram—the energy of the state and the magnitude of the $e_g^* - t_{2g}$ orbital energy difference (Δ_o) and Δ_o plotted on the x axis—are each divided by what is known as the Racah B parameter which, for any given case, expresses the amount of interelectronic repulsion. The specifying of the B parameter allows one diagram quantitatively to cover all examples for a given d configuration.

Both Orgel and Tanabe-Sugano diagrams are used (i) to interpret the spectra of transition metal complexes, and (ii) to estimate the appropriate value of Δ_o. The latter quantity is the basis of the establishment of the spectrochemical series. Orgel diagrams are more easily understood and a general treatment of them follows.

Orgel Diagrams

All Orgel diagrams depend for their understanding on the principles outlined above and, in particular, on the details of the d^1 case. The ground state of the d^2 configuration, ^3F, splits into the three states, $^3T_{1g}$, $^3T_{2g}$ and $^3A_{2g}$, in an octahedral environment, which correlate with the configurations, t_{2g}^2, $t_{2g}^1(e_g^*)^1$ and $(e_g^*)^2$ respectively. Such information allows the slopes of the correlation lines to be inferred from those for one t_{2g} electron and one e_g^* electron. The line for the ground term, $^3T_{1g}$ (t_{2g}^2), will have a slope which is $-2 \times (^2/_5)\,\Delta_o = -(^4/_5)\Delta_o$, because of there being two t_{2g} electrons. The line for the first excited state, $^3T_{2g}$ ($t_{2g}^1(e_g^*)^1$), (one electron in each orbital), has a slope given by:

$$\text{Slope}(^3T_{2g}) = -(^2/_5)\Delta_o + (^3/_5)\Delta_o = +(^1/_5)\Delta_o \qquad (10.22)$$

The line for the second excited state, $^3A_{2g}$ $(e_g^*)^2$, has a slope given by:

$$\text{Slope }(^3A_{2g}) = +2 \times (^3/_5)\Delta_o = +(^6/_5)\Delta_o \qquad (10.23)$$

These lines are shown in Fig. 10.20 together with the line representing the variation with Δ_o of the $^3T_{1g}$ state which is derived from the atomic ^3P state. The ^3P state is unaffected by the value of Δ_o and is shown with a positive slope which represents the increasing difference between the atomic ^3P and ^3F states as the metal–ligand interaction increases.

Since there are two $^3T_{1g}$ states there is the possibility that they may interact, causing stabilization of the lower state and destabilization of the upper one. The extent of the interaction depends upon the difference in energy between the two states. The effects of the interaction causes both $^3T_{1g}$ state lines to be curved as shown in Fig. 10.20. There are no such complications with the $^3T_{2g}$ and $^3A_{2g}$ states and this is important since the difference between them amounts to $(^6/_5)\Delta_o - (^1/_5)\Delta_o = \Delta_o$ at any value of Δ_o. This means that Δ_o can be estimated as the difference in energy between the two transitions: $^3T_{1g} \rightarrow {}^3T_{2g}$ and $^3T_{1g} \rightarrow {}^3A_{2g}$, which may be the first and second transitions or the first and third transitions (in terms of increasing energy, decreasing wavelength), depending upon whether Δ_o is large enough to allow the $^3A_{2g}$ state to lie above the higher of the two $^3T_{1g}$ states.

The two diagrams in Figs 10.18 and 10.20 may be extended in a relatively simple way to cover almost all the cases of d electron configurations in either octahedral or tetrahedral complexes. The d^1 case in a tetrahedral coordination is the opposite of that in octahedral coordination in that the e* orbitals have lower energy than do the t_2. The value of Δ_t, the energy difference between the t_2 and e* levels has the value $(^4/_9)\Delta_o$ for a given ligand as may be confirmed by carrying out an angular overlap calculation for the two symmetries.

Exercise. Using the angular overlap approximation, show that $\Delta_t = (^4/_9)\Delta_o$

The ground state for the T_d environment is ^2E, the excited state being $^2T_2$. As Δ_t increases the ^2E state is stabilized with the line having a slope of $-(^3/_5)\Delta_t$. The excited state is similarly destabilized with a slope of $+(^2/_5)\Delta_t$. The d^9 (O_h) case is very similar to that of the d^1 (T_d) configuration. The ground state, ^2E_g $(t_{2g}{}^6(e_g{}^*)^3)$, is associated with a slope of stabilization given by:

$$\text{Slope}(^2E_g) = -6 \times (^2/_5)\Delta_o + 3 \times (^3/_5)\Delta_o = -(^3/_5)\Delta_o \tag{10.24}$$

The excited state, $^2T_2$ $(t_{2g}{}^5(e_g{}^*)^4)$, has a slope of destabilization given by:

$$\text{Slope}(^2T_2) = -5 \times (^2/_5)\Delta_o + 4 \times (^3/_5)\Delta_o = +(^2/_5)\Delta_o \tag{10.25}$$

Note that the Orgel diagrams for the d^9 (O_h) and d^1 (T_d) cases are (apart from the scale and the absence of the *g* subscripts in the T_d) the inverse of that for the $d^1(O_h)$ configuration. The splittings in O_h and T_d environments of atomic states (the ingredients for the appropriate Orgel diagrams) and whether the d configurations are d^n or d^{10-n} have the following general relationships:

$$d^n\ (O_h) = d^{10-n}\ (T_d) = \text{inverse of } d^{10-n}\ (O_h) \tag{10.26}$$

The above relationship may be understood in terms of the 'hole' formalism which, for example, considers the d^9 case as a positive 'hole' in a d^{10} configuration. The arguments which apply to one negative electron in constructing the d^1 diagrams apply in reverse to a positive hole in giving the d^9 diagram.

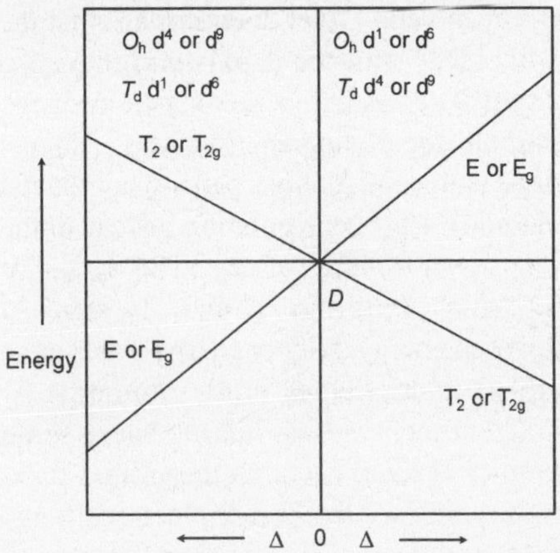

Fig. 10.22 A combined Orgel diagram applicable to octahedral and tetrahedral complexes with one, four (high spin), six (high spin) or nine d electrons

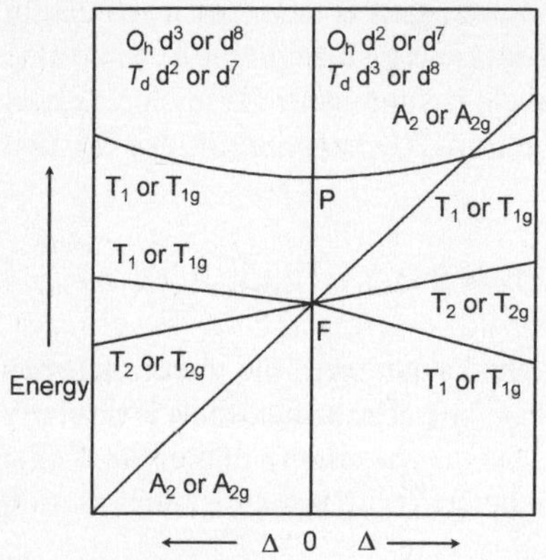

Fig. 10.23 A combined Orgel diagram applicable to octahedral and tetrahedral complexes with two, three, seven (high spin), or eight d electrons

The majority of spectra may be interpreted via the diagrams in Figs 10.22 and 10.23, which are based upon the above generalizations. The cases involving spin pairing induced by high values of Δ_o are considered in the next section. For octahedral complexes the value of Δ_o may be obtained as the energy of the lowest energy (longest wavelength) transition in the following cases: $d^{1,3,4,6,7\text{(high spin)},8 \text{ and } 9}$. For the $d^{2,7\text{(low spin)}}$ cases the value of Δ_o is obtained as the difference in energy between the first and third bands.

The spectrum of the hexaaquanickel(II) complex ion in aqueous solution may be interpreted by the left-hand part of the diagram in Fig. 10.23. The three d-d transitions are, in order of increasing energy (and increasing frequency expressed as wavenumber, cm^{-1}), ${}^3A_{2g} \rightarrow {}^3T_{2g}$ (8500 cm^{-1}), ${}^3A_{2g} \rightarrow {}^3T_{1g}$ (centred at 13500 cm^{-1}), and ${}^3A_{2g} \rightarrow {}^3T_{1g}$ (25300 cm^{-1}). The second and third transitions are from the ground state (${}^3A_{2g}$) to the ${}^3T_{1g}$ states derived from the F and P atomic states (in a spherical ligand field) respectively.

Tanabe-Sugano Diagrams

Simplified versions of Tanabe-Sugano diagrams for 3, 4, 5, 6, 7 and 8 d electrons are shown in Fig. 10.24. Those for 1 and 9 d electrons are too similar to the appropriate Orgel diagrams to include. The main differences between Tanabe-Sugano and Orgel diagrams are that the former have the lowest energy electronic state coincident with the Δ axis and they also indicate the abrupt change from high spin to low spin states as the value of Δ exceeds the magnitude necessary to cause spin-pairing. For example, the d^4 diagram includes a change of ground state multiplicity as Δ_o increases. At large values of Δ_o spin pairing is induced causing the ground state of the complex to be based upon the t_{2g}^4 configuration instead of the $t_{2g}^3(e_g^*)^1$ configuration which is more stable at lower values of Δ_o. The $t_{2g}^3(e_g^*)^1$ configuration produces an atomic ground state, ${}^5D$, which splits into ${}^5E_g$ and ${}^5T_{2g}$ states in an octahedral environment, the former being the lower in energy. Higher energy states arise from triplet and singlet atomic states and retain those multiplicities in O_h symmetry. At sufficiently high values of Δ_o (indicated on the diagram by a vertical line) to cause some electron pairing the ${}^3T_{1g}$ state (derived from the atomic ${}^3H$ state) becomes the ground state. At these high values of Δ_o the spectrum of a d^4 complex consists of triplet-triplet transitions. The singlet states and higher triplet states are omitted from the simplified diagram. They have no relevance to the observed spectra of complexes.

In the midst of all the colourful chemistry of the transition elements there is the almost colourless chemistry of the commonest (II) oxidation state of manganese. Manganese(II) ions have a d^5 configuration and are normally of high spin $t_{2g}^3(e_g^*)^2$ so that the ground states are ^6S. Any electronic transition must alter the multiplicity of the state and is therefore spin forbidden. The very pale colours of Mn^{II} compounds are produced by orbitally and spin forbidden transitions.

The most useful chemical information that can be drived from a d–d spectrum is the value of Δ, the ligand field splitting energy. For octahedral complexes the general rules are that (i) Δ is equal to the energy of the one and only transition observed for d^1, d^4, d^6 and d^9 configurations, (ii) Δ is equal to the energy of the lowest transition of the three which are observable for d^3 and d^8 configurations, and (iii) Δ for d^2 and d^7 configurations is given by the difference in energy between the third and highest energy transition and the lowest one. In the latter case the third d–d transition is sometimes overlain by a charge transfer transition of far higher intensity.

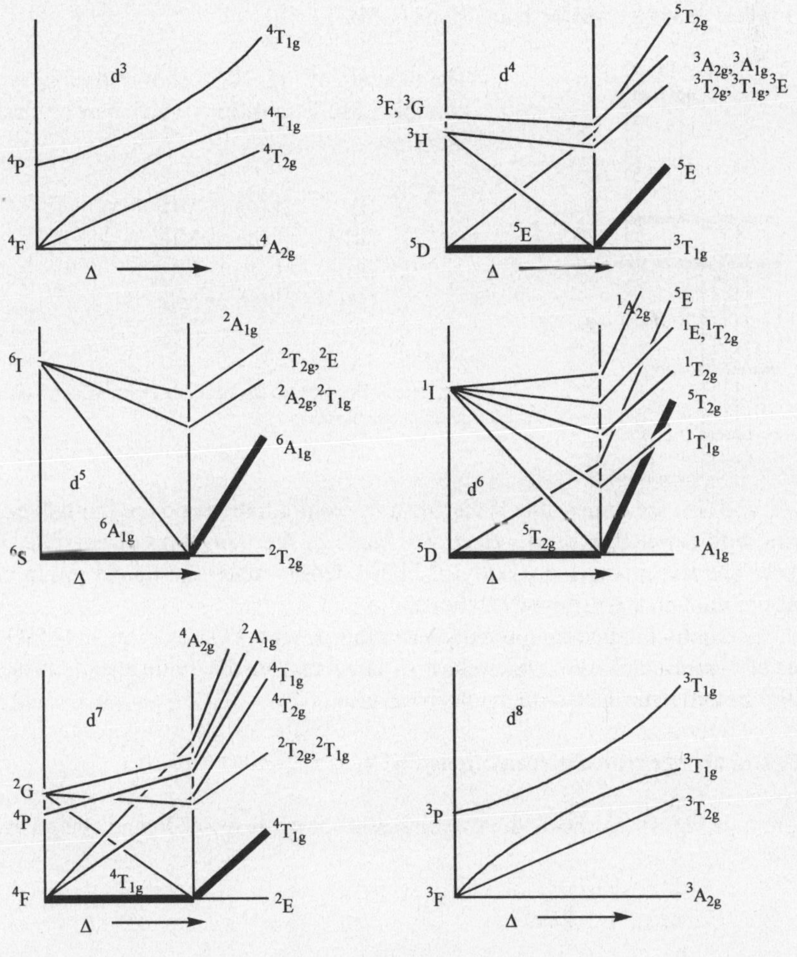

Fig. 10.24 Tanabe-Sugano diagrams for d^3 to d^8 configurations; the base line in all cases is the ground state

Charge transfer spectra

The term, *charge transfer*, implies the movement of an electron from one part of a complex to another. Charge transfer transitions occur as the result of the absorption of a photon of suitable energy to cause such a movement of an electron. The movement of charge may be either (i) from an orbital largely localized on the ligands to one largely localized on the metal (LMCT) or (ii) the reverse process (MLCT).

In a complex there are bonding σ orbitals largely composed of ligand group orbitals with only a minor contribution from any metal orbitals. In addition there is the possibility of there being π ligand orbitals which may interact with the t_{2g} orbitals of the metal to give bonding and anti-bonding combinations. In such a case the t_{2g} orbitals become anti-bonding (with respect to the metal-ligand interaction). If the ligands possess anti-bonding π orbitals interaction with the t_{2g} metal orbitals causes the stabilization of the latter.

Ligand-metal charge transfer transitions (LMCT)

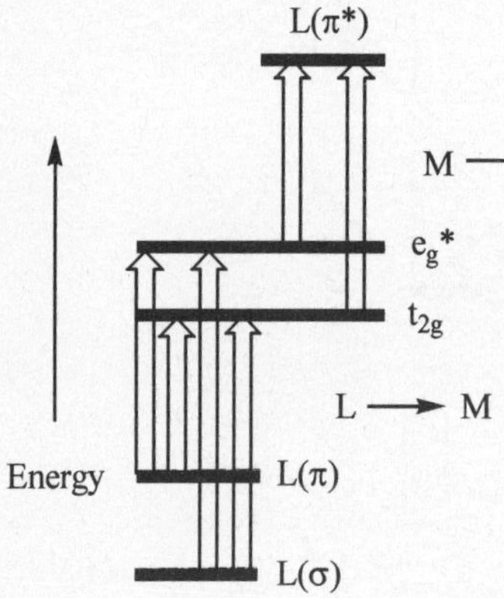

The diagram of Fig. 10.25 shows the four types of possible LMCT transitions which may be classified as follows.

(I) L(σ) → M(t_{2g})
(ii) L(σ) → M(e_g*)
(iii) L(π) → M(t_{2g})
(iv) L(π) → M(e_g*)

Fig. 10.25 The various charge transfer transitions which can occur in complexes

Classes (i) and (iii) are impossible if there is a t_{2g}^6 configuration and are limited to those complexes with fewer than six t_{2g} electrons. None of the transitions is possible for d^{10} complexes. The transitions are usually of high intensity since the ligand group orbital combinations contain *u*-type representations.

The highly coloured compounds containing the ions, VO_4^{3-}, CrO_4^{2-} and MnO_4^-, are examples of d^0 metal centres to which electron transfers from the oxide ligands cause their respective characteristic spectra in the visible region.

Metal-ligand charge transfer transitions (MLCT)

The diagram of Fig. 10.25 shows the two classes of possible metal-ligand charge transfer transitions:

(i) M(e_g*) → L(π*)

(ii) M(t_{2g}) → L(π*)

which may occur. These transitions are available in those complexes in which there is π back bonding (unsaturated ligands). The existence of class (i) depends upon the anti-bonding e_g* orbitals being occupied in the ground state. As the π group orbitals contain *u*-type representations the MLCT transitions are allowed and are observed to have high intensities.

The detailed interpretations of charge transfer spectra are complicated and are not pursued any further in this book. The complications arise from two sources. One is that the ligand orbitals (π and/or π*) may form group orbitals so that several states are produced. The other is that some of the charge transfer bands may be obscured by π−π* transitions between various ligand orbitals. Examples of CT spectra follow.

The hexaiodoosmium(IV) ion, OsI_6^{2-}, exhibits intense absorptions at 862-538 nm, 373-281 nm and 224 nm, which are thought to be π-t_{2g}, π-e_g* and σ−e_g* LMCT transitions respectively. The tris-1,10-phenanthrolineiron(II) ion, $[Fe(phen)_3]^{2+}$, exhibits intense absorptions at 520 nm, 320 nm and 290 nm. They are MLCT transitions. Any higher energy MLCT transitions are obscured by the two intra-ligand π−π* transitions of the ligand system which are observed at 229 and 263 nm.

10.5 MAGNETIC PROPERTIES OF COMPLEXES

The operation of molecular orbital theory allows the prediction of the number of unpaired electrons to be made for any complex. The actual number for any d configuration depends upon the value of Δ_o (for octahedral complexes) relative to the pairing energy, *P*, which represents the interelectronic repulsion energy which has to be overcome for electron pairing to occur. If $P > \Delta_o$ pairing will not occur and a high spin complex results. A low spin complex is formed if $\Delta_o > P$. For octahedral complexes there are significant differences observable for the d^{4-7} configurations. High spin and low spin complexes have different d-d absorption spectra as has been dealt with above. The other area of experimentation in which differences are observable is the determination of magnetic moments of such complexes.

If all the electrons in a compound are paired up in molecular orbitals the compound is *diamagnetic*. Compounds with unpaired electrons are *paramagnetic*. They are attracted towards the most intense part of an applied magnetic field (diamagnetic compounds being slightly rejected by a magnetic field) so that a sample of a paramagnetic substance has an apparently higher mass when weighed in the presence of a magnetic field. The magnetic moment, m, of a compound depends upon the number of unpaired electrons (with parallel spins) and whether their orbital contributions are quenched or not. The orbital contribution may be quenched by an octahedral environment. The orbital contribution arises from the availability of another orbital (to the one containing the electron) which the electron may occupy (without violating the Pauli exclusion principle) by a rotatory motion. A single electron in one of the t_{2g} orbitals can undertake such motion and a t_{2g}^1 configuration would be expected to give an orbital contribution to the magnetic moment of the complex. A t_{2g}^2 configuration would also give an orbital contribution. A t_{2g}^3 configuration would not be able to contribute orbitally to the magnetic moment since rotation from one orbital to another by one electron would violate the Pauli principle. By similar arguments it may be decided that t_{2g}^4 and t_{2g}^5 configurations contribute, but the t_{2g}^6 one does not. High spin configurations, $t_{2g}^4(e_g)^2$ and $t_{2g}^5(e_g)^2$ can also contribute to the orbital fraction of the magnetic moment. The magnetic moment, μ, of a compound is given by the equation:

$$\mu = g\mu_B(S(S+1) + L(L+1))^{1/2} \tag{10.27}$$

where g is the *magnetogyric* or *gyromagnetic* ratio, μ_B is the *Bohr magneton* ($eh/4\pi m$), and S and L are the total spin and angular momentum quantum numbers of the state (m is the rest mass of the electron). g has the value:

$$g = 1 + \left(\frac{J(J+1) - L(L+1) + S(S+1)}{2J(J+1)}\right) \tag{10.28}$$

where J is the appropriate value for the interaction of L and S for the state.

For a fully orbitally quenched situation (i.e where L is effectively zero so that J is equal to S) the value of g reduces to 2 (strictly 2.00023 because of a relativistic effect) and this, together with the absence of the L term from equation (10.28) produces the spin-only formula:

$$\mu = 2\mu_B(S(S+1))^{1/2} \tag{10.29}$$

and, since S is given by $n/2$, n being the number of unpaired electrons, the equation may be modified to:

$$\mu = \mu_B(n(n+2))^{1/2} \tag{10.30}$$

The expected spin-only values for the magnetic moments of complexes with one to five unpaired electrons are shown in Table 10.10 (in terms of Bohr magnetons).

Table 10.10 Spin-only magnetic moments

Number of unpaired electrons	**Magnetic moment (μ/μ_B)**
1	1.73
2	2.83
3	3.87
4	4.90
5	5.92

Some typical values for the spin-only theoretical and observed magnetic moments for a range of ions with d^{0-10} configurations are given in Table 10.11.

The values of the observed magnetic moments given in Table 10.11 are those obtained at room temperatures. There are two important influences upon the value of the magnetic moment of an ion. One is the operation, in some cases, of spin-orbit coupling resulting in an orbital contribution. If there are orbital contributions to the magnetic moment they may be dealt with by modifying equation (10.30) to read:

$$\mu = \left(1 - \frac{\lambda}{\Delta_o}\right)\mu_B[n(n+1)]^{1/2} \tag{10.31}$$

where λ is the *spin-orbit coupling parameter* and Δ_o is the ligand field splitting energy (for the O_h). The value of λ for a given complex may be either positive (if the electron shell is less than half full) or negative (if the electron shell is more than half full). The orbital contribution is distinguishable from the spin-only contribution by its temperature dependence, the spin-only moment being independent of temperature.

Table 10.11 A comparison of spin-only and observed magnetic moments for d^{0-10} O_h ions

Ion	d	t_{2g}	e_g*	n	μ(spin-only)	μ(observed)
Sc^{III}	0	0	0	0	0	0
Ti^{III}	1	1	0	1	1.73	1.73
V^{III}	2	2	0	2	2.83	2.75-2.85
Cr^{III}	3	3	0	3	3.87	3.7-3.9
Cr^{II},Mn^{III}	4	3	1	4	4.9	4.75-5.0
Mn^{III}	4	4	0	2	2.83	3.18
Mn^{II}	5	3	2	5	5.92	5.65-6.1
Mn^{II}, Fe^{III}	5	5	0	1	1.73	1.8-2.5
Fe^{II}	6	4	2	4	4.9	5.1-5.7
Co^{III}	6	6	0	0	0	0
Co^{II}	7	5	2	3	3.87	4.3-5.2
Co^{II}	7	6	1	1	1.73	1.8
Ni^{II}	8	6	2	2	2.83	2.8-3.5
Cu^{II}	9	6	3	1	1.73	1.7-2.2
Zn^{II}	10	6	4	0	0	0

Thc other important influence upon the value of μ is temperature. The Curie law, which relates the variations of the molar susceptibility (χ) with temperature (χ is proportional to μ^2), followed by normal paramagnetic substances, may be written as:

$$\chi = C/T \tag{10.32}$$

where C is a proportionality factor (the Curie constant). It applies to magnetically 'dilute' cases where there is no interaction between the individual magnetic ions. If long-range interaction between the magnetic ions does occur then two further cases arise.

Below a certain temperature, known as the Curie temperature or Curie point, there is a large increase in susceptibility as the temperature decreases. The individual magnetic ions, at temperatures below the Curie point, become aligned in the same direction rather than being dispersed in random directions as is the case for a normal paramagnetic substance. The parallel
alignment of individual ions causes the great increase in susceptibility at temperatures below the Curie point, the solid substance acquiring a permanent magnetism. Such behaviour is known as *ferromagnetism*. Chromium(IV)oxide is an example of a ferromagnetic substance and is used in the manufacture of magnetic data storage media. The associated behaviour known as *ferrimagnetism* occurs when there are two different magnetic species, the alignments of which are in opposition. An example of ferrimagnetism is magnetite, Fe_3O_4, which contains Fe^{II} (d^6, high spin) and Fe^{III} (d^5). The two magnetic species are individually aligned, but in opposite directions, thus reducing the overall magnetic susceptibility of the oxide.

Opposite behaviour below another critical temperature known as the Néel point where there is a sudden decrease in susceptibility as the temperature decreases is known as anti-ferromagnetism. This is brought about by the alignment of the individual magnetic ions with respect to a single direction in such a way that there is a pairing or cancelling out of their effects.

10.6 REDUCTION POTENTIALS

This section is a continuation of Section 7.2 applied to the redox properties of the M^{3+} and M^{2+} ions of the elements V to Co.

Not all the first row transition elements have the same pairs of stable oxidation states. The elements from V to Co are chosen for this example because they all have (II) and (III) states for which reliable experimental data are available. The data and the calculated E° values are shown in Table 10.12. The reduction enthalpies are calculated by using the equation:

$$\text{Reduction enthalpy} = -\Delta_{hyd}H^\circ(M^{3+}) - I_3(M) + \Delta_{hyd}H^\circ(M^{2+}) + 439 \qquad (10.33)$$

Table 10.12 E° data for the elements V to Co

	V	Cr	Mn	Fe	Co
E°(observed)	-0.26	-0.42	1.5	0.77	1.92
$-\Delta_{hyd}H^\circ(M^{3+})$	4408	4626	4596	4487	4713
I_3(M)	2828	2987	3248	2957	3232
$-\Delta_{hyd}H^\circ(M^{2+})$	1896	1926	1863	1959	2080
E°(calculated)	-1.27	-1.58	0.79	0.01	1.66

The trends in the terms $\Delta_{hyd}H^\circ(M^{3+})$, I_3(M) and $\Delta_{hyd}H^\circ(M^{2+})$ are shown in Fig. 10.26. They may be understood from a consideration of the changes in electronic configurations. Across the sequence of elements, V to Co, there is a general reduction in atomic and ionic sizes as the increasing nuclear charge becomes more effective. Applied to hydration enthalpies this factor implies that they would be expected to become more negative as the ionic size decreases across the series. Superimposed on this trend is a destabilization as electrons are added which occupy the anti-bonding e_g* orbitals, i.e. of the M^{2+} ions, Cr^{2+} $(e_g*)^1$ and Mn^{2+}, Fe^{2+} and Co^{2+} (all $(e_g*)^2$ are destabilized to some extent, the anti-bonding electrons causing the metal-water bonding to be weakened, with consequent increase in size.

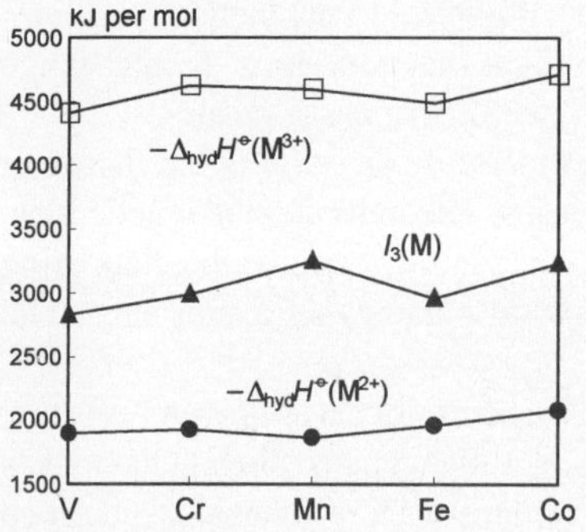

Fig. 10.26 Trends in the standard hydration enthalpies for the M^{II} and M^{III} ions and the third ionization enthalpies for the elements V to Co

The same consideration applies to the M^{3+} ions Mn^{3+} $(e_g*)^1$ and Fe^{3+} $(e_g*)^2$ which have less negative hydration enthalpies as a consequence. The higher stability (more negative hydration enthalpy) of the Co^{3+} ion arises because the d^6 configuration is low-spin $t_{2g}^{\,6}$ with consequent stronger metal-water bonding. The trends in the third ionization energies are understandable as a general increase across the series, but with the values for iron and cobalt being lower than expected because the electron removal occurs from a doubly occupied orbital, those from vanadium, chromium and manganese occurring from singly occupied orbitals.

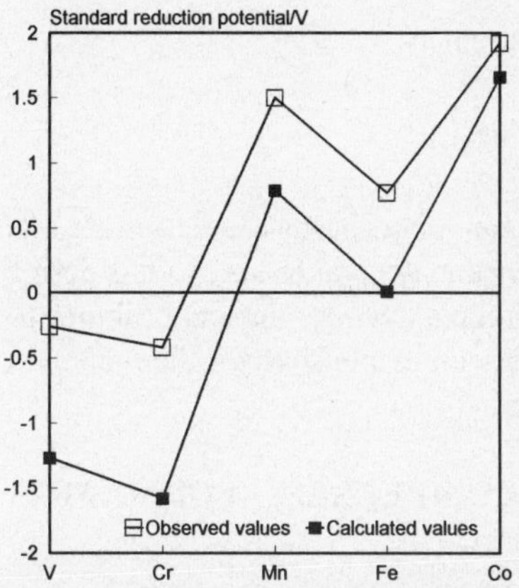

Fig. 10.27 A comparison of the observed and calculated values for the M^{III} to M^{II} standard reduction potentials for the elements V to Co

The observed and calculated values of the reduction potentials for the M^{3+}/M^{2+} couples are compared in Fig. 10.27. The trends in the calculated values are similar to those in the observed data. In the calculations the omission of the entropy terms has caused some large differences between the observed and calculated values for the M^{3+}/M^{2+} potentials.

It is clear from the above examples that quite simple calculations can identify the major factors which govern the values of E° for any couple and that major trends may be rationalized.

The effects of complexation upon E° values

The above discussions concern aquated ions but the value of E° is very dependant upon the ligands surrounding any metal ion centre. Some typical values are given for Fe^{III}/Fe^{II} systems in Table 10.13.

Table 10.13 Some iron(III)/iron(II) standard reduction potentials showing the effects of different ligands

Redox couple	E°/V
$[Fe(phen)_3]^{3+}$/ $[Fe(phen)_3]^{2+}$	1.14
$[Fe(H_2O)_6]^{3+}$/ $[Fe(H_2O)_6]^{2+}$	0.77
$[Fe(CN)_6]^{3-}$/ $[Fe(CN)_6]^{4-}$	0.36
$[Fe(C_2O_4)_3]^{3-}$/ $[Fe(C_2O_4)_3]^{4-}$	0.02
$[Fe(edta_3)]^{-}$/ $[Fe(edta)]^{2-}$	-0.12

The variations in the value of E° for the Fe^{III}/Fe^{II} systems can be rationalized in terms of the interactions between the metals and their ligands. Compared to the aquated system the phen (phen = 1,10-phenanthroline) couple has a higher potential which is mainly due to the relatively greater stabilization of the Fe^{II} state. The lower charge on Fe^{II} facilitates the occurrence of better back bonding compared to the higher charged Fe^{III} complex. Both complexes have very large values for their formation constants because of the strength of the bonding. Cyanide ion, being negatively charged, interacts more strongly with the higher oxidation state, Fe^{III}, although both ions are stabilized considerably by the π-acid back bonding behaviour of the ligands. The effect of the oxalate ($C_2O_4^{2-}$, dithioate ion) ligand is electrostatic and preferentially stabilizes the iron(III) complex. There is no back bonding, but there is the chelate effect which stabilizes the iron(III) preferentially. This is taken to an extreme when edta is the ligand.

Another striking example is the effect on the E° values for the Co^{III}/Co^{II} couple of replacing water by ammonia ligands. The values of E° for the two half-reactions:

$$[Co(H_2O)_6]^{3+} + e^- \rightleftharpoons [Co(H_2O)_6]^{2+}$$

$$[Co(NH_3)_6]^{3+} + e^- \rightleftharpoons [Co(NH_3)_6]^{2+}$$

are 1.84 and 0.11 V respectively. The complexation by ammonia ligands causes a great loss of oxidizing power of the Co^{III} state. This may be explained in general terms by considering the bonding in the (III) state to be stronger with the ammonia ligands which contributes to a lowering of the effective nuclear charge so that the electron responsible for the reduction is attracted less well than in the case of the hydrated ion.

10.7 THERMODYNAMIC STABILITY OF COMPLEXES; FORMATION CONSTANTS

The thermodynamic stability of a complex is conventionally expressed as the magnitude of the equilibrium constant appropriate to the formation reaction from the metal atom or ion and the ligands of which the complex is composed. The formation reaction may be generalized as:

$$M^{n+}(aq) + xL^{m-}(aq) \rightleftharpoons ML_x^{(n-xm)+}(aq) \qquad (10.34)$$

The oxidation state of the metal, M, may be positive, zero or (rarely) negative. The charge on the ligand, L, is either negative or zero. Some ligands exist in acid solution as positive ions, e.g. HL^+, with the neutral ligand being protonated. The displacement of coordinated water molecules from $M^{n+}(aq)$ when M-L bonds are formed is not indicated in equation (10.34). The water molecule which enters the bulk solution does not affect the equilibrium constant for the reaction because the activity of the liquid water is normally taken to be unity.

The equilibrium constant appropriate to reaction (10.34) may be written as:

$$K_x = \frac{[ML_x^{(n-mx)+}]}{[M^{n+}][L^{m-}]^x} \qquad (10.35)$$

where the square brackets are the usual indication of the equilibrium activities of the respective species having been divided by the activity of the species in their standard states ($a^{\circ} = 1$) so as to make the equilibrium constant dimensionless.

In practice so very few activity coefficients are known (or even measurable for the concentrations normally used) that actual concentrations are used in equations like (10.35). It should be realised that the vast number of published formation constants (the equilibrium constants for the formation) of complexes are not true thermodynamic equilibrium constants. For that reason some caution should be used in drawing too fine a conclusion about their absolute magnitudes and, in particular, in the discussion of relatively small differences between values for different complexes.

Note that the term 'formation constant' is synonymous with 'stability constant'. Some treatments refer to 'instability constants' which are the reciprocals of the stability or formation constants. They refer to the reverse of the formation reactions - the dissociation reactions of complexes which give their constituent metal ions in hydrated form and aquated ligands.

The formation constant as described by equation (10.35) is the *overall equilibrium constant*, $\beta_x (= K_x$ in this case). Overall formation constants (βs) are related to the values of *step-wise formation constants* (Ks), which relate to the successive additions of ligands to the central metal ion. The step-wise addition of the ammonia molecule as a ligand to copper(II) ion may be taken as an example.

The first step is written as:

$$Cu(H_2O)_4^{2+}(aq) + NH_3(aq) \rightleftharpoons Cu(NH_3)(H_2O)_3^{2+}(aq)$$

with K_1 given by:

$$K_1 = \frac{[Cu(NH_3)(H_2O)_3]^{2+}}{[Cu(H_2O)_4][NH_3]} \qquad (10.36)$$

The second step is the addition of the second ligand to the product of the first step:

$$Cu(NH_3)(H_2O)_3^{2+}(aq) + NH_3(aq) \rightleftharpoons Cu(NH_3)_2(H_2O)_2^{3+}(aq)$$

with the second step-wise formation constant being given by:

$$K_2 = \frac{[Cu(NH_3)_2(H_2O)_2^{2+}]}{[Cu(NH_3)(H_2O)_3^{2+}][NH_3]} \qquad (10.37)$$

The complex which is normally formed when copper(II) ion, in aqueous solution, is treated with an excess of ammonia is the tetraamminecopper(II) ion, $[Cu(NH_3)_4]^{2+}$. There are four step-wise constants which determine the value of the overall formation constant. The step-wise constants (K values) may be combined to give cumulative (or overall) formation constants, denoted by β, so that, in the case of the first two steps of the formation of $[Cu(NH_3)_2(H_2O)_2]^{2+}$, $\beta_1 = K_1$, but $\beta_2 = K_1 \times K_2$. This may be seen from the operation of carrying out the multiplication of the first two step-wise constants as given by equations (10.36) and (10.37) to give:

$$\beta_2 = K_1K_2 = \frac{[Cu(NH_3)(H_2O)_3^{2+}]}{[Cu(H_2O)_4^{2+}][NH_3]} \cdot \frac{[Cu(NH_3)_2(H_2O)_2^{2+}]}{[Cu(NH_3)(H_2O)_3^{2+}][NH_3]}$$

$$= \frac{[Cu(NH_3)_2(H_2O)_2^{2+}]}{[Cu(H_2O)_4][NH_3]^2} \tag{10.38}$$

which is the cumulative formation constant for the first two stages of the reaction:

$$Cu(H_2O)_4^{2+}(aq) + 2NH_3(aq) \rightleftharpoons Cu(NH_3)_2(H_2O)_2^{2+}(aq)$$

Likewise, it can be written that:

$$\beta_3 = K_1K_2K_3 \tag{10.39}$$

and

$$\beta_4 = K_1K_2K_3K_4 \tag{10.40}$$

such a procedure being continued as far as is necessary to describe the *cumulative constants* in terms of the *step-wise* ones. Both types of constant are published in data books and in scientific papers and some care has to be exercised in distinguishing between the two. One common error in the publication of the values of Ks and βs is that their values are sometimes given as the logarithm (to base 10) without actually specifying that the quoted numbers are logarithms of the actual values of the K or β in question. This arises because of the very large magnitudes of most formation constants. The K and β values concerned in the step-wise production of the tetraamminecopper(II) ion are shown in Table 10.14.

Table 10.14 The K and β values for the four ammine complexes of copper(II)

i	K_i	β_i
1	1.48×10^4	1.48×10^4
2	2.75×10^3	4.07×10^7
3	8.32×10^2	3.38×10^{10}
4	1.32×10^2	4.47×10^{13}

The values in the table refer to the standard conditions of 298 K and 1 atm pressure. Although the β values increase as the number of ligand molecule increases the K values decrease. The observation is a general one.

In the case considered above the subscript of the β value is equal to the number of metal ligand interactions at each stage in the formation of the complex. This is not always the case. If a bidentate ligand is used (such as diaminoethane, $H_2NCH_2CH_2NH_2$) which can form two linkages (via the nitrogen atoms) to a metal ion then the relationships between the K and β values are the same as in the Cu^{2+}-ammonia case but now the subscripts refer to the number of ligands bonded to the metal ion and not to the number of metal-donor atom linkages. In the extreme case of one multi-dentate ligand being used to complex with a metal ion there could be just one value for the formation constant, K_1 being identical to β_1. That may be the case for the interaction of metal ions with EDTA (ethylene diamine tetraacetic acid; formal name, diaminoethane-N,N,N′,N′-tetraethanoic acid, $(HOOCCH_2)_2N(CH_2)_2N(CH_2COOH)_2$, which has two nitrogen atoms and four oxygen atoms that may act as donors) or with more complex ligands such as protein molecules.

Factors affecting stability constants

The magnitude of any particular stability constant (or formation constant) is governed by several factors, some depending upon the nature of the central metal ion and others upon the nature of the ligands.

The nature of the metal ion

Variations in the magnitude of stability constants for a series of complexes of metals ions with a common set of ligands may be discussed in terms of the relative sizes and charges of the metal ions.

(i) Effect of ionic size

Table 10.15 contains the values of the ionic radii (crystal radii), the standard enthalpies of hydration and the decadic logarithms of the stability constants for the formation of $[M(en)(H_2O)_4]^{2+}$ complexes for the (II) oxidation states of the elements from Mn to Zn. The radius and formation constant data are shown graphically in Fig. 10.28.

Table 10.15 Thermodynamic data for M^{II}(en) complexes

M	r_i/pm	$-\Delta_{hyd}H^{\circ}$/kJ mol^{-1}	$\log_{10}K_1$
Mn	83	1863	2.7
Fe	78	1959	4.3
Co	75	2080	5.9
Ni	69	2121	7.6
Cu	73	2121	10.7
Zn	74	2059	5.9

The crystal radius used for copper(II) is the one accepted for octahedral coordination. The Jahn-Teller effect is important in the d^9 case and is probably responsible for the extraordinary stability of the $[Cu(en)(H_2O)_4]^{2+}$ complex. The trends in the values of $\Delta_{hyd}H^{\ominus}$ and log K_1(en) are those which would be expected from the variations in the radii of the central metal ions. The smaller ions produce the more stable complexes.

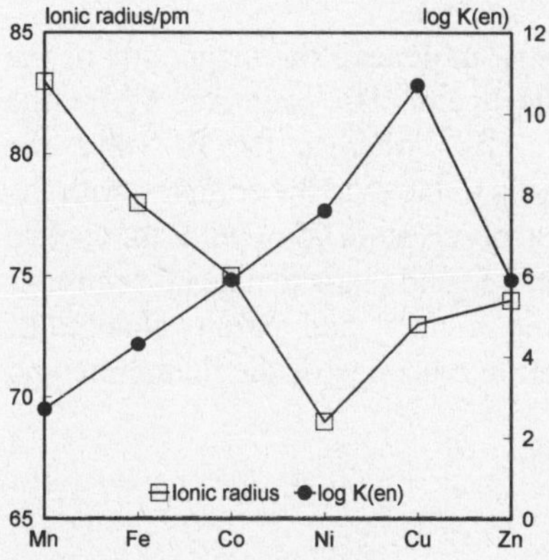

Fig. 10.28 The variations of ionic radii and the log K_1 values for the M^{II}(en) complexes of the elements Mn to Zn

The general shape of the graph of the log K_1 values for the diaminoethane(en) 1:1 complexes is an example of the general trend for the stability constants for any particular ligand. The general trend in stability for the complexes of the M^{II} cases:

$$Mn < Fe < Co < Ni < Cu > Zn$$

is known as the Irving-Williams series.

(ii) Effect of the oxidation state of the central metal ion

As the charge on the central metal ion increases the stability constant for a given metal with the same ligand environment would be expected to show an increase. One example suffices to make the general point; the β_6 values for the complexes, $[Fe(CN)_6]^{4-}$ and $[Fe(CN)_6]^{3-}$, are 10^{34} and 10^{44} respectively, implying a very large increase in stability as the formal charge on the metal ion increases from +2 to +3.

The nature of the ligand

There are three main effects of the nature of the ligand upon the stability constants of complexes.

(i) The donor atom

The effect of the donor atom upon complex stability is mainly summarized by the spectrochemical series, which is the order of ligands in terms of the differences in energy between the e_g* and t_{2g} orbitals produced for a given central metal ion. Bearing in mind that the series takes into account not only σ bonding but π bonding and anti-bonding effects it is to be expected that a large e_g*-t_{2g} energy difference would be associated with a large value for the stability constant of the complexes of any particular metal ion and a variety of ligands. There are minor deviations from such a generalization which apply when metal-ligand π bonding is possible. That kind of bonding produces an increase in stability (compared to a purely σ bonded complex) but reduces the magnitude of the e_g*-t_{2g} energy difference.

(ii) Denticity and the chelate effect

The *denticity* of a ligand is its number of donor atoms. In general the magnitude of the stability constant increases with the denticity of the ligand. For example, K_1 for the $[Cu(en)(H_2O)_2]^{2+}$ complex is observed to be 5 x 10^{10} whereas the β_2 value for $[Cu(NH_3)_2(H_2O)_2]^{2+}$ is only 5 x 10^7 —a factor of 10^3 less stable than the complex with the bidentate ligand with the same donor atoms. This general observation is known as the chelate effect, which is related to ring formation when a chelate ligand binds to a metal centre.

The chelate effect is due to differences in the enthalpy and entropy changes of formation of the complexes. The data for the formation reactions of the diammine and diaminoethane (en) complexes are given below.

Reaction	$\Delta H^{\ominus}$/kJ mol^{-1}	$\Delta G^{\ominus}$/kJ mol^{-1}
$[Cu(H_2O)_4]^{2+} + 2NH_3 \rightleftharpoons [Cu(NH_3)_2(H_2O)_2]^{2+} + 2H_2O$	-46.4	-43.9
$[Cu(H_2O)_4]^{2+} + en \rightleftharpoons [Cu(en)(H_2O)_2]^{2+} + 2H_2O$	-54.4	-61.1

The respective $\Delta G^{\ominus}$ values indicate the greater stability of the Cu^{II}-en combination over the $Cu^{II}-(NH_3)_2$ complex by 17.2 kJ mol^{-1}. The greater stability of 17.2 kJ mol^{-1} is made up from a greater exothermicity (difference in $\Delta H^{\ominus}$ values) of 8.0 kJ mol^{-1} and an advantage of 9.2 kJ mol^{-1} from the difference in the $T\Delta S^{\ominus}$ terms (at 298 K). There is a small decrease in entropy in the formation of the diammine complex compared to a comparatively large increase in entropy of formation of the diaminoethane complex. This difference may be understood in terms of there being no change in the number of species taking part in the reaction with ammonia whereas there is an increase in the number of species in the reaction with en. The translational freedom afforded to one of the water molecules in the en reaction is mainly responsible for the entropy increase observed.

The greater exothermicity of the chelate formation reaction is due mainly to a change in the relative solvation energies of one diaminoethane molecule compared to the two water molecules it replaces. The en molecule, with its two CH_2 groups, is not as hydrophilic (hydrocarbon groups are hydrophobic, i.e. do not form hydrogen bonds with water) as ammonia and does not interact as strongly with the water solvent. The two water molecules it displaces are very hydrophilic (!) and their entry into the bulk solution is a major factor in determining the enthalpy advantage of the chelation reaction.

(iii) Ring size

In general a five-membered ring (metal plus appropriate ligand atoms) confers a greater stability to a complex than a six (or more) membered ring. Table 10.16 contains data for the Mn^{II} complexes with ethandioate (oxalate, $C_2O_4^{2-}$), propandioate (malonate, $O_2CCH_2CO_2^{2-}$) and butandioate (succinate, $O_2C(CH_2)_2CO_2^{2-}$) ions. Here and in many other examples the five membered ring is preferred to any other number.

Table 10.16 $\log_{10}K_1$ values for Mn^{II} complexes

Ligand	$\log_{10}K_1$
Ethandioate (oxalate)	2.93
Propandiaoate (malonate)	2.30
Butandioate (succinate)	1.26

10.8 KINETICS AND MECHANISMS OF REACTIONS

The sign of $\Delta G^{\ominus}$ for a reaction is an indication of its feasibility. A negative value of $\Delta G^{\ominus}$ indicates that the position of equilibrium favours the products. Although that criterion may be fulfilled there are *kinetic barriers* which may cause a reaction to be extremely slow. This section is restricted to an account of ligand substitution reactions. Redox reactions and their mechanisms are dealt with in the succeeding section.

Ligand substitution reactions

There are three classes of ligand substitution reactions:
(i) *Associative* (A) reactions in which the rate determining step is that which involves the association of the two reactants - the complex and the incoming ligand. The first stage of the reaction is the production of an intermediate (not a transition state) where the metal centre has a coordination number one more than it has in the initial state.
(ii) *Interchange* (I) reactions which are single stage processes not involving an intermediate of any kind. In this kind of process there is a synchronous interchange so that, as the incoming ligand approaches the complex, the outgoing ligand leaves. It has been necessary to define two sub-sections of interchange reactions. One is the I_a process in which the rate depends upon the nature of the incoming group as well as its concentration. In such a case bond formation in the transition state is important in determining the magnitude of the activation energy. The other sub-section consists of I_d processes in which the rate is independent of the nature of the incoming group but may depend upon its concentration. The activation energy is largely determined by bond breaking in the transition state. In such a process the incoming ligand enters the outer sphere of the complex as the leaving group dissociates (hence I_d) from the central atom.
(iii) *Dissociative* (D) reactions in which there is a detectable intermediate with a coordination number which is lower than that of the reactant complex.

The above classification derives from the ideas of Langford and Gray and is in general use in this branch of kinetic investigation.

Only two general examples of ligand substitution reactions are dealt with in this book; (i) square planar platinum(II) complexes and (ii) octahedral complexes.

Ligand substitution reactions of platinum(II) complexes

Platinum(II) complexes are in general square planar and conform to the associative interchange (I_a) mechanism. The incoming ligand is able to approach the complex without experiencing much steric hindrance. There are complications with the participation of the solvent. The observed rate laws for such reactions as:

$$PtL_3X + Y \rightarrow PtL_3Y + X$$

in which X represents the outgoing ligand, L represents ligands which remain attached to the metal centre (spectator ligands) and Y is the incoming ligand, have the form:

$$\text{Rate} = k_{obs}[PtL_3X] \qquad (10.41)$$

This is the case where the reactions are carried out under *pseudo-first order* conditions with an excess of Y (compared to the concentration of the platinum complex). Under such conditions the change in the concentration of Y is minimal and [Y] can be regarded as a constant term. The observed pseudo-first order rate constant, k_{obs}, is found (by carrying out reactions with different concentrations of Y) to have the form:

$$k_{obs} = k_S + k_Y[Y] \qquad (10.42)$$

in which k_S and k_Y are rate constants. Whenever the observed rate constant has a form such as this the interpretation is that there must be two competing pathways for the observed reaction. The expression for k_{obs}, equation (10.42), may be placed into equation (10.41) and then multiplied out to give:

$$\text{Rate} = (k_S + k_Y[Y])[PtL_3X] = k_S[PtL_3X] + k_Y[Y][PtL_3X] \qquad (10.43)$$

the two terms representing the rates of the competing pathways. The second term is that which would be expected for a simple bimolecular association between the two reactants. The first term is independent of the concentration of the incoming ligand and may be thought to be typical of a dissociative process. When experiments are carried out with different solvents it becomes clear that the first term represents a pathway in which k_S is determined by the nature of the solvent. The rate determining step does not involve the incoming ligand and is a slow process in which X is possibly replaced by a solvent molecule, S:

$$PtL_3X + S \rightarrow PtL_3S + X$$

the solvent molecule being itself replaced by a faster (and therefore not rate limiting) step:

$$PtL_3S + Y \rightarrow PtL_3Y + S$$

Ligand substitution reactions of octahedral complexes

The discussion of these reactions is limited to the replacement of coordinated water molecules by water molecules (exchange reactions) or by other ligands. The mechanism of reactions of six coordinate complexes is dominated by the dissociative (D) or dissociative interchange (I_d) processes. It is very improbable that an associative process with the production of a seven coordinate intermediate (or transition state) would be able to compete with the dissociative alternatives.

An important possibility in these reactions is the formation of an outer-sphere complex between the six coordinate complex and the incoming ligand:

$$ML_5X + Y \rightarrow ML_5X,Y$$

in which the incoming ligand, Y, replaces one of the water molecules in the outer sphere or solvation shell of the complex, ML_5X. The second stage of the process is an I_d type of ligand substitution between the inner sphere and outer sphere:

$$ML_5X,Y \rightarrow ML_5Y,X$$

The outgoing ligand, X, then equilibrates with the bulk solution phase, thus completing the substitution reaction.

$$ML_5Y,X + H_2O \rightarrow ML_5Y + X$$

If an equilibrium is established between the complex, C (= ML_5X) , and the incoming ligand, Y, to give an outer sphere complex, OS:

$$C + Y \rightleftharpoons OS \tag{10.44}$$

the equilibrium constant, K_{OS}, is given by:

$$K_{OS} = \frac{[OS]}{[C][Y]} \tag{10.45}$$

If the rate determining step is the interchange reaction, the rate will be first order in [OS]:

$$\text{Rate} = k_I[OS] \tag{10.46}$$

where k_I is the rate constant for the interchange reaction. Placing the expression for [OS] from equation (10.45) into equation (10.46) gives:

$$\text{Rate} = k_I K_{OS}[C][Y] \tag{10.47}$$

implying a second order process.

The total concentration of the complex undergoing substitution, T, is given by:

$$T = [C] + [OS]$$

$$= [C] + K_{OS}[C][Y] \tag{10.48}$$

and solving this equation for [C] gives:

$$[C] = \frac{T}{1 + K_{OS}[Y]} \tag{10.49}$$

Substitution in equation (10.47) gives:

$$\text{Rate} = \frac{k_I K_{OS} T[Y]}{(1 + K_{OS}[Y])} \tag{10.50}$$

Equation (10.50) reduces to one implying second order kinetics if K_{OS} has a value such that $K_{OS}[Y] >> 1$ so that:

$$\text{Rate} = k_I K_{OS} T[Y] \tag{10.51}$$

(under these conditions $T \sim [C]$).

For reactions carried out under pseudo-first order conditions where $[Y] >> T$ the observed rate constants are given by:

$$k_{obs} = \frac{k_I K_{OS}[Y]}{(1 + K_{OS}[Y])} \qquad (10.52)$$

which may be inverted to give:

$$\frac{1}{k_{obs}} = \frac{1}{k_I K_{OS}[Y]} + \frac{1}{k_I} \qquad (10.53)$$

in which form a plot of $1/k_{obs}$ against $1/[Y]$ gives a straight line with a slope of $1/k_{obs}K_{OS}$ and an intercept of $1/k_I$. Determination of the slope and intercept from experimental data gives values for both k_I and K_{OS}.

Rates of water molecule replacement

The above formulation for the kinetics of ligand replacement reactions at an octahedral centre are modified in the case of water replacement reactions taking place in aqueous solution because of the very high concentration of the incoming ligand which is 55.5 *M*. This has the consequence that $K_{OS}[H_2O] >> 1$ so that $k_{obs} \sim k_I$.

The observed rate constants for water exchange in a series of transition metal hexaaqua-complexes are given in Table 10.17. All the aqua-complexes are of the high spin variety (water does not bond particularly strongly with +2 transition elements). There is an enormous variation in the water exchange rates of these complexes ranging from the relatively very slow V^{II} to the very fast Cr^{II} and Cu^{II} cases.

Table 10.17 Observed rate constants for water exchange reactions

Central metal	d	$\log(k_I/s^{-1})$
Ca^{II}	0	8.5
V^{II}	3	1.8
Cr^{II}	4	9.0
Mn^{II}	5	7.5
Fe^{II}	6	6.5
Co^{II}	7	6.4
Ni^{II}	8	4.4
Cu^{II}	9	9.0
Zn^{II}	10	7.5

The reason for this variation cannot be based upon any consideration of ionic size. There is a variation of ionic radius from the largest (Ca^{2+}, $r_i = 100$ pm) to the smallest (Cu^{2+}, $r_i = 57$ pm) which is irrelevant compared to the variation in rate constant from the highest

($Cr^{II} = Cu^{II}$, $k_i = 10^9\ s^{-1}$) to the lowest (V^{II}, $k_i = 63\ s^{-1}$) - a factor of 1.58×10^7. In the Cu^{II} case there is no doubt a participation of the Jahn-Teller effect which tends to ensure that the axial ligands are very weakly bound. In all the cases except for Ca^{II} and V^{II} there are e_g* anti-bonding electrons present which weaken the metal-ligand binding and in the V^{II} case the symmetrical filling of the t_{2g} orbitals (non-bonding) is consistent with the low rate of exchange.

The above data give rise to the classification of complexes as being either inert or labile with respect to ligand substitution. These terms belong to descriptions of kinetic behaviour and must not be confused with thermodynamic terms such as stable or unstable which are used when discussing the stability or otherwise of complexes with respect to their constituent metal and ligands.

The conjugate base mechanism of ligand substitution reactions

The replacement of a chloride ion ligand by a hydroxide ion in a cobalt(III)-ammine complex:

$$Co(NH_3)_5Cl^{2+} + OH^- \rightarrow Co(NH_3)_5OH^{2+} + Cl^-$$

is observed to be a second-order reaction, the rate law being:

$$\text{Rate} = k_2[Co(NH_3)_5Cl^{2+}][OH^-] \quad (10.54)$$

where k_2 is a second-order rate constant. The rate law could be interpreted in terms of an associative (A) or an associative interchange (I_a) process. Such processes are relatively unusual for six-coordinate complexes of cobalt(III). The expected dissociative (D) or dissociative interchange (I_d) processes may be assigned to reactions of ammine complexes if it is considered that they participate in the formation of their conjugate bases. The chloropentamminecobalt(III) ion forms its conjugate base according to the reaction:

$$Co(NH_3)_5Cl^{2+} + OH^- \rightleftharpoons Co(NH_3)_4(NH_2)Cl^+ + H_2O$$

the equilibrium constant being K_{CB}. The conjugate base has not been isolated and the equilibrium constant must have a very small value. The rate determining step of the ligand substitution reaction consists of the replacement of the chloride ion ligand of the conjugate base by a hydroxide ion in a dissociative manner according to the equations:

$$Co(NH_3)_4(NH_2)Cl^+ \rightarrow Co(NH_3)_4(NH_2)^{2+} + Cl^-$$

$$Co(NH_3)_4(NH_2)^{2+} + H_2O \rightarrow Co(NH_3)_5(OH)^{2+}$$

The formation of the conjugate base reduces the overall charge on the Co^{III} complex and facilitates the dissociation of the negative chloride ion. If the dissociative step is rate determining, the rate of the reaction is given by:

$$\text{Rate} = k_D[Co(NH_3)_4(NH_2)Cl^+] \quad (10.55)$$

where k_D is the first order dissociation rate constant. The concentration of the conjugate base is given by:

$$k_D[Co(NH_3)_4(NH_2)Cl^+] = K_{CB}[Co(NH_3)_5Cl^{2+}][OH^-] \quad (10.56)$$

so that equation (10.55) may be written in the form:

$$\text{Rate} = k_D K_{CB}[Co(NH_3)_5Cl^{2+}][OH^-] \quad (10.57)$$

with the product $k_D K_{CB}$ representing the second-order rate constant, k_2.

10.9 KINETICS AND MECHANISMS OF REDOX PROCESSES

Redox reactions between transition metal complexes

There are three classes of electron transfer reactions to be discussed:
(i) one electron exchange processes between metal centres involving different oxidation states of the same metal (exchange reactions),
(ii) one electron redox processes between complexes of two different metal centres, and
(iii) redox processes involving the transfer of more than one electron.

Exchange reactions

The classical example of this type of process is the scrambling of the ^{55}Fe (radioactive, decays by electron capture with a $t_{1/2}$ of 2.73 y) and ^{56}Fe isotopes between the (II) and (III) oxidation states:

$$^{55}Fe^{II} + {}^{56}Fe^{III} \rightarrow {}^{55}Fe^{III} + {}^{56}Fe^{II}$$

which is followed by selectively precipitating the iron(III) from solution at various times after the initial mixing of the two isotopes. The reaction is thermoneutral in that $\Delta H^{\ominus}$ is zero. There is a slight entropy increase as the two isotopes become scrambled between the two oxidation states. The activation energy of the reaction is 41.4 kJ mol^{-1}, the bimolecular rate constant at 273 K being 0.87 L mol^{-1}s^{-1}. The rate constants of ligand water replacement in the two iron species (Fe^{II}, Fe^{III}) are 3×10^6 s^{-1} and 3×10^3 s^{-1} respectively, which become 5.4×10^4 L mol^{-1} s^{-1} and 54 L mol^{-1} s^{-1} when divided by the molarity of water (55.5 *M*) to convert them into bimolecular rate constants. Both those rate constants are far greater than that observed for the exchange process and this precludes the so-called inner sphere mechanism in which one ion loses a ligand by dissociation and a bridged complex is formed with the other ion which facilitates electron transfer. The Fe^{II}, Fe^{II} exchange reaction takes place by the outer sphere mechanism in which both reactant complexes retain their inner coordination spheres.

Another factor which would rule out the formation of a bridged intermediate complex is that the water molecule has only one pair of non-bonding electrons of high enough energy to participate in bonding to a metal centre. There are two main factors which cause the reaction to have a reasonably high activation energy. These are (i) the overcoming of the electrostatic repulsion between the two ions, and (ii) the 'Franck-Condon' restriction which permits electron transfer only when the two ions have identical Fe–O distances. The Franck-Condon Principle is important in electronic spectroscopy and indicates that 'the vertical transition between any two potential energy curves (section 3.3) is the most probable'. It arises from the recognition that electronic transitions occur in times which are far shorter than the time for a molecular vibration to occur. Applied to exchange processes it means that, in the transition state, the metal-ligand distances must be identical for the two participants, otherwise the first law of thermodynamics would be violated! The ground state ions would, if an electron were to be transferred, become vibrationally excited states which could relax to their respective ground states with the production of energy. The smaller Fe^{III}–O bonds must therefore be stretched and the larger Fe^{II}–O bonds must contract (both by vibrational excitation) before the electron may be transferred.

The mechanism whereby the two ions approach each other so that their primary solvation shells are within appropriate Van der Waals radii is known as the *outer sphere mechanism.* In this mechanism the inner spheres (the primary coordination spheres) of the two reactants remain intact throughout the process.

The most convincing evidence which supports the outer sphere mechanism is for the reaction:

$$Fe(CN)_6^{4-} + Fe(phen)_3^{3+} \rightarrow Fe(CN)_6^{3-} + Fe(phen)_3^{2+}$$

which has a very high rate of exchange. All four participating complexes undergo ligand hydrolysis at very low rates. The inner sphere mechanism is thus precluded. The rate constants for isotopic scrambling in the couples: $Fe(CN)_6^{3-}/Fe(CN)_6^{4-}$ and $Fe(phen)_3^{3+}/Fe(phen)_3^{2+}$ are 3×10^2 and $> 10^5$ L mol^{-1} s^{-1} respectively, again indicating that the outer sphere mechanism must apply.

One–Electron Redox Processes

When one electron is transferred between oxidation states of different elements there is no Frank-Condon restriction in operation. The mechanism of any one process may be either outer
sphere or inner sphere. The classical work of H.Taube (Nobel prize winner for Chemistry, 1983) established the criteria for the operation of the *inner sphere mechanism*.

The reaction:

$$Cr(H_2O)_6^{2+} + Co(NH_3)_5Cl^{2+} + 5H_3O^+ \rightarrow Cr(H_2O)_5Cl^{2+} + Co(H_2O)_6^{2+} + 5NH_4^+$$

has a rate constant of 6×10^5 L mol^{-1} s^{-1} at 298K. The Cr^{II} reactant has d^4 configuration and is expected to be substitutionally labile, the rate constant for water exchange being 10^9 s^{-1}. The cobalt(III) complex has a d^6 configuration and is thus substitutionally inert and has a hydrolysis rate constant of 1.7×10^{-7} s^{-1}. The mechanism of the above reaction is thought to involve the replacement of a water ligand of the Cr^{II} complex by the chloride ion ligand of the Co^{III} complex to form a *bridged activated complex* in which the electron transfer may take place:

$$Cr(H_2O)_6^{2+} \rightarrow Cr(H_2O)_5^{2+} + H_2O$$

$$Cr(H_2O)_5^{2+} + Co(NH_3)_5Cl^{2+} \rightarrow (H_2O)_5Cr^{II}ClCo^{III}(NH_3)_5^{4+} \leftrightarrow (H_2O)_5Cr^{III}ClCo^{II}(NH_3)_5^{4+}$$

Both formulations of the bridged activated complex contribute to the transition state and when in the right hand form there is a dissociation such that the chloride ion is retained by the Cr^{III} (the d^3 configuration causing substitutional inertness), the other product is (initially) the substitutionally labile, d^7, $Co^{II}(NH_3)_5^{2+}$ ion which loses all five of its ammonia ligands to the bulk solution with the addition of six water ligands. There is no other way in which the chloride ion could be transferred between the Cr^{II} and Co^{III} complexes. If the Co^{III} complex released the chloride ion into the bulk solution (which is a very slow process) the labile Cr^{II} complex would not be likely to react with it. If an electron were to be transferred by an outer sphere process then the chloride would appear in the bulk solution (released by the labile Co^{II} and would be unable to react with the inert Cr^{III} which would appear as the hexaaquo-complex). Confirmatory evidence for the inner sphere mechanism is furnished by the observation that if the reaction is carried out in the presence of free $^{36}Cl^-$ (radioactive, $t_{1/2} = 3.01 \times 10^5$ y, β^- and β^+emissions) there is no incorporation of the radioactive chloride ion by the Cr^{III} product.

It should be noted that the chloride ion does possess a suitable pair of non-bonding electrons which allows it to act as a bridging ligand. The reaction of $[Co(NH_3)_6]^{3+}$ with $[Cr(H_2O)_6]^{2+}$ is a very slow process, with a rate constant of 10^{-3} L mol^{-1} s^{-1} - a factor of 6 x 10^8 times slower than the reaction involving a chloride ligand, the ammonia molecule being unable to participate in bridge formation. The mechanism must be of an outer sphere type.

Multiple Electron Redox Reactions

The basic rule for so-called *non-complementary* reactions in which the changes in oxidation state of the oxidizing and reducing agents are different is that the mechanism will consist of the smallest number of one-electron steps. In the reaction:

$$Cr^{VI} + 3Fe^{II} \rightarrow Cr^{III} + 3Fe^{III}$$

the observed rate law for the production of Cr^{III} is:

$$\frac{d[Cr^{III}]}{dt} = \frac{k[Cr^{VI}][Fe^{II}]^2}{1 + \left(\dfrac{k'[Fe^{II}]}{[Fe^{III}]}\right)} \qquad (10.58)$$

where k and k' are rate constants.

This may be explained by the three one-electron steps:

$$Cr^{VI} + Fe^{II} \rightarrow Cr^{V} + Fe^{III} \qquad \text{(Rate constant} = k_1) \qquad (10.59)$$

$$Cr^{V} + Fe^{II} \rightarrow Cr^{IV} + Fe^{III} \qquad \text{(Rate constant} = k_2) \qquad (10.60)$$

$$Cr^{IV} + Fe^{II} \rightarrow Cr^{III} + Fe^{III} \qquad \text{(Rate constant} = k_3) \qquad (10.61)$$

with the reverse of reaction (10.59) being important (rate constant given by k_{-1}). The mechanism involves two reactive intermediates—Cr^V and Cr^{IV}—whose respective concentrations would attain steady state values consistent with the equations:

$$d[Cr^V]/dt = k_1[Cr^{VI}][Fe^{II}] - k_{-1}[Cr^V][Fe^{III}] - k_2[Cr^V][Fe^{II}] = 0 \qquad (10.62)$$

and

$$d[Cr^{IV}]/dt = k_2[Cr^V][Fe^{II}] - k_3[Cr^{IV}][Fe^{II}] = 0 \qquad (10.63)$$

Digression on steady state theory

It is a normal practice, when dealing with reaction mechanisms, to assume that there is a *rate-determining step* which is the slowest reaction in the sequence. Other steps in the mechanism are assumed to proceed at higher rates and the overall rates of production and reaction of reactive intermediates (these are identifiable species, not to be confused with transition states) are assumed to be equal, so leading to a steady state concentrations of the intermediates. The steady state assumption can be written as:

$$\frac{d[\text{reactive intermediate}]}{dt} = 0 \qquad (10.64)$$

The equation for the rate of production of Cr^{III}—the final chromium product—is:

$$d[Cr^{III}]/dt = k_3[Cr^{IV}][Fe^{II}] = k_2[Cr^V][Fe^{II}] \qquad (10.65)$$

and the unknown $[Cr^V]$ may be derived from equation (10.62):

$$[Cr^V] = \frac{k_1[Cr^{VI}][Fe^{II}]}{k_{-1}[Fe^{III}] + k_2[Fe^{II}]} \qquad (10.66)$$

the final expression for the overall rate being:

$$\frac{d[Cr^{III}]}{dt} = \frac{k_1k_2[Cr^{VI}][Fe^{II}]^2}{1 + \left(\dfrac{k_2[Fe^{II}]}{k_{-1}[Fe^{III}]}\right)} \qquad (10.67)$$

which is consistent with the observed rate law with $k = k_1k_2$, and $k' = k_2/k_{-1}$.

10.10 PROBLEMS

1. Calculate the ligand field splitting parameter, Δ, in $[Ni(DMF)_6]^{2+}$ given that the longest wavelength d–d bands in the spectra of $[Ni(bipy)_3]^{2+}$ and $[Ni(bipy)(DMF)_4]^{2+}$ have the wavelengths of absorption maxima at 791 nm and 985 nm respectively. Estimate λ_{max} for the corresponding band in $[Ni(DMF)_6]^{2+}$.

2. Give reasoned explanations for the following observations.
(a) Whereas the spectrum of $[Fe(MeCN)_6]^{2+}$ has a single d–d band at 917 nm (absorption coefficient = 0.66 $m^2\ mol^{-1}$), the spectrum of $[Fe(CN)_6]^{4-}$ has d–d bands at 270 nm and 320 nm (ϵ = 50–350 $m^2\ mol^{-1}$) and a much weaker band at 420 nm ($\epsilon = 10^{-2}\ m^2\ mol^{-1}$).
(b) The absorption spectrum of *cis*-$[Co(en)_2F_2]^+$ has bands at 360 nm and 510 nm, whereas the *trans*-isomer has weaker bands at 360, 440 and 570 nm. The spectrum of $[CoF_6]^{3-}$ has two bands at 690 nm and 877 nm.
(c) The wavelengths, λ_{max}, of an intense band ($\epsilon \sim$ 200–800 $m^2\ mol^{-1}$) of the following complex anions are:

Complex	$[RuCl_6]^{3-}$	$[RuCl_6]^{2-}$	$[OsCl_6]^{2-}$	$[OsI_6]^{2-}$
λ_{max}/nm	309	406	383	537

(d) The molar absorption coefficients, ϵ, of the sharp visible bands of $[Mn(H_2O)_6]^{2+}$ lie in the range 0.002–0.004 $m^2\ mol^{-1}$, whereas a solution of $[Mn(H_2O)_6]^{3+}$ has a very broad band at 480 nm with an ϵ value of ~5 $m^2\ mol^{-1}$.
(e) $[Fe(phen)_3]^{2+}$ has an absorption band at 510 nm (ϵ = 1120 $m^2\ mol^{-1}$) plus absorption bands in the UV region, whereas $[Fe(phen)_3]^{3+}$ has an absorption band at 590 nm (ϵ = 100 $m^2\ mol^{-1}$) and similar bands in the UV region.
(f) The major band in the spectrum of the pink $[Co(H_2O)_6]^{2+}$ has a maximum at 513 nm (ϵ = 0.5 $m^2\ mol^{-1}$). On the addition of of HCl (conc.) the solution turns blue and is more intensely coloured with absorption maxima at 625, 670 and 700 nm ((ϵ ~ 35 - 60 $m^2\ mol^{-1}$).
(g) In the series $[Cr(NH_3)_6]^{3+}$, $[Cr(NH_3)_5CN]^{2+}$ and $[Cr(NH_3)_4(CN)_2]^+$ the wavelengths of the longest spin-allowed bands are 464, 451 and 437 nm respectively, whereas in the series $[Cu(NH_3)_4(H_2O)_2]^{2+}$, $[Cu(NH_3)_5(H_2O)]^{2+}$ and $[Cu(NH_3)_6]^{2+}$ the wavelengths of the only d–d band are 590, 640 and 575 nm respectively.

3. (a) On the basis of the following standard reduction potentials in acid solution, predict which species can oxidize Fe^{II} to Fe^{III} and which can reduce Fe^{III} to Fe^{II}.

Reaction	E°/V
$Ti^{IV} + e^- \rightarrow Ti^{3+}$	0.01
$Fe^{3+} + e^- \rightarrow Fe^{2+}$	0.77
$Cr_2O_7^{2-} + 14H^+ + 6e^- \rightarrow 2Cr^{3+} + 7H_2O$	1.33
$Sn^{IV} + 2e^- \rightarrow Sn^{2+}$	0.15

(b) 3.48 g of an iron oxide were dissolved in 250 cm^3 of dilute acid. A 25 cm^3 aliquot of the solution was titrated with a Ti^{3+} solution (0.096 mol dm^{-3}) of which 31.34 cm^3 were required. Another 25 cm^3 aliquot was treated with an Sn^{2+} solution sufficient to complete the reaction and the solution was then titrated with a 1/60 mol dm^{-3} solution of $K_2Cr_2O_7$ of which 45.10

cm^3 were required. Calculate the formula of the iron oxide.

4. (a) What conclusions may be drawn from the following effective magnetic moments, μ, measured at room temperature?

Complex	Magnetic moment, μ/μ_B
$[(CH_3)_4N]_2[MnCl_4]$	5.9
$K_4[Mn(CN)_6]$	2.2
$K_2[Mn(IO_3)_6]$	3.82
$K_3[Mn(C_2O_4)_3]$	4.81
$K_3[Mn(CN)_5OH]$	2.92
$[Fe(CN)_6]^{3-}$	2.3
$[Fe(H_2O)_6]^{2+}$	5.3
$[Fe(H_2O)_6]^{3+}$	5.9
$[FeCl_4]^{2-}$	5.4

(b) Calculate the spin-only magnetic moments for the species given below and compare them with their observed values.

Species	$[Cr(H_2O)_6]^{3+}$	$[Ni(H_2O)_6]^{2+}$	$[Co(H_2O)_6]^{2+}$
μ/μ_B	3.79	3.27	5.0

SELF-ASSEMBLY: STOP PRESS

The self-assembly of complexes is referred to in Chapter 1 as one of the areas of inorganic chemistry which is of great current interest. It is not treated further in the book as it is a topic more appropriate to specialist courses, but one recent example deserves mention. This is the complex cage/cluster (Mingos and Vilar, to be published) which is formed from Ni^{2+}(aq) and the compound (LH), $NH_2C(NH)NHCSNH_2$, in the presence of bromide ion. A cage is formed $Ni_6L_8Br^{3+}$ consisting of an octahedral arrangement of six Ni^{II} in which the two apical ions are both bonded to four sulfur donor atoms giving each of them a square planar environment. The apical Ni centres are nearly, but not quite, co-planar with the sulfur atoms.

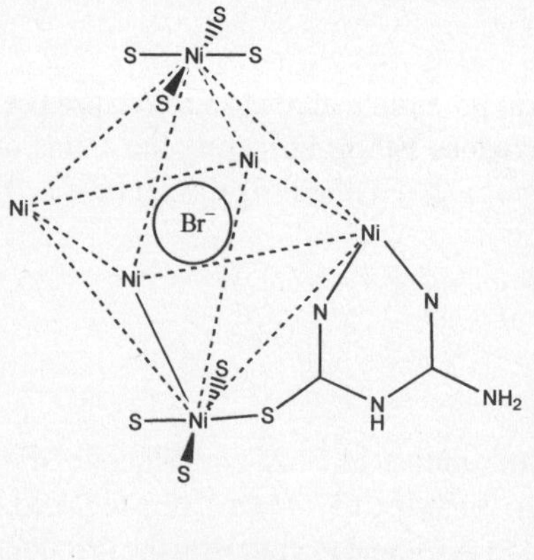

The other ends of the eight ligands, L^- use two nitrogen donors each to provide the Ni species in the octahedral plane with square planar coordination. The six Ni centres are thus 4-coordinate and more-or-less square planar, in keeping with expectations for a d^8 configuration. The bromide ion lies at the centre of the cage and is presumably responsible for the apical Ni centres being out-of-plane. The bromide ion is replaceable with F^- and Cl^-, but not by I^-, NO_3^- or ClO_4^- which are too large. The structure is shown in the diagram in which only one ligand is drawn and in which some of the hydrogen atoms of the ligand are omitted for clarity.

11

Chemistry of the d and f Block Metals

11.1 INTRODUCTION

The d block contains the elements of Groups 3–12 in which the d orbitals are progressively filled in each of the four long periods. The f block elements are those in which the 4f and 5f orbitals are progressively filled in the latter two long periods. They are formal members of Group 3. The names transition metals and inner transition metals are often used to refer to the d and f blocks respectively. However, two groups among the d block groups differ in many respects from the others. The Sc, Y, Lu triad invariably form M^{3+} ions with noble gas configurations. The Zn, Cd, Hg triad in Group 12 form M^{2+} ions which have a full d^{10} configuration, whereas the remaining d block metals exhibit variable valency when the d sub-shell is incomplete. Of the group 12 elements only Hg has a stable +1 state in the form of the Hg_2^{2+} cation. These metals are conveniently dealt with together. On the other hand, the 4f metals, called the lanthanides, behave like the Sc triad and both are discussed together with the 5f metals which are referred to as the actinides. The actinides differ from the lanthanides in exhibiting much more variable valency than the latter elements.

11.2 CHARACTERISTICS OF THE METALS IN GROUPS 4 TO 11

Because the d orbitals project to the periphery of an atom more than the other orbitals (i.e. s and p), they are more influenced by the surroundings as well as affecting the atoms or molecules surrounding them. In some respects, ions of a given d^n configuration (n = 1–9) have similar magnetic and electronic spectral properties.

Successive ionization energies do not increase as steeply as in the main group elements. As a result of this, the loss of a variable number of electrons is not energetically prohibitive. This means that the formation of compounds in which the valency or oxidation state of the metal is variable. Ionic compounds are formed by those elements in low oxidation states. Variable valency is also found in covalent compounds especially with F and/or O, where the maximum oxidation number can be exhibited. In these compounds a variable number of metal orbitals are used for bonding. Fig. 11.1 shows the trends in the first three ionization energies of the first row transition elements together with first two ionization energies of calcium which are included for comparison.

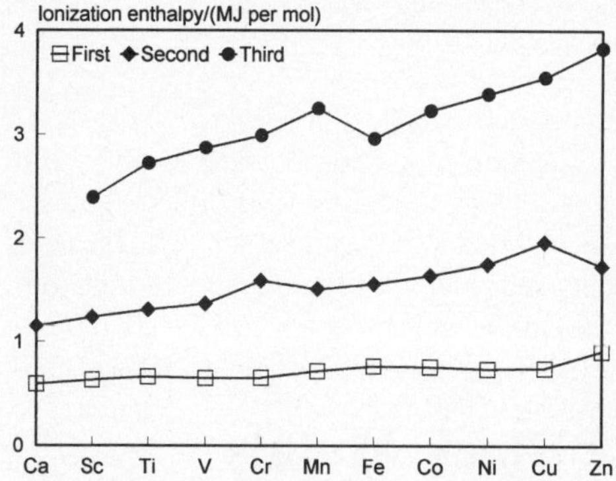

Fig. 11.1 The 1st, 2nd and 3rd ionization enthalpies of the first transition series and the 1st and 2nd values for calcium

The irregular trend in the first ionization energy of the 3d metals, though of little chemical significance, can be accounted for by considering that the removal of one electron alters the relative energies of the 4s and 3d orbitals so that the uni-positive ions have d^n configurations with no 4s electrons. There is thus a reorganization energy accompanying ionization with some gains in exchange energy as the number of d electrons increases and from the transference of s electrons into d orbitals. There is the generally expected increasing trend in the values as the effective nuclear charge increases with the value for Cr being lower because of the absence of any change in the d configuration and the value for Zn being relatively high because it represents an ionization from the 4s level. The lowest common oxidation state of these metals is +2. To form the M^{2+} ions from the gaseous atoms, the sum of the first and second ionization energies is required in addition to the enthalpy of atomization for each element. The dominant term is the second ionization energy which shows unusually high values for Cr and Cu where the d^5 and d^{10} configurations of the M^+ ions are disrupted, with considerable loss of exchange energy. The value for Zn is correspondingly low as the ionization consists of the removal of an electron which allows the production of the stable d^{10} configuration. The trend in the third ionization energies is not complicated by the 4s orbital factor and shows the greater difficulty of removing an electron from the d^5 Mn^{2+} and d^{10} Zn^{2+} ions superimposed upon the general increasing trend.

Another characteristic of these metal ions or complexes is their catalytic role in homogeneous reactions. This is related to the variable valency and depends upon the facility by which one oxidation state changes to the next. One electron changes are important and much more probable than others.

The highest oxidation state corresponds to the group number up to Group 7 when all the two s and the five d electrons participate in the bonding. The oxidation state of +8 is achieved with Os in OsO_4. Low oxidation numbers of 0, −1, −2 or even −3 are found when a complex compound has ligands capable of π acceptor character in addition to the σ bonding.

The formation of coloured and paramagnetic species, another characteristic of the transition metals (discussed in Chapter 10), is related to the splitting of the d orbitals in the field of surrounding ligands.

11.3 TRENDS IN THE M^{II} – M REDUCTION POTENTIALS

Table 11.1 contains the thermochemical parameters related to the transformation of the solid metal atoms to M^{2+} ions in solution and the standard reduction potentials. The observed values of $E^{\ominus}$ and those calculated using the data of Table 11.1 and equation (7.46) with n =

2 are compared in Fig. 11.2.

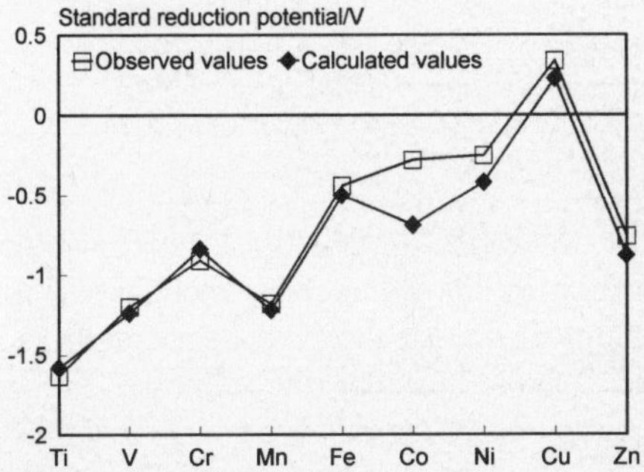

Fig. 11.2 Observed and calculated values for the standard reduction potentials ($M^{2+} \rightarrow M$) of the elements Ti to Zn

The unique behaviour of Cu, having a positive E°, accounts for its inability to liberate H_2 from acids. Only oxidizing acids (nitric and hot concentrated sulfuric) react with Cu, the acids being reduced. The high energy to transform Cu(s) to Cu^{2+}(g) is not balanced by its hydration enthalpy. The general trend towards less negative E° values across the series is related to the general increase in the sum of the first and second ionization energies. It is interesting to note that the values of E° for Mn, Ni and Zn are more negative than expected from the general trend.

Table 11.1 Thermochemical data (all in kJ mol^{-1}) for the first row transition elements and the standard reduction potentials for the reduction of M^{II} to the element

Element(M)	$\Delta_a H^{\circ}$(M)	I_1	I_2	$\Delta_{hyd}H^{\circ}(M^{2+})$	E°/V
Ti	469	661	1310	-1866	-1.63
V	515	648	1370	-1895	-1.2
Cr	398	653	1590	-1925	-0.91
Mn	279	716	1510	-1862	-1.18
Fe	418	762	1560	-1958	-0.44
Co	427	757	1640	-2079	-0.28
Ni	431	736	1750	-2121	-0.25
Cu	339	745	1960	-2121	0.34
Zn	130	908	1730	-2059	-0.76

The stability of the half-filled d sub-shell is Mn^{2+} and the completely filled d^{10} configuration in Zn^{2+} are related to their E° values, whereas E° for Ni is related to the highest negative $\Delta_{hyd}H^{\circ}$, associated with high LFSE. The Ti^{2+} ion is unstable in aqueous solution except as the oxocation TiO^2 (or possibly $Ti(OH)_2{}^{2+}$).

11.4 LATIMER AND VOLT-EQUIVALENT DIAGRAMS

(*Educ. Chem.*, **1**, 123, 1964)

Under standard acidic conditions the most stable oxidation states of the first row transition elements are: Sc^{III}, Ti^{IV}, V^{III}, Cr^{III}, Mn^{II}, Fe^{III}, Co^{II}, Ni^{II}, Cu^{II}, and Zn^{II}, as the hydrated cations except for TiO^{2+}(aq). Although the ion Ti^{4+} of noble gas configuration is only found in the ionic oxide Ti O_2, Ti O^{2+} ions are stable in solution and these are examples of the efficiency of O in stabilizing the highest oxidation state. Ti^{3+} solutions are readily oxidized by air and should be kept and used under an inert gas atmosphere. The diagram for Ti is not shown because it has no other interesting features. However, the Latimer and volt-equivalent diagrams for V and Mn are interesting and are shown below and in Fig. 11.3 for acid solutions together with the diagram of Mn in alkaline solution. The diagrams for Cr are contained in Section 7.4.

$$MnO_4^- \xrightarrow{0.56} MnO_4^{2-} \xrightarrow{2.26} MnO_2 \xrightarrow{0.95} Mn^{3+} \xrightarrow{1.51} Mn^{2+} \xrightarrow{-1.18} Mn$$

$$MnO_4^- \xrightarrow{0.56} MnO_4^{2-} \xrightarrow{0.6} MnO_2 \xrightarrow{-0.2} Mn(OH)_3 \xrightarrow{0.1} Mn(OH)_2 \xrightarrow{-1.55} Mn$$

$$VO_2^+ \xrightarrow{1.0} VO^{2+} \xrightarrow{0.36} V^{3+} \xrightarrow{-0.25} V^{2+} \xrightarrow{-1.2} V$$

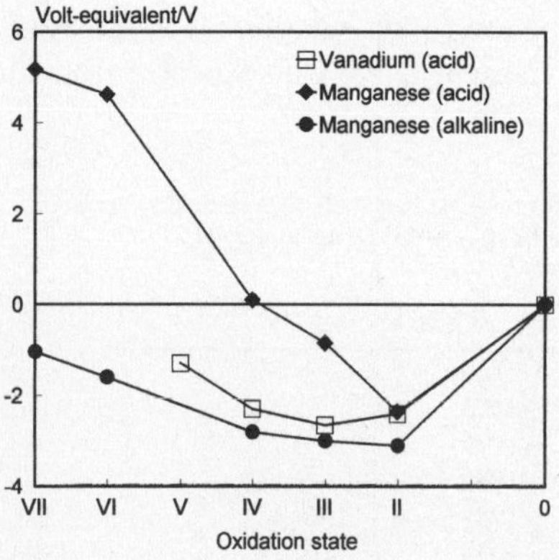

Fig. 11.3 The volt-equivalent diagrams for V and Mn under 1 *M* acid conditions and that for Mn with 1 *M* OH^-

The lowest points in the volt-equivalent diagrams show that V^{3+}, Cr^{3+} and Mn^{2+} are the most stable oxidation states in acid solution. The highest oxidation states of the three elements have the ionic forms: VO_2^+, $Cr_2O_7^{2-}$ and MnO_4^-, which correspond to the group number. Again the role of O in stabilizing these states is evident. The relative oxidizing power of these species can be compared by comparing the slopes of the lines joining them to the minima of the plots. Thus the oxidizing power increases in the series: $VO_2^+ < Cr_2O_7^{2-} < MnO_4^-$, emphasizing the increased stability of the lower species to which they are reduced. Beyond Mn, as the number of valence electrons exceeds seven, a high oxidation state corresponding to the group number requires unpairing of electron spins and cannot be achieved. The possibility of comproportionation is used for preparative purposes, e.g.

$$MnO_4^- + 4Mn^{2+} + 8H^+ \rightarrow 5Mn^{3+} + 4H_2O \qquad (11.3)$$

Similarly Cr^{2+} reduces $Cr_2O_7^{2-}$ to Cr^{3+} and V^{2+} reduces VO_2^+ to VO^{2+}. The latter reaction is useful for the determination of the air-sensitive V^{2+} solution. The two relatively high points in the Mn diagram in acid solution are +3 and +6. Both oxidation states are subject to disproportionation:

$$2Mn^{3+} + 2H_2O \rightarrow Mn^{2+} + MnO_2 + 4H^+ \qquad (11.4)$$

$$3MnO_4^{2-} + 4H^+ \rightarrow 2MnO_4^- + MnO_2 + 2H_2O \qquad (11.5)$$

MnO_2 is, however, stable to disproportionation being on a relatively low point. Both Mn^{3+} and MnO_4^- are powerful oxidants which are useful in titrimetric analysis. MnO_2 can also act as a heterogeneous catalyst (in the thermal decomposition of $KClO_3$ to KCl and O_2) and as a homogeneous catalyst in the decomposition of H_2O_2 to H_2O and O_2. The relative kinetic stability of H_2O_2 contrasts the negative $\Delta G^{\ominus}$ of its decomposition.

The diagram of Mn in alkaline solution shows the expected reduced oxidising power of MnO_4^- and the stability of MnO_4^{2-} in this medium. The two minima at $Mn(OH)_2$ and MnO_2 again demonstrate the stability of these two compounds.

Comparing the slope of the line joining MnO_4^- to Mn^{2+} in acid solution with the lines joining VO_2^+ to either VO^{2+}, V^{3+} or V^{2+}, it can be seen that acidified MnO_4^- can oxidize any of the latter three vanadium ions to VO_2^+. This is the basis of titrimetric determination of solutions of lower V ions.

The aqueous chemistry of Fe, Co, Ni, Cu and Zn is dealt with in the sections concerning the variations in the $E^{\ominus}$ values for $M^{III} \rightarrow M^{II}$ and $M^{II} \rightarrow M$ reductions (Section 11.4).

Problem 11.1

Given the following $E^{\ominus}$ values in acid solution for the reactions:

$$Mn^{3+} + e^- \rightarrow Mn^{2+} \qquad E^{\ominus} = 1.51 \text{ V}$$

$$Mn\,O_2 + 4H^+ + e^- \rightarrow Mn^{3+} + H_2O \qquad E^{\ominus} = -0.95 \text{ V}$$

Calculate $E^{\ominus}$ for the disproportionation of Mn^{3+} in acid solution and indicate the sign of $\Delta G^{\ominus}$ for the reaction. See how this is in agreement with the mid-point in the line joining MnO_2 and Mn^{2+} in Fig. 11.3.

Answer Reversing the last equation and adding the first equation gives:

$$2Mn^{3+} + 2H_2O \rightarrow Mn^{2+} + Mn\,O_2 + 4H^+ \qquad E^{\ominus} - 1.51 - (0.95) = 0.56 \text{ V}$$

This means $\Delta G^{\ominus}$ will be negative as is clear from the points for Mn^{3+} and the mid-point on the line joining MnO_2 and Mn^{2+}.

11.5 RELATIVE STABILITY OF M^{II}/M^{III} HALIDES

Thermodynamic considerations account for the relative stability of MX_2/MX_3 solids, where lattice energies are used instead of $\Delta_{hyd}H^{\ominus}$ values as is the case of M^{2+}/M^{3+} ions in aqueous solution. This topic is dealt with as the following problem.

Problem 11.2

(a) Given that the lattice energies, L, of $MnCl_2$ and $MnCl_3$ are 2540 and 5544 kJ mol^{-1} respectively and $D(Cl_2) = 248$ and $E_{ea}(Cl) = -364$ kJ mol^{-1} respectively, calculate $\Delta H^{\ominus}$ and $\Delta G^{\ominus}$ for the reaction:

$$MnCl_3(s) \rightarrow MnCl_2(s) + ½Cl_2(g)$$

given that I_3(Mn) = 3249 kJ mol^{-1} and $\Delta S^\ominus$ for the reaction is 27.6 J K^{-1} mol^{-1}.

(b) Plot the following thermodynamic quantities for the elements Sc to Zn against atomic number

M	Sc	Ti	V	Cr	Mn	Fe	Co	Ni	Cu	Zn
δL	2498	2640	2736	2916	-	2787	2879	2887	2967	3105
I_3	2389	2653	2828	2987	-	2960	3232	3394	3554	3833
$\Delta G^\ominus$	314	184	109	126	-	29	−155	−305	−389	−536

($\delta L = L(MCl_3) - L(MCl_2)$; $\Delta G^\ominus$ for $MCl_3 \rightarrow MCl_2 + ½Cl_2$)

Plot δL and $-I_3$ using the same scale and $\Delta G^\ominus$ using another scale. Comment on the trends in the three plots.

Answer

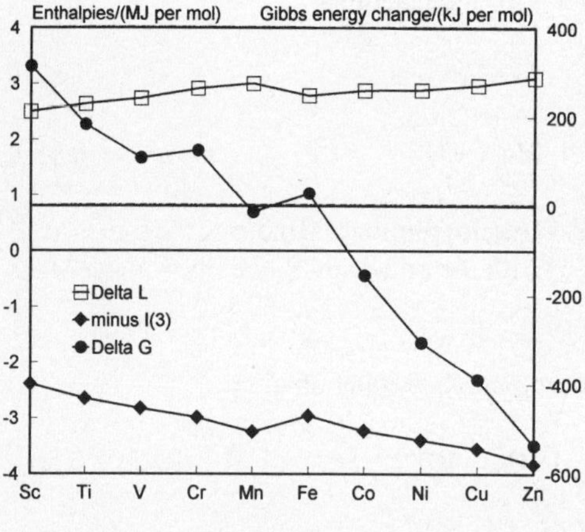

(a) $\Delta H^\ominus$ = 3004 − 3249 + 364 − 124 = −5 kJ mol^{-1}

$\Delta G^\ominus$ = −5 − [(298 × 27.6)/1000] = −13.2 kJ mol^{-1}

(b) The general increase in I_3 along the series reflects the increased Z_{eff}. The low value for Sc reflects the stability of Sc^{3+} which has a noble gas configuration. The highest value for Zn is due to the removal of an electron from the stable d^{10} configuration of Zn^{2+}. The comparatively high value for Mn shows that Mn^{2+} (d^5) is particularly stable, whereas the comparatively low value for Fe shows the extra stability of Fe^{3+}(d^5). The plot for δL is roughly a mirror image of I_3. The plot for $\Delta G^\ominus$ for the decomposition is roughly parallel to I_3. The comparatively low value for V is related to the stability of V^{2+} (half-filled t_{2g} level). MCl_3 for M = Co, Ni, Cu and Zn are apparently unstable.

11.6 HIGHER OXIDATION STATES

Halides of the 3d elements

Table 11.2 shows the stable halides of the 3d series of transition metals. The highest oxidation numbers are achieved in Ti tetrahalides, in VF_5 and CrF_6 The +7 state for Mn is not represented but $MnOF_3$ is known and beyond Mn no metal has a trihalide except Fe and CoF_3. The ability of F to stabilise the higher oxidation state is due to either: higher L for the higher compound as in the case of CoF_3, or higher bond energy terms for the higher covalent compounds e.g. VF_5 and CrF_6. These are evidently covalent since they are not represented by simple ions. The covalence of CrF_4 and especially CrF_5 is demonstrated by their low melting points which are much lower than the ionic CrF_3 and CrF_2.

Table 11.2 Formulae of halides of groups 4 to 12

Ox.No.									
+6			CrF_6						
+5		VF_5	CrF_5						
+4	TiX_4	VX^I_4	CrX_4	MnF_4					
+3	TiX_3	VX_3	CrX_3	MnF_3	FeX^I_3	CoF_3			
+2	TiX^{III}	VX_2	CrX_2	MnX_2	FeX_2	CoX_2	NiX_2	CuX^I_2	ZnX_2
+1								CuX^{III}	

Key: $X = F \rightarrow I$; $X^I = F \rightarrow Br$; $X^{II} = F, Cl$; $X^{III} = Cl \rightarrow I$

Even in the case of Cr^{III}, departure from ionic character with the increase in size and polarizability of the halide is evident from the decrease in melting point from the fluoride to the chloride to the bromide. The same trend is shown by Co^{II} halides. The ability of O to stabilize the higher oxidation state is also demonstrated in the oxohalides. Although V^V is represented only by VF_5, the oxohalides VOX_3 are known where X = F, Cl or Br. Another feature of fluorides is their instability in the low oxidation states e.g. VX_2 (X = Cl, Br or I) and the same applies to CuX. On the other hand, all Cu^{II} halides are known except the iodide. In this case, Cu^{2+} oxidises I^- to I_2:

$$Cu^{2+} + 2\,I^- \rightarrow CuI(s) + {}^1/_2\, I_2 \qquad (11.10)$$

However Cu^I is unstable in solution and undergoes disproportionation:

$$2Cu^+ \rightarrow Cu^{2+} + Cu \qquad (11.11)$$

The stability of $Cu^{2+}(aq)$ rather than $Cu^+(aq)$ is due to the much more negative $\Delta_{hyd}H^\circ$ of $Cu^{2+}(aq)$ than Cu^+, which more than compensates for the second ionization energy of Cu.

Oxides of the 3d elements

Table 11.3 lists the known oxides in the series.

Table 11.3 Oxides of the 3d metal series

Ox.No.										
+7					Mn_2O_7					
+6				CrO_3						
+5			V_2O_5							
+4		TiO_2	V_2O_4	CrO_2	MnO_2					
+3	Sc_2O_3	Ti_2O_3	V_2O_3	Cr_2O_3	Mn_2O_3	F_2O_3				
					Mn_3O_4	Fe_3O_4	Co_3O_4			
+2		TiO	VO	(CrO)	MnO	FeO	CoO	NiO	CuO	ZnO
+1									Cu_2O	
Group	3	4	5	6	7	8	9	10	11	12

All the metals except Sc form MO oxides and these have the NaCl structure. Whereas TiO and VO have metallic appearance and electronic conductivity, the other oxides are poor conductors although ZnO and Li^+-doped or heated NiO are semiconductors. Fig.11.4 shows plots of $-\Delta_f H^\circ$ and lattice enthalpies for the known MO compounds.

The effect of increasing Z_{eff} across the series explains the trend in L and the nearly mirror image trend in $-\Delta_f H^\circ$ for these oxides. It is worth noting that L for MnO is noticeably lower than the expected trend, which shows that Mn^{2+} in high spin O_h symmetry has zero LFSE and hence has a larger radius than expected.

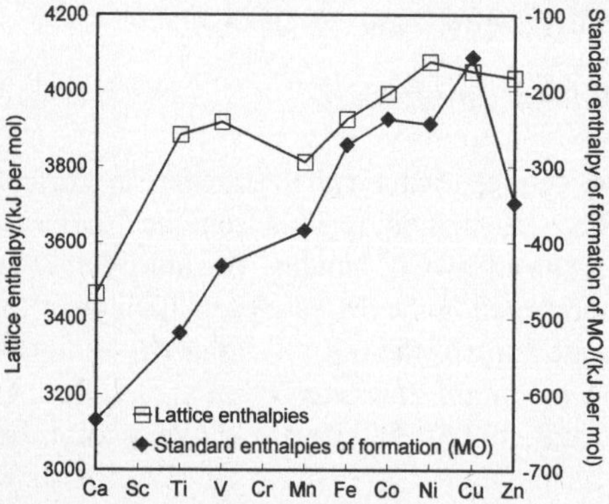

Fig. 11.4 Plots of the lattice enthalpies and the standard enthalpies of formation of the M^{II} oxides of the elements Ca to Zn

The highest oxidation number in the oxides in Table 11.3 coincides with the group number and is attained in Sc_2O_3 to Mn_2O_7. Beyond Group 7, no higher oxides of Fe above Fe_2O_3 are known, although ferrates (VI):$(FeO_4)^{2-}$ are formed in alkaline media but they readily decompose to Fe_2O_3 and O_2. Besides the oxides in the table, oxocations stabilize V^V as VO_2^+, V^{IV} VO^{2+} and Ti^{IV} as TiO^{2+}. The ability of O to stabilize these high oxidation states exceeds F in this respect. Thus the highest Mn fluoride is MnF_4 whereas the highest oxide is Mn_2O_7. The ability of O to form multiple bonds to metals explains its superiority. In the covalent oil Mn_2O_7, each Mn is tetrahedrally surrounded by O's including a Mn–O–Mn bridge. The tetrahedral $[MO_4]^{n-}$ ions are known for V^V, Cr^{VI}, Mn^V, Mn^{VI} and Mn^{VII}.

The numerous nonstoichiometric Ti oxides are not included in Table 11.3 because of the wide range of compositions. Many of these as well as V_2O_4 exhibit characteristic transition temperatures of change from insulator or semiconductor to electronic conductor. In the low temperature form of V_2O_4, V–V bonds are evident from the V–V distances. Metal-metal bonds are also found especially in Ti_xO ($x = 2, 3, 6$).

As the oxidation number of a metal and its electronegativity increase, ionic character decreases. In the case of Mn, Mn_2O_7 is a covalent (very unstable) green oil. Even CrO_3 and V_2O_5 have low melting points. In these oxides, acidic character is predominant. Thus Mn_2O_7 gives $HMnO_4$ and CrO_3 gives H_2CrO_4 and $H_2Cr_2O_7$. V_2O_5 is, however, amphoteric though mainly acidic and it gives VO_4^{3-} as well as VO_2^+ salts. In V, there is a gradual change from the basic VO to the less basic V_2O_3 and the amphoteric V_2O_4, which dissolves in acids to give VO^{2+} salts. The ill-characterized CrO is basic but Cr_2O_3 and $Cr(OH)_3$ (or hydrous Cr_2O_3) are amphoteric.

Problem 11.3

(a) Calculate L (CuO) using the following standard thermochemical data (in kJ mol^{-1} at 298 K):

$$\Delta_a H^\circ(\text{Cu}) = 339,\ I_1(\text{Cu}) = 745,\ I_2(\text{Cu}) = 1960,\ \Delta_a H^\circ(\text{O}) = 248,$$
$$\Delta H^\circ(\text{O,g}\rightarrow\text{O}^{2-}\text{,g}) = 702,\ \Delta_f H^\circ(\text{CuO}) = -155.$$

Given these data, your calculated L and the following information:

$$L(\text{Cu}_2\text{O}) = 3285,\ \Delta_f H^\circ(\text{Cu}_2\text{O}) = -167,$$

discuss the relative stability of the two oxides of copper

(b) Calculate the standard electrode potential in acid solution

(pH = 0) for: $Cu^+(aq) + e^- = Cu(s)$

given that in acid solution:

	E°/V
$Cu^{2+}(aq)+2e^- = Cu(s)$	0.34
$Cu^{2+}(aq)+e^- = Cu^+(aq)$	0.15

Then calculate the standard free energy change (in kJ mol^{-1}) and the equilibrium constant for the disproportionation:

$$2Cu^+(aq) = Cu^{2+}(aq) + Cu(s)$$

(c) Comment on the relative stability of the two oxidation states of Cu in solids and in acidic solution and suggest reasons for the observed behaviour.

Answer

(a) $L(\text{CuO}) = 155 + 339 + 745 + 1960 + 248 + 702 = 4149$ kJ mol^{-1}

Although $L(\text{Cu}_2\text{O})$ is lower, $\Delta_f H^\circ(\text{Cu}_2\text{O})$ is more negative than $\Delta_f H^\circ(\text{CuO})$ and can be taken to represent greater stability.

(b) $E^\circ(Cu^+(aq.) + e^- = Cu(s)) = (0.34 \times 2) - 0.15 = 0.53$ V

E° for disproportionation = $(0.53 \times 2) - (0.34 \times 2) = 0.38$V

$\Delta G^\circ = -2 \times 96500 \times 0.38/1000$ kJ mol^{-1}

$= -73.34$ kJ mol^{-1}

$73340 = 8.314 \times 298 \times \ln K$

$\ln K = 29.6,\ K = 7.175 \times 10^{12}$

(c) The more negative $\Delta_{hyd}H^\circ$(for Cu^{2+}) than for Cu^+ (larger ion with a single positive charge) compensates the extra ionization energy required to form Cu^{2+} compared to Cu^+. $\Delta_{hyd}S^\circ$ has to be taken into account as well.

11.7 STABILIZATION OF LOW OXIDATION STATES

A metal in a zero or negative oxidation state has all its valence electrons plus 1, 2 or 3 more. If ligands form coordinate bonds with the metal, they donate a pair of electrons to the metal for each ligand. This accumulation of negative charge on a metal which is essentially of low electronegativity must be relieved by a drift of electrons from the metal to available vacant antibonding orbitals of the ligands. For an octahedral complex, its t_{2g} orbitals are of π symmetry. Hence t_{2g} electrons can be back donated to vacant π^* orbitals of the ligands.

Alternatively stated, when ligands have a powerful π acceptor ability i.e. have vacant π* orbitals not much higher than the metal d orbitals. the π overlap will lower the energy of π orbitals of mainly metal character and raise the energy of the π* orbitals of mainly ligand character, resulting in an increase in Δ_o. Hence such ligands occupy a position near the top of the spectrochemical series. A typical such ligand is CO or substituted phosphanes (commonly called phosphines) PR_3. This type of interaction is described as synergic since the σ donation from the ligand reduces its charge (δ+) and hence enhances its acceptor ability, the reverse is expected from the back bond. This can be represented by

$$M \underset{\pi}{\overset{\sigma}{\rightleftarrows}} L$$

The mere possibility of the formation of stable carbonyls of metals, which have high $\Delta_a H^\circ$ shows that the M–CO bonds are strong, which is unexpected from a ligand of low polarity. Confirmation of this comes from the C—O vibration frequencies and force constants, which increase in the series:

$$Ti(CO)_6^{2-} < V(CO)_6^{-} < Cr(CO)_6 < Mn(CO)_6^{+}$$

The decrease in the negative charge or increase in the positive charge hinder the drift of π electrons from the metal to CO, thus increasing the strength of the C–O bond. The reverse is observed for the M–C bond which becomes weaker across the series as found for (M–C) frequencies. In all cases, the C–O frequency is lower than in CO(g). Evidence of the strength of the M–C bond is also provided by the bond length which increases as the bonds become weaker. Additionally other ligands, which may be incapable of π accepting or weaker or stronger π acceptors were found to alter the C–O frequency in mixed carbonyls. Thus $(dien)Cr(CO)_3$ has a lower ν(CO) than $Cr(CO)_6$ whereas $(PF_3)Ni(CO)_3$ has a higher ν (CO) than $Ni(CO)_4$. Unlike dien ($H_2N(CH_2)_2NH(CH_2)_2NH_2$), which is incapable of π acceptance, PF_3 is a more powerful π acceptor than CO. Estimates of M–C bond polarity indicate a low bond moment, confirming the synergic mechanism. N_2 is isoelectronic with CO although it is non polar. The first compound containing N_2 as a ligand, $[Ru(NH_3)_5N_2]X_2$, was reported in 1965. It has Ru–N–N linear bonds, and its ν(N–N) is lower than that of N_2(g). Similar arguments as above confirm the synergic mechanism. Other compounds in which this mechanism is operative include the nitrosyls of transition metals e.g. $Cr(NO)_4$ in which σ donation from N to Cr and π back donation from the metal to NO is shown by similar evidence as mentioned above. It is worth noting that NO is an odd electron molecule and its bond order = 2½.

11.8 ISOMERISM IN TRANSITION METAL COMPLEXES

Besides geometrical and optical isomerism encountered in organic compounds, novel types are encountered in complexes. It is convenient to discuss these together with the geometry of the complexes, starting with the more common coordination numbers (CN) of 4 and 6. Two geometries are known for the former i.e. tetrahedral (T_d) and square planar (D_{4h}). For ML_6 complexes, the 6 ligands are usually arranged octahedrally (O_h).

Geometrical isomerism

In square planar complexes *cis-* and *trans-* isomers are possible when two different ligands are coordinated in pairs: $[Pt(NH_3)_2Cl_2]$ where the Cl's are adjacent to each other (*cis*) or at opposite corners (*trans*) as shown in Fig. 11.5. The same is possible when a chelating ligand has two different coordinating atoms e.g. H_2NCH_2COO (glycinate). When the 2N's are adjacent in a Pt(II) square planar complex (*cis-)* it will have a dipole moment but when they are at opposite corners (*trans*), the dipole moment is zero. Three geometric isomers are possible when 4 different ligands are coordinated to the metal e.g. $[Ptpy(NO_2)(NH_3)(NH_2OH)]$. Either NO_2, NH_3 or NH_2OH are *trans* to pyridine (py).

Fig. 11.5 The square planar structures of *cis-* and *trans*-diamminedichloroplatinum(II)

In octahedral complexes where 2 ligands A and 4 ligands B are in the inner coordination sphere of the metal, the two A ligands may be adjacent (*cis*) or opposite (*trans*), e.g. $[Co(NO_2)_2(NH_3)_4]^+$. However in $[MA_3B_3]$ complexes, the three A or B ligands may be on one face of the octahedron (*fac*) or two A or B are *trans* to each other and the third is adjacent to both (*mer* from meridional) as shown in Fig. 11.6. An example of this is $[Co(NH_3)_3(NO_2)_3]$.

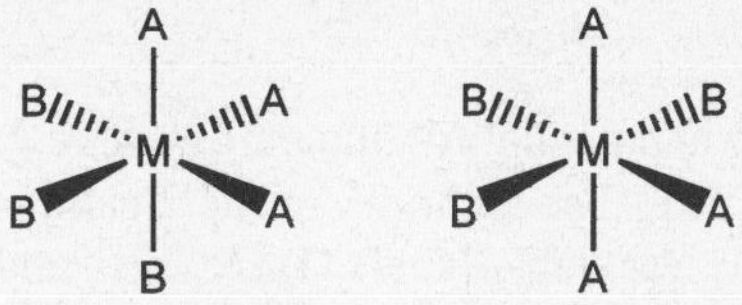

fac-MA_3B_3 *mer*-MA_3B_3

Fig. 11.6 Diagrams of the *fac-* and *mer*-isomers of the octahedral complex MA_3B_3

Cis- and *trans-* isomers are also observed when 2 bidentate and 2 unidentate ligands are coordinated to the metal ion e.g. $[Cr(H_2O)_2(C_2O_4)_2]^-$, the 2 chelate rings are adjacent to each other (*cis*) or opposite to each other (*trans*). Another example is $[Co(en)_2Cl_2]^+$(en = $H_2N(CH_2)_2NH_2$). In these cases the *cis* isomer can be resolved into 2 optical isomers.

Optical isomerism

Octahedral complexes containing three chelate rings e.g. $[Cr(C_2O_4)_3]^{3-}$ or $[Co(en)_3]^{3+}$ in which both lack an improper axis of rotation can exist in two enantiomeric forms as shown in Fig. 11.7 and are optically active or chiral. Their solutions alter the plane of polarization of polarized light. A 50/50 racemic mixture does not alter the plane. Tetrahedral complexes, which do not have geometric isomers, may be chiral (i.e. optically active) if all four ligands are dissimilar, e.g. $[(C_5H_5)Fe(CO)(PR_3)(COCH_3)]$.

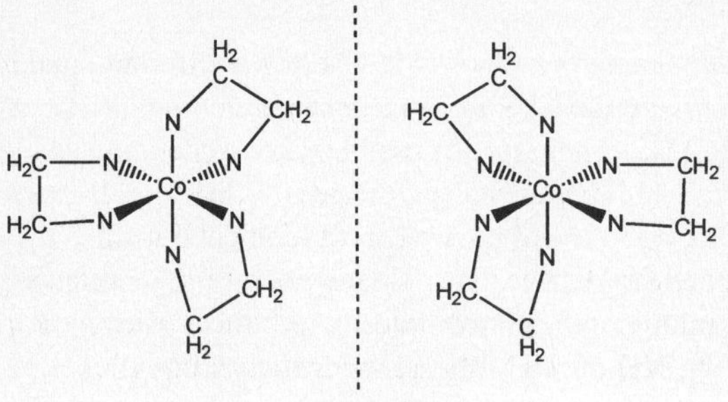

Fig. 11.7 The two optically active isomers of $[Co(en)_3]^{3+}$, the two hydrogen atoms on each of the nitrogen donor atoms are omitted

Ionization isomerism is encountered when two different anionic ligands exchange their position in the inner and outer coordination sphere of a complex e.g. $[Co(NH_3)_5(SO_4)]Br$ (precipitates AgBr with Ag^+) and $[Co(NH_3)_5\ Br]SO_4$ (precipitates $BaSO_4$ with Ba^{2+}).

Hydration isomers have water molecules in the inner coordination sphere or as water of crystallization e.g. violet $[Cr(H_2O)_6]Cl_3$, pale green $[Cr(H_2O)_5Cl]Cl_2.H_2O$ and dark green $[Cr(H_2O)_4Cl_2]Cl.2H_2O$. $AgNO_3$ solution precipitates three moles of AgCl, two moles of AgCl and one mole of AgC1 from the three isomers respectively, per mole of complex. A solution of the last can be separated into *cis* and *trans* cations.

Linkage isomers have at least one ligand with 2 different potential coordinating atoms e.g. $[Co(NH_3)_5NO_2]^{2+}$ (nitro) and $[Co(NH_3)_5ONO]^{2+}$ (nitrito). The chlorides of the 2 cations differ in colour and in I.R. spectra. The latter is thermally less stable. $[Cr(H_2O)_5(NCS)]^{2+}$ and $[Cr(H_2O)_5(SCN)]^{2+}$ can be distinguished by their visible spectra, since the 2 anions occupy different positions in the spectrochemical series.

Coordination isomerism is possible when two different ligands can exchange in the coordination sphere of two different metals e.g. $[Co(NH_3)_6][CrC_2O_4)_3]$, $[Co(C_2O_4)_3]$ $[Cr(NH_3)_6]$.

Coordination-position isomerism is different in that ligands occupy different positions in a binuclear complex e.g. $[Cl_2Pt(\mu\text{-}Cl_2)(PEt_3)_2]$ where the two 2 Cl ligands may be on one side (*cis*) or on opposite sides (*trans*), or may be both on either side: $[(PEt_3)ClPt(\mu\text{-}Cl_2)Pt(PEt_3)Cl]$.

Polymerization isomerism is observed when two compounds have the same ratio of metal and ligand but one has double the RMM e.g. $[Pt(NH_3)_2Cl_2]$ and $[Pt(NH_3)_4][PtCl_4]$.

Ligand isomerism is encountered when two isomers of a ligand are coordinated with the same metal e.g. $[Co(pn)_2Cl_2]$ and $[Co(tn)_2Cl_2]$ (pn = 1,2 diaminopropane, tn = 1,3-isomer).

11.9 SOME TRENDS IN COMPLEXES OF TRANSITION METALS

The few examples of coordination number (CN) = 2, giving the linear [LML], are found in complexes of M^I ions (M = Group 11 metal) or Hg^{II} complexes (of the same d^{10} configuration).

In tetrahedral complexes, LFSE's are small, and ligands having a weak L.F. tend to coordinate with small metal ions such as Zn^{2+} or Co^{2+} (no or small LFSE).

On the other hand, the other geometry with C.N.4, is the square. This is generally found in metal ions with d^8 configuration, when spin pairing produces diamagnetic complexes. M(II) ions (M = group 10 metal) or Au(III) tend to form such complexes. In the case of Ni(II), strong field ligands e.g. CN or chelating ligands favour this geometry. For the heavier metals, even Cl ligands give square planar $[MCl_4]^{n-}$ ($n = 2$ for Pd or Pt, $n = 1$ for Au).

The most common C.N.6 is represented by numerous octahedral complexes: $[M(H_2O)_6]^{n+}$ ($n = 2$ or 3 for the majority of the 3d metals). In the case of Cu(II), however, Jahn Teller effect leads to a distortion, accentuated in $[Cu(NH_3)_5(OH_2)]^{2+}$. The predominance of octahedral low spin Co(III) complexes is ascribed to the high LFSE. Although CoF_3 is an odd example of a binary Co(III) compounds, complex formation stabilizes this oxidation number. Although C.N.5 is less common, two limiting geometries are found: trigonal bipyramidal, TBP, or square pyramidal SP. Both geometrics in $[Ni(CN)_5]^{3-}$ are found in the same compound $[Cr(en)_3)][Ni(CN)_5].1.5H_2O$. In complexes with π acceptor ligands, each of these geometries may be encountered: $[M(CO)_5]^{n-}$ ($n = 0$ or 1 for Fe or Mn), both TBP; $[Ru(NO)_2Cl(PPh_3)_2]^+$(SP).

C.N.8 is the next most common after C.Ns 6 and 4, and is favoured with the larger and heavier M ions especially with small ligands. The 2 common geometries are the trigonal dodecahedron or square antiprism. Examples include $[M(CN)_8]^{4-}$ (M = Mo, W) and $[TaF_8]^{3-}$ representing the two structures respectively.

11.10 METAL–METAL BONDING (*J. Chem. Educ.*, 60, 713 (1983))

Although M–M bonding is found among the metals of the 3d series, especially in bi-and polynuclear carbonyls and in $M_2(CH_3CO_2)_4.2H_2O$ (M = Cr, Cu), M–M bonding is more extensive in the 4d and 5d metals. M–M distances and magnetic measurements are useful indicators of M–M bonding.

In the case of binuclear carbonyls or mixed cyclopentadienyl/carbonyls or nitrosyls, single σ bonds, with or without bridges, are formed between two metal atoms. For example, in $Mn_2(CO)_{10}$, orbitals from each Mn overlap to give the M–M σ bond (M may be Mn and/or Re). The diamagnetism, the M–M distance and the estimated M–M bond energy are all in agreement with a single bond.

Multiple M–M bonds

Although the formal oxidation number of Cr in the above complex is +2 i.e. d^4 configuration, the red colour, relative stability and diamagnetism all suggest Cr–Cr multiple bonds. Here and in the corresponding Cu complex, each metal is coordinated to one water molecule, 4 O's from the 4 acetate ligands and is bound to the other metal atom by a multiple bond.

It is observed that the heavier 4d and 5d metals have more numerous multiple M–M bonds. In these, the d orbitals of two metal atoms overlap: the two d_{z^2} orbitals give σ and σ* orbitals (taking the *z* axis as the bond axis), the two d_{xz} orbitals give π and π* orbitals, the same for the d_{yz} orbitals but the $d_{x^2-y^2}$ orbital overlap produces δ and δ* orbitals. Since this latter overlap is the least efficient, the δ and δ* orbitals will not be far apart energetically. Since M.O. energies are proportional to the overlap integrals, the energy sequence of the orbitals will be:

$$\sigma < \pi \ll \delta < \delta^* \ll \pi^* < \sigma^*$$

The bond order is calculated as usual from the number of e^- pairs in bonding orbitals—pairs in antibonding orbitals. Thus in $[Re_2Cl_8]^{2-}$ of ox. no. +3, each Re has 4 valence electrons and the 8 electrons in the ion give the configuration: $\sigma^2\pi^4\delta^2$ (with vacant antibonding orbitals). Hence the bond order is four. The very short Re---Re distance and the eclipsed configuration of the ions are in agreement with this order (the σ and pair of π bonds are insensitive to angle of internal rotation but δ overlap is efficient when the two $ReCl_4$ groups are eclipsed). Since the δ m.o. is weakly bonding and the δ* is weakly antibonding, the loss or gain of 2 electrons transforms the bond order to 3.

Fractional bond orders

In $[Mo_2Cl_8]^{4-}$, Mo has d^4 configuration and the eight electrons occupy the orbitals as shown above. In $[Mo_2(SO_4)_4]^{3-}$, the two Mo atoms have d^3–d^4 configuration giving $\sigma^2\pi^4\delta^1$ and a bond order 3.5, but in $[Mo_2(HPO_4)_4]^{2-}$ there are two d^5 Mo atoms and the ion has $\sigma^2\pi^4$ configuration and a bond order of three. The change in the Mo---Mo distance: 211, 217 and 223 pm respectively is in agreement with the bond order allocated. The loss or gain of an electron in the Re compound also gives fractional bond orders.

Metal–metal bond energies are not easily measured but estimates show that the Re–Re energy term in $[Re_2Cl_8]^{2-}$ is appreciably higher than in $Re_2(CO)_{10}$.

M–M bonds are also found in polynuclear complexes. In $[(ReCl_3)_3]$, each Re is bonded to the other two by M–M bonds. Other metal clusters may contain 4 or 6 metal atoms in the cluster but they are also bound to other ligands or bridging groups. However in a group of metal cluster cations or anions, formed by p block metals, the atoms are only bound together without other ligands e.g. Pb_5^{2-}, Bi_9^{5+} and they are referred to as **Zintl ions.**

11.11 THE 18 ELECTRON RULE (*J. Chem. Educ.*, 46, 411 (1969))

Sidgwick noticed that in many complexes when the number of valence electrons of the metal is added to the electron pairs donated from the ligands, the electron shell becomes equal to that of a noble gas. In an octahedral complex, when the metal has 6 valence electrons and the 6 ligands donate 2 each, the total will be 18 electrons. In ordinary complexes, sometimes called Werner complexes, the 18 electrons may be not reached or even exceeded depending on the metal and its oxidation state. However, in most carbonyls (except those in which > 4 metal atoms are in the polynuclear complex), mixed cyclopentadienyl/carbonyls and nitrosyl carbonyls the rule is obeyed. The counting assumes that:

each terminal CO or N≡N supplies 2 electrons to the metal valence shell;
each bridging CO provides one electron;

each M–M bond involves 1 electron from each;

a linear nitrosyl terminal group contributes 3 electrons;

a bent nitrosyl supplies 1 electron; and

a cyclopentadienyl ligand provides all its five π electrons (or if the metal ox.no. is considered, it supplies 6 electrons).

The alternate monomeric/dimeric carbonyls of the 3d series and of the cyclopentadienyl/carbonyls of the same series and the series: $Cr(NO)_4$, $Mn(CO)(NO)_3$, $Fe(CO)_2(NO)_2$, $Co(CO)_2(NO)$ and $Ni(CO)_4$; and $V(CO)_5(NO)$, $Cr(CO)_6$; $Mn(CO)_4(NO)$, $Fe(CO)_5$ are in agreement with the rule.

It can be argued that in O_h symmetry, if the 6 ligands are strong σ donors and strong π acceptors, Δ_o is large and there will be nine molecular orbitals of low energy, including the t_{2g} set, which will accommodate 18 electrons. Otherwise stated, to fill all the available valence orbitals of the metal will allow back bonding to the ligands, which relieves the accumulation of negative charge on the metal.

Although CN^- and CO are isoelectronic, the 18 electron rule is not applicable to cyano-complexes. To emphasize this it is noticeable in the series $[M(CN)_6]^{n-}$, where M can be one of several 3rd row 3d transition metals, which have different d configurations and n might vary even when M represents the same metal. Thus, in $[Fe(CN)_6]^{n-}$ when $n = 4$ the electron count is $6 + (2 \times 6) = 18$, but when $n = 3$ it is $5 + (2 \times 6) = 17$. In $[Mn(CN)_6]^{3-}$ the count is $4 + (2 \times 6) = 16$.

11.12 INTRODUCTION TO METALLIC CLUSTERS

Cluster chemistry is a rapidly expanding branch of inorganic chemistry as is pointed out in Chapter 1, but a detailed account is beyond the scope of this volume. Some clusters of the p block elements are discussed in Chapter 9 together with an introduction to the Wade-Mingos electron-counting rules which rationalize their cage geometries. The similarity between simple transition metal clusters and boranes and carboranes was emphasized by Mingos. Thus $B_6H_6^{2-}$ and $Co_6(CO)_{14}^{4-}$ both have eleven antibonding orbitals (see Chapter 9). On the basis of similar arguments the structures of many transition metal clusters can be classified as *closo*, *nido* or *arachno*, depending on whether the electron count is $14n + 2$, $14n + 4$ or $14n + 6$ respectively, where n is the number of cluster atoms. Thus, $Os_5(CO)_{15}^{2-}$ has an electron count of 5×8 (from 5Os) plus 30 (from 15CO) plus 2 (for the charge) making a total of $72 = (14 \times 5) + 2$. The structure is based upon a *closo*-trigonal bipyramid with 5 vertices. Similarly $Rh_6(CO)_{16}$ has an electron count of 6×9 (from 6Rh) plus 32 (from 16CO) making a total of $86 = (14 \times 6) + 2$. The structure is based on a *closo*-octahedron. $Ru_5C(CO)_{15}$ has an electron count of 5×8 (from 5Ru) plus 4 (from C) plus 30 (from 15CO) making a total of $74 = (14 \times 5) + 4$. The molecule has a *nido*-structure (square pyramid) based upon the geometry of the octahedron (with an extra vertex), with one vertex absent. The ion $[Re_4(CO)_{16}]^{2-}$ has an electron count of 4×7 (from 4Re) plus 32 (from 16CO) plus 2 (for the charge) making a total of $62 = (4 \times 14) + 6$. The structure of the ion is based upon an octahedron (i.e. the structure with 4 + 2 vertices), but with two vertices absent giving an *arachno* (butterfly or folded square) structure.

For compounds in which the transition elements obey the 18 electron rule there are three extreme forms in which compounds with ≥3 transition metal atoms can exist. For n metal atoms a chain structure is possible if there are $16n + 2$ valence electrons. The reasoning behind this formula is that to form single metal–metal bonds to two neighbours each metal atom would derive one electron from each neighbour making a requirement of 16*n* valence electrons to form an unending chain. The two extra electrons are those needed to end the chain with two single bonds to other atoms (e.g. H). The alternative method of ending a chain is ring formation to occur, in which case 16*n* valence electrons are sufficient. Chains and rings are sometimes excluded from a definition of clusters which are then deemed to be more regular three-dimensional arrangements of metal atoms with, in most cases, their ligand groups.

Any metal atom can be regarded as having the possibility of being three-connected (i.e. bonded to three neighbouring metal atoms) and the resulting 3-connected polyhedron would have *n* vertices and $^3/_2n$ edges (i.e. metal–metal bonds). The number of valence electrons required for such structures is given by $N_v = 18n - (2 \times {}^3/_2n) = 15n$ for compounds in which the transition metals obey the 18 electron rule. In many cases clusters are formed in which the 18 electron rule cannot be obeyed as there are insufficient electrons to form the 2c2e bonds that would be necessary.

Considering the treatment of the main group cluster, $B_6H_6^{2-}$, it was shown that there are seven bonding orbitals, i.e. $n+1$ orbitals where *n* is the number of vertices of the regular polyhedraon—a general rule also applicable to transition metal clusters. In such clusters the metal atoms are bonded, for instance, to three CO ligands (either terminally or in a bridging manner), but still supply one radial and two tengential orbitals per atom to the cluster skeleton. This means that for a cluster with *n* vertices there will be 6*n* bonding orbitals (e.g. concerned with the σ and π bonding in each $M(CO)_3$ unit) plus $n+1$ bonding orbitals for the skeleton making a total of $7n + 1$ orbitals. Thus, $14n + 2$ electrons are required to give a cluster in which the metal atoms all participate in M–M bonding to the same extent. The production of clusters with symmetries less that those of the parent *closo*-structures are described above.

There are many more complicated clusters which are not dealt with in this book, but the one reported in Chapter 1 deserves some discussion. The cluster has the formula $[Au_{13}Cl_2(PMe_2Ph)_{10}]^{3+}$ and is based on an icosahedral arrangement of twelve gold atoms, the thirteenth gold atom being at the centre of the icosahedron as an interstitial atom. There are 162 valence electrons representing $12n + 18$ electrons based upon the $n = 12$ vertices.

A normal *closo*-icosahedral cluster would possess $14n+2=170$ valence electrons, but the presence of the interstitial Au atom provides 9 bonding orbitals for the cluster skeleton instead of the $n+1$ orbitals of a cluster with no interstitial atom. Thus, the bonding orbitals of the interstitial cluster become 6*n* (from the ligand interactions) + 9, making the valence electron requirement $12n + 18$. In a formal sense this means that 18 electrons are associated with the interstitial Au atom which would be Au^{7-} surrounded by the icosahedral cluster skeleton of twelve Au^{I} species. That gives the cluster a formal charge of +5 which is reduced to +3 by the presence of the two chloride ion ligands. Clearly the molecular orbital situation is more complex, but more in keeping with the Pauling electroneutrality principle. Clusters of greater and greater nuclearity are being synthesized and as the nuclearity increases the clusters become more like the metals whose atoms form the cluster. The high nuclearity clusters are considerably electron deficient in the sense that there are fewer electrons than are needed to form all the individual metal–metal bonds and this must mean that electron density is removed from the peripheral atoms and the ligand groups. This factor is possibly the main reason that the high nuclearity structures do not behave catalytically like the finely divided metal would.

11.13 SPECIAL FEATURES OF THE 4d AND 5d SERIES

Although the metals in these two series share many of the characteristics of the 3d series, the heavier metals exhibit distinctive characteristics, summarised as follows:

Metallic and ionic radii

As a result of the lanthanide contraction, the metallic, covalent and cationic radii of the 5d metals are nearly equal to those of the 4d atoms. In fact the metallic radii of Zr and Hf are equal. The same applies to Pd and Pt and to Ag and Au. As expected these radii are longer than those of the higher 3d metals in each group, and the metallic radii are slightly longer than the covalent radii. The ionic radii of the 4d and 5d metals are not always available for the oxidation states which are the same as the 3d metals. In the relatively small 5d atoms, as a result of increased Z, the s electrons accelerate as they approach the nucleus as implied by the relativity theory, leading to an increase in electron mass and hence a decrease in orbital size. This effect is thought to add to the effect of the lanthanide contraction.

Ionization energies

Although I decreases down a group in the s or p blocks, this is not always the case in the d block. Although I_1 for a 4d metal may be greater than or less than I_1 for the 3d metal in the same group: I for the later 5d metals from group 6 onwards are greater than I_1 of either 3d or 4d member of the same group. This is again due to the above-mentioned relativistic effect and leads to the noble character of these metals and even Tl and Pb have higher I_1 than Al or Sn.

Stability of higher oxidation states

The higher oxidation states exhibited by the 3d metals is exceeded in 4d or 5d metals of groups 7 or higher. None of the 3d metals from a tetroxide but MO_4 (M = Ru, Os) are known. Although the highest fluoride of Mn is MnF_4, MF_n (M = Tc, Re with $n = 5$ or 6) and the highest ReF_7 is the group maximum. Similarly MF_5 and MF_6 are known for the heavier members of groups 6-8. Even OsF_7 and AuF_5 have been reported for these noble metals. Evidently covalent bonding is predominant in these compounds. This may be related to the high electronegativities which increase with an increase in oxidation number. The high oxoanions are not easily reduced unlike the strong oxidising oxoanions of Mn, Cr, V and Fe.

Low oxidation states in aqueous solution

Unlike the stable Cr^{3+}, Mn^{2+}, Fe^{3+} and the later M^{2+} ions in the 3d series, the 4d and 5d metals do not generally form stable cationic species. This is related to the relatively high I_{1+2} and $I_{1\to3}$. Even when aqueous cations are known e.g. $[Ru(H_2O)_6]^{2+}$, these are low spin although H_2O is a weak ligand. Cationic complexes are not common except for Ru, Rh, Pt and Pd, although anionic oxo-or halo-complexes are known.

High Δ_o and L.F.S.E. values

For a series of complexes, Δ_o generally increases down a group e.g. $M(CO)_6$ where Δ_o increases in M = Cr < Mo < W. Accordingly spin-paired complexes are common e.g. $[Ru(NH_3)_6]^{n+}$ (n = 2 or 3) although NH_3 is not a particularly strong ligand.

Exceptional magnetic behaviour

The magnetic moments are often less than what is expected from the spin-only moments. Some compounds have a complex magnetic behaviour.

Technetium, Tc, and promethium, Pm

The only 4d metal which has no stable isotopes is technetium. Chemical considerations indicated a gap in the Periodic Table between Mo (Group 6) and Ru (Group 8). Technetium shows similarities to Mn and especially to Re in Group 7. It has as many as 22 synthetic isotopes of which the metastable ^{99m}Tc is very useful in radio-diagnosis. The only lanthanide with no stable isotopes is promethium. Its existence could not be predicted from chemical considerations because of the small differences in properties of the lanthanides in general, but characterization of Nd (Z = 60) and Sm (Z = 62) by their X-ray spectra exhibited the gap in the Periodic Table which is now filled by the synthetic element, Pm.

11.14 SPECIAL FEATURES OF THE ZINC GROUP (GROUP 12)

Since the d subshell is filled in the atoms of this group and since the two s electrons are preferentially lost, the M^{2+} ions formed by Zn, Cd and Hg have the stable d^{10} configuration. For Zn and most Cd compounds, this is the only stable oxidation state. Because the elements of group 12 do not exhibit typical properties of transition metals (e.g. coloured paramagnetic ions) some authors regard them as main group elements. If transition elements are those in which the appropriate d shell is being populated, the elements of group 12 are transition elements. In Hg, and to a less extent Cd, the +1 oxidation state is attained, not as M^+ ionic compounds but in $(M-M)^{2+}$ compounds, which are common for Hg but rare for Cd and rarer for Zn. Although E° (the standard reduction potential of the (II) state to the metal) is negative for Zn (−0.76 V) and Cd (−0.4 V) (the latter is below Zn in the electrochemical series), E° is positive for Hg (0.86 V). This is one example of the noble character of Hg. This derives from a combination of the higher value of the first two ionization energies (Zn, Cd, Hg: 906 & 1733, 868 & 1631, 1007 & 1810 kJ mol^{-1}) and a lower value of the standard hydration enthalpy (Zn, Cd, Hg: −2059, −1828, −1845 kJ mol^{-1}) which offset the lower standard enthalpy of atomization (Zn, Cd, Hg: 131, 112, 59 kJ mol^{-1}). This leads to a tendency of covalent bonding in many of its compounds. Even Zn, and to a greater extent Cd, compounds depart from purely ionic character, since their ions (of d^{10} configuration) are more polarizing than group 2 ions (Z_{eff} is higher for Group 12). The first ionization energy reaches a maximum in each transition period at Group 12. However I_{1+2} is lower than I_{1+2} of group 11. However I_{1+2} increases in the series: Cd < Zn < Hg. Another characteristic of the metals is the relatively low melting points which decrease in the series Zn > Cd > Hg, the latter being the only liquid among the metals. This reflects the weaker metal–metal bonds in the metals, which can be ascribed to the absence of contributions from d orbitals.

The ion, Hg^{2+}, and to an extent, Cd^{2+}, are soft acceptors and tend to favour coordination with soft donors e.g. I^-. Thus $[MI_4]^{2-}$ (M = Cd, Hg) are formed by dissolving MI_2 in excess iodide solution. Complex and covalent compound formation is more pronounced than observed in group 2. Among the covalent compounds of Hg, the linear $Hg(CH_3)_2$ is similar to the linear $Hg(CN)_2^-$. Although $M(CH_3)_2$ (M = Cd, Zn) are also linear, most Zn and Cd complexes are tetrahedral or octahedral e.g. $[M(CN)_4]^{2-}$ (M = Zn, Cd) and $[Zn(NH_3)_6]^{2+}$ (the latter is found only in solids). Among the unique Hg compounds, those in which it is directly bound to N are remarkable. The reaction of HgO (yellow) with NH_3 solution gives $Hg_2NOH.2H_2O$ and with HCl, pure $Hg_2NCl.H_2O$ is obtained. $HgNH_2Cl$ has NH_2–Hg–NH_2–Hg chains. Numerous organo–Hg compounds are known. RZnX (R = alkyl, X = halide) resembles Grignard reagents and is more readily prepared than the Cd compound. All 3 metals of the group form R_2M.

Zn^{2+} aq. is precipitated by alkali as $Zn(OH)_2$ which is amphoteric dissolving in excess alkali to give $[Zn(OH)_4]^{2-}$. On the other hand, $Cd(OH)_2$ similarly precipitated, is more basic than the Zn compound and is insoluble in alkali. $Hg(OH)_2$ is a very weak base. The departure of Hg compounds from ionic character is clear from the partial dissociation:

$$HgCl_2(aq) \rightarrow HgCl^+ + Cl^- \qquad (11.15)$$

11.15 GROUP 3 AND THE f BLOCK METALS

The arguments about the details of which elements form the true members of Group 3 and which are either lanthanides or actinides are to be found in Chapter 3. Conventionally, and about which there is no controversy, Group 3 contains scandium and yttrium—the first members of the 1st and 2nd transition series (although some would question the latter phrase). Because of their electronic and chemical similarities lanthanum and lutetium have competing claims to be the first member of the 3rd transition series and in this book lutetium gets the vote. The lanthanide elements are best regarded as the fifteen metals from lanthanum to lutetium (see Chapter 3 for their electronic configurations). Likewise, the fifteen elements from actinium to lawrencium are to be regarded as the actinides.

The elements are metals with low values for their electronegativity coefficients and highly negative values for their standard reduction potentials from the aqueous III state to the element. The electronegativity coefficients are Sc 1.2, Y 1.1, and the values for the lanthanides vary between 1.0 – 1.1. These values should be compared with those for Cs 0.9, Ba 1.0 to realize that the metals are highly electropositive and are to be expected to have appropriate properties.

The metals react easily with water to give solutions containing the +3 ions. They react with hydrogen to give saline hydrides mainly with the general formula LnH_3 and which contain the hydride ion. Their oxides have the general formula Ln_2O_3. Europium also forms the II oxide EuO and cerium forms the IV oxide CeO_2. All the elements form all four III halides and Sm, Eu, and Yb form all four II halides. Nd, Dy and Tm form the II chlorides, bromides and iodides.

Scandium, Y and the lanthanides form M^{3+} ions (of noble gas configuration) rather than lower M^+ or M^{2+} ions because the high ionization energies, $I_{1\rightarrow 3}$, are compensated by the high L values of the ionic solids or highly negative $\Delta_{hyd}H^{\circ}$ when ions are formed in solution, a similar situation to the behaviour of group 2 metals. There are a few exceptions in the case of the lanthanides, some of which form +2 and/or +4 ions, but these are referred to below. However, the M^{3+} ions are smaller than the M^{2+} ions of group 2 and hence they have a higher φ and show departure from ionic character.

In the lanthanides, Ln^{3+} and Ln^{III} compounds are the predominant species. However, occasionally +2 or +4 ions in solution or in solid compounds are obtained. This and some irregularities in electronic configuration arise from the extra stability of empty, exactly half-filled or filled f sub-shell. Thus La has the configuration $[Xe]6s^25d^14f^0$ but the following elements have vacant d orbitals, filling the 4f orbitals singly with electrons of the same spin until Eu which has 7 f electrons. In Gd, the configuration $[Xe]6s^25d^14f^7$ shows the stability of f^7. From Gd, the extra electrons fill the f orbitals but now pairing spins until the f sub-shell is filled in Yb. In the last metal, Lu, the configuration is $[Xe]6s^25d^14f^{14}$.

The formation of Ce^{IV} is favoured by its noble gas configuration, but it is a strong oxidant reverting to the common +3 state. Eu^{2+} is formed by losing the two s electrons and its f^7 configuration accounts for the formation of this ion. However, Eu^{2+} is a strong reductant reverting to the common +3 state. Similarly Yb^{2+}, which has f^{14} configuration is a reductant. Tb^{IV} has a half-filled f^7 configuration and is an oxidant.

By contrast to the lanthanides, the chemistry of the early actinides Th – Am shows similarities to that of the corresponding elements in the transition groups 4 – 9, in particular with regard to the range of oxidation states and the most stable oxidation states. For example, uranium exhibits the oxidation states III, IV, V and VI and has VI as its most stable state. The corresponding element in the 3rd transition series, W, shows similar chemistry. These and other similarities led to the early actinide elements being classified as members of the 4th transition series at one time.

In aqueous solution all the elements form the Ln^{3+} aquated ions, with Ce forming a +4 ion and the elements Sm, Eu and Yb forming +2 ions. In all cases the +3 ion is the more stable state. The standard reduction potentials of the elements Sc, Y and La are compared in Table 11.4 which also gives the calculated values derived from the equation:

$$\Delta_{red}H^{\circ} = \Delta_{hyd}H^{\circ}(Ln^{3+}) - I_3 - I_2 - I_1 - \Delta_{at}H^{\circ}(Ln) + 3 \times 439 \qquad (11.16)$$

which is a modification of equation (7.46).

Table 11.4 Accepted and calculated values for the standard reduction potentials ($Ln^{3+} \rightarrow Ln$) for some Group 3 elements

Ln	E° Accepted value/V	E° Calculated value/V
Sc	-2.08	-2.22
Y	-2.37	-2.55
La	-2.52	-2.47

The closeness of the calculated values to the accepted ones indicates that the terms of equation (11.16) are those mainly responsible for the determination of the values.

Unlike the 3d metals, the lanthanides show certain characteristics associated with their f orbitals which are efficiently shielded metal ion surroundings. Hence, f orbitals do not contribute significantly to the bonding in complexes. As a result, the strengths of the lanthanide–ligand bonds have little directional preference unlike the 3d metals. The coordination numbers for the lanthanide compounds and complexes are 6–9, with typical shapes of trigonal prism, square antiprism and dodecahedron. The higher coordination numbers are not unexpected since the lanthanide ions are larger than the corresponding 3d ions in spite of the lanthanide contraction.

Another characteristic of the lanthanides is the sharp line-like electronic absorption spectra associated with the fact that the intra-f orbital transitions are forbidden and that the ligand field effect on the f orbitals is slight. The effective magnetic moments of Ln^{3+} ions shows two maxima at Pr and Dy, the latter being highest at > 10 μ_B. There is a minimum at Sm. The spin-only calculated values of the magnetic moments are of little significance since they show one maximum at Gd. There is quite clearly a large and dominant contribution from spin-orbital interactions.

The actinides

The actinides are discussed as the fifteen elements from Ac to Lr, thus avoiding the argument about whether Ac or Lr is the first actinide element or the first member of the fourth transition series.

There are differences in chemistry between the first seven and the latter eight elements in that the first seven exhibit variable oxidation states from II up to VI in some cases, whereas the elements from Cm onwards behave much like the lanthanides in having almost solely the oxidation state III in their compounds and in solution.

Because of the radioactivity of these elements, most of which have been artificially produced, their chemistry has been less studied than the lanthanides. Although the naturally occurring elements and the earlier members have relatively long half-lives, the later members have values ranging from days to 3 minutes for lawrencium ($Z = 103$). This and the high activity render their study more difficult. Although the general decrease in the size of the atoms and M^{3+} ions follows the same pattern as in the lanthanides, some radii are not known accurately. Again the +3 oxidation state is found in all the members, there is a tendency in the earlier members to have a range of oxidation numbers. As an example, +3, +4, +5 and +6 states are known for U, whose chemistry has been extensively studied not only because of its availability but also because of its use in nuclear power stations. However, the later members exhibit the oxidation states found in the lanthanides. The main difference between the 4 f and 5 f orbitals is that the latter are available for bonding and this accounts for the range of oxidation states, which can reach +7 in Np. The larger radii of the actinides allow the attainment of high coordination numbers in complexes. For example, $[UO_2(NO_3)_2(OH_2)_4]$ has a coordination number of 8, with NO_3 acting as a bidentate ligand. This is used for separating U from other metals by solvent extraction. The chlorides of U show the variation in character from ionic solids: UCl_3, UCl_4 to the molecular U_2Cl_{10} and UCl_6. The highest fluoride UF_6 is used for separating the U isotopes. In solution, U(VI) exists as the linear ion, UO_2^{2+}.

The irregularities in the electronic configurations of the actinides, like those in the lanthanides, are related to the stabilities of the f^0, f^7 and f^{14} occupancies of the 5f orbitals. Thus, the configurations of Am and Cm are: $[Rn]5f^7 7s^2$ and $[Rn]5f^7 6d^1 7s^2$. Although the 5f orbitals resemble the 4f in their angular part of the wavefunction, unlike the 4f orbitals which are 'buried' and hence their electrons are rarely ionized or shared, the electrons in the 5f orbitals are more available for chemical interactions. This is particularly the case for the earlier actinides which show some resemblance to the 3d metals.

In the earlier actinides, the maximum oxidation number increases from +4 in Th to +5, +6 and +7 respectively in Pa, U and Np, but decreases in the succeeding metals. The actinides resemble the lanthanides in many respects, e.g. in having more compounds in the +2 oxidation state than in the +4 state. However, the +3 and +4 ions tend to hydrolyse and polymerize readily.

The differences between the earlier and later actinides are apparent in their electronic spectra and magnetic properties. The line-like absorption spectra of the later actinides, e.g. Cm^{3+}, resemble the spectra of the lanthanides, whereas the earlier actinides, e.g. U^{III}, have broader bands in their spectra in addition to to some very narrow ones. The broadening reflects the greater metal–ligand interaction expected from the availability of thcir 5f electrons. The magnetic properties of the actinides are more complex than those of the lanthanides. Although the variation of the magnetic susceptibility of the actinides with the number of unpaired 5f electrons is roughly parallel to the corresponding results for the lanthanides, the latter have higher values.

Lanthanide and actinide contractions

Along the lanthanide and actinide series *Z* gradually increases since the f electrons do not efficiently screen the extra protons added in the nucleus, the f orbitals being more diffuse than the d orbitals. As a result, there is a gradual decrease in the size of atoms or M^{3+} ions across the series. This is referred to as the lanthanide or actinide contraction.

In spite of this effect, the additional f electrons do not have a major role in controlling the chemical behaviour especially in the lanthanides since the 4 f orbitals are "buried" in the atoms or ions unlike the d orbitals. Hence the Ln^{3+} ions or Ln^{III} compounds show remarkable similarity to each other and to Y. The chemical similarity makes it difficult to separate individual lanthanides. However, they can be separated using solvent extraction or ion-exchange resins. The relative size of Ln^{3+} ion determines the equilibria in either the solvent mixture or the resin/solution surface.

The lanthanides, together with Y resemble the alkaline earth compounds e.g. they form insoluble fluorides but soluble halides. Although their carbonates and hydroxides are insoluble, the latter are not amphoteric. $E^{\ominus}$ values do not vary regularly or significantly along the series. Compared to group 2 metals, the lanthanides form more and more stable complexes and organometallics e.g. $[M(acac)_3]$ (acac = $OC(CH_3)CH{=}C(CH_3)O^-$) and $M(\eta^5\text{-}C_5H_5)_3$, the latter is a cyclopentadienide. Because of the reluctance of the f orbitals to be involved in bonding, ligand field effects are much less important than in the d transition metals. However, it is worth noting that the seven f orbitals are split in O_h symmetry into: one labelled a_{2u}, three labelled t_{2u} and the highest energy three labelled as t_{1u}, the energy sequence is based on electrostatic considerations. Hence the latter are $^3/_7\,\Delta_o$ above the level of the spherically symmetric orbitals and the three t_{2u} are $^1/_7\,\Delta_o$ below that level while the a_{2u} is $^6/_7\,\Delta_o$ below it. However, the stability trends of the lanthanide complexes rarely exhibit the double-humped trend found for the 3d complexes.

The lanthanide and actinide contractions, as well as being the main factor which allows the separation of the +3 species in solution, have extended effects on the sizes, and therefore the properties, of the elements succeeding them in their respective periods.

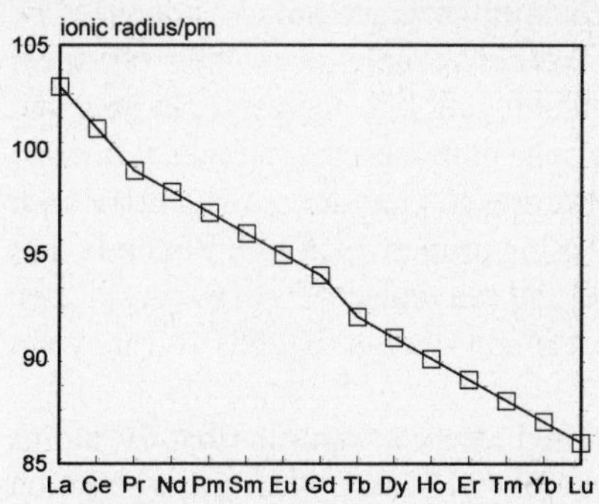

Fig. 11.8 The ionic radii of the lanthanide +3 ions showing the lanthanide contraction

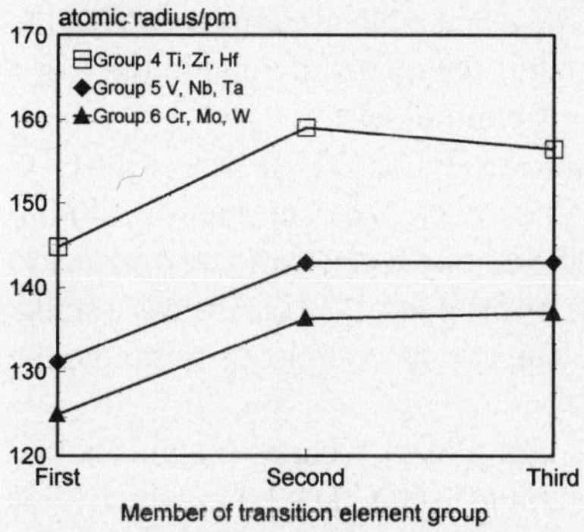

Fig. 11.9 The atomic radii of some transition elements showing the effects of the lanthanide contraction

The lanthanide contraction is the more important because the elements succeeding the actinides are much less interesting chemically at the present time. Nevertheless, the actinide contraction is expected to exert a similar effect as does the lanthanide contraction. A plot of the Ln^{3+} ionic radii is shown in Fig. 11.8 the atomic radii for some of the members of the transition series are shown in Fig. 11.9 as examples of how the lanthanide contraction contributes to the relatively small members of the sixth period. It is considered that the lanthanide contraction accounts for about half the effect of reducing the sizes of the 3rd series of transition elements, the other main contributor being the relativistic effect discussed in Chapter 3.

The 6d transition series

Elements with Z values above 102 have been synthetically produced and currently the maximum value of Z is 112. These artificial elements have very short half-lives which makes the study of their chemistry difficult. With advances in radiochemistry, it was established that these elements form the early members of the 6d, fourth transition series.

11.16 INTRODUCTION TO ORGANOMETALLIC COMPOUNDS

Although CO cannot be called organic, metal carbonyls, which have been discussed earlier are usually considered as organometallic compounds, whereas complexes like: $[M(C_2O_4)_3]^{3-}$, $[M(en)_3]^{3+}$ or $[M(acac)_3]$ are not usually classified as organometallic since the metal is not directly bound to C in these complexes. On the other hand, in complexes containing CN^- as a ligand, the metal is coordinated to carbon, but such complexes are not considered to be organometallic. They resemble the very common Werner complexes in their chemical behaviour. Cyanide ion can be regarded as a pseudo-halide ion. It is interesting to note that CN^- is isoelectronic with CO and both are near the high end of the spectrochemical series and both exert strong ligand field effects causing spin-pairing in complexes. The difference between the two ligands lies with the stronger π-accepting properties of CO. PF_3 and other substituted phosphines are strong π-acceptors like CO and can replace CO in its complexes. Complexes containing both CN^- and CO may also contain organic ligands and they are sometimes considered as organometallics.

In organometallics, the hapticity notation is used where η^n signifies that *n* C atoms of the ligand are within bonding distance from the metal. Alkyl ligands η^1 are more common among the p or s metals than among d metals, although the M−C bond energy of the latter is not particularly weak. Another difference between the d and p block elements is that the E-C bond energy increases down a d block group, which is the opposite trend in the p or s block. This could be related to the trends in χ in the different blocks.

The alkene complexes η^2 include the earliest known $[PtCl_3(C_2H_4)]^-$ where the C=C is at right angles to the coordination plane of Pt and 3Cl's. A synergic mechanism, similar to that discussed previously (11.8) is proposed where the ethene π electrons are donated to the Pt and the latter is involved in a back bond to vacant π* orbitals of C_2H_4. The most stable compounds are those with one or 2 chelating dienes. The alkyne complexes resemble the alkene complexes.

A similar bonding is proposed for the allyl η^3 complexes, where a metal is bound to 2 allyl ligands or one allyl and other ligands. An example is $(\eta^3C_3H_5)_2Ni$ in which Ni is sandwiched between the two allyl ligands, which may be in staggered or eclipsed forms.

There are several classes of cyclic polyene ligands: $C_4H_4(\eta^4)$, $C_5H_5(\eta^5)$, $C_6H_6(\eta^6)$, $C_7H_7^+(\eta^7)$, C_8H_8 or cot (η^8). The best representative is ferrocene $(\eta^5\text{-}C_5H_5)_2Fe$. A molecular orbital diagram shows that 12 electrons occupy the lower bonding molecular orbitals (mainly of ligand character) whereas the higher molecular orbitals (of metal character) are: e_2 (derived from d_{xy} and $d_{x^2-y^2}$), a_1 (derived from d_{z^2}) and e_1* (derived from d_{xz} and d_{yz}), arranged in increasing energy. In ferrocene 6 electrons occupy e_2 and a_1 orbitals. In the Co and Ni complexes 1 and 2 electrons occupy e_1* (singly in the latter). In V and Cr complexes, the e_2 and a_1 are filled according to Hund's rules. The metallocenes have the metal sandwiched between the two rings which may be staggered or eclipsed. The most stable benzene compound is $(\eta^6\text{-}C_6H_6)_2Cr$, which is diamagnetic similar to ferrocene.

The m.os which are mainly of ligand character are filled by 12 electrons. Hence, the electron count is 15, 16, 17, 18, 19 and 20 for $(\eta^5\text{-}C_5H_5)_2M$ for M = V, Cr, Fe, Co and Ni respectively. The electronic configurations described are consistent with the effective magnetic moments for these metallocenes, giving the number of unpaired electrons as 3, 2, 0, 1 and 2 respectively. In the metallocenes of Fe, Co and Ni the metal−ring distance increases along the series, although the atomic or ionic radii of the metals decrease. This observation is in agreement with the occupation of the antibonding e_{1g} orbitals singly and doubly in the Co and Ni compounds respectively. It is noteworthy that a C_5H_5 group may be bonded to a metal in a η^3 or η^5 manner. Some cyclopentadiene complexes have one or up to four rings attached to the metal. When the metal is bound to two rings plus 1, 2 or 3 other non-bonding π ligands, the ring−M−ring angle is appreciably greater than 180°. The most stable benzene compound, $(\eta^6\text{-}C_6H_6)_2Cr$, has an electron count of 18 as does ferrocene. In numerous complexes the metal is sandwiched between two different rings or between one ring and 2–4 CO ligands. Some complexes have a metal–metal bond in addition to C_5H_5 and CO groups which may be terminal and/or bridging.

Ring and sandwich compounds exhibit fluxional behaviour (stereochemical non-rigidity) in which, depending upon the temperature, there is movement of the rings with respect to their general position in the molecule. For example, the complex $Ru(\eta^4\text{-}C_8H_8)(CO)_3$ has a room temperature n.m.r. proton spectrum which consists of a single sharp line indicating that all eight hydrogen atoms have equivalent positions. At lower temperatures the signal broadens and eventually splits into four peaks as would be expected for four pairs of protons with different environments.

The more recent organometallic compounds contain M=C and M≡C bonds, named carbenes and carbynes respectively. The former are sometimes distinguished as being either Fischer or Schrock carbenes. In the former carbenes a σ pair of electrons is donated to a vacant metal orbital forming a σ bond and a vacant π* orbital (p_z) of the carbene accepts π electrons from the metal. The bond may be considered to be a double bond, although metal–C distances suggest a fractional bond order. The bonding in Schrock carbenes may be considered to consist of two carbene orbitals which are singly occupied and which overlap with two metal orbitals, each containing one electron. These carbenes are nucleophilic unlike the electrophilic Fischer carbenes. Fischer also reported complexes containing an M≡C bond. It is suggested that an electron pair in an sp orbital of C is donated to the metal in addition to a bond formed by pairing an electron from the metal and a p electron from carbon. An interesting compound contains W bound to CH_2CMe_3, $CHCMe_3$ and $CCMe_3$ representing single, double and triple metal–ligand bonds respectively. This is confirmed by the W–C distances which decrease along the series described. The ultimate bond between a metal and carbon is that where C itself is bonded to a metal, e.g. as in the square pyramidal $C[Fe(CO)_3]_5$. Such molecules are termed carbido-complexes. The carbyne ligand may be considered as a 3 electron donor: the sp pair of electrons form a σ bond with the metal and the odd electron of the carbyne pairs with a metal electron in a π bond.

11.17 SOME APPLICATIONS OF d AND f BLOCK ELEMENTS

Iron and steels are the most important construction materials. Their production is based on the reduction of iron oxides, the removal of impurities and the addition of carbon and alloying metals such as Cr, Mn and Ni. Some compounds are manufactured for special purposes such as TiO_2 for the pigment industry and MnO_2 for use in dry battery cells. The battery industry also requires Zn and Ni/Cd. The elements of group 11 are still worthy of being called the coinage metals, although Ag and Au are restricted to collectors items and the contemporary UK 'copper' coins are copper-coated steel. The 'silver' UK coins are a Cu/Ni alloy. Many of the metals and/or their compounds are essential catalysts in the chemical industry. V_2O_5 catalyses the oxidation of SO_2 in the manufacture of sulfuric acid. $TiCl_4$ with $Al(CH_3)_3$ forms the basis of the Ziegler catalysts used to manufacture polyethylene (polythene). Iron catalysts are used in the Haber process for the production of ammonia from N_2/H_2 mixtures. Nickel catalysts enable the hydrogenation of fats to proceed. In the Wacker process the oxidation of ethyne to ethanal is catalysed by $PdCl_2$. Nickel complexes are useful in the polymerization of alkynes and other organic compounds such as benzene. The photographic industry relies on the special light-sensitive properties of AgBr.

Transition metal compounds in homogeneous catalysis

A catalyst is a substance which lowers the Gibbs energy of activation of a reaction, leading to a more rapid transformation of the reactants into the products. When a catalyst does not form a separate phase from the reaction mixture, it is termed homogeneous. In many cases, a reaction may be catalysed by a homogeneous or a heterogeneous catalyst, the latter reactions being usually those occurring on the surface of a solid catalyst. The advantage of using a homogeneous catalyst rather than a heterogeneous one is that the former usually allows the reaction to proceed at lower temperatures and pressures than the latter. In addition, homogeneously catalysed reactions are more specific, thus giving purer products. Numerous transition metal complexes or organometallic compounds are used as catalysts.

Wilkinson's catalyst is the compound $(Ph_3P)_3RhCl$ which is square planar Rh^I with sixteen valence electrons. It catalyses the hydrogenation of ethene and other alkenes at near normal atmospheric pressure. Although the detailed mechanism of its reactions is not known for certain, the initial reaction is the addition of H_2 to the compound or a solvated complex, $(Ph_3P)_3RhClS$ (where S is a solvent molecule). The reaction is termed oxidative addition in which the oxidation state of Rh increases from +1 to +3 and its coordination number increases from 4 to 6. The next step in the reaction is the formation of an ethene complex by the replacement of the solvent molecule. The rate-determining step is thought to be hydride migration to form the 5-coordinate
$(Ph_3P)_3RhHCl(CH_2CH_3)$. The fast following steps are: solvent re-coordination, a second hydride migration to complete the formation of the alkane and the detachment of the alkane. The original solvated complex is regenerated, completing the catalytic cycle. The last two stages in the cycle is regarded as reductive elimination. Other Rh^I compounds containing chiral phosphines have been used to produce optically active products.

In the hydroformylation reaction, represented by the equation:

$$RCH_2CH_2 + CO + H_2 \rightarrow RCH_2CH_2CHO$$

$Co_2(CO)_2$ can be used as a catalyst. It is assumed that a pre-equilibrium produces $HCo(CO)_4$. The cycle includes the loss of one CO, followed by the addition of the alkene forming a compound with an alkene–Co bond. Hydride migration and addition of CO gives a compound with an alkene–Co bond and on reaction with CO, a C=O group is inserted between the Co and the alkyl group. The final stage is the regeneration of $HCo(CO)_4$ by reaction with dihydrogen and the release of the aldehyde product.

In the Wacker process, alkenes are oxidized to aldehydes with the same number of carbon atoms. The ion $[PtCl_4]^{2-}$ is the homogeneous catalyst used. The catalytic cycle is complex and includes a Cu^{II}/Cu^I step in which dioxygen participates.

A new catalyst (Ghaffar and Parkins, *Tetrahedron Letters*, **36**, 8657, 1995) for the hydrolysis of nitriles (RC≡N) to amides ($RCONH_2$) is a formal platinum(II) compound, $[(Me_2POHOPMe_2)PtH(PMe_2OH)]$. Apart from its catalytic activity, the catalyst has the advantage that the hydrolysis reaction stops at the amide stage, other conventional hydrolyses proceeding inconveniently to the acid RCOOH.

11. 18 PROBLEMS

1. Calculate the standard entropy changes ΔS° at 298 K and 1 atmosphere of the following reactions given the standard entropies of the species shown below.

$$2C(s) + O_{2}(g) = 2CO(g)$$
$$C(s) + O_2(g) = CO_2(g)$$
$$Ca(s) + Cl_2(g) = CaCl_2(s)$$
$$^{4}/_{3}Al(s) + O_2(g) = ^{2}/_{3}Al_2O_3(s)$$

Species	Al(s)	Al_2O_3(s)	C(s)	CO(g)	CO_2(g)	Ca(s)	$CaCl_2$(s)
S°……………	28.3	57.0	5.7	197.9	214.0	41.6	114.0
Species	Cl_2(g)	O_2(g)					
S°……………	223.0	205.0					

Comment on the relative magnitudes of the ΔS° values calculated.

2. Calculate the standard free energy change ΔG°, for the reaction:
$2Hg(s, red) = 2Hg(1) + O_2(g)$ at 298 K, using the standard entropies
S° (in J $K^{-1}mol^{-1}$ at 298 K):
S°(HgO, red) = 72.0, S°(Hg,1) = 77.4, $S^{\circ}(O_2)$ = 205,
and taking ΔH° for the reaction as 91 kJ mol^{-1} at 298 K
Calculate the temperature abovc which the compound is thermally unstable. Comment on the difference between your calculated value and the value at 750 K (from Fig. ???).

3. (a) 0.535 g of a diamagnetic complex were heated with sodium hydroxide solution liberating ammonia which was neutralised by 12.0 cm^3 of 1 *M* hydrochloric acid.
The cobalt oxide left reacted with acidified iodide solution according to:

$$Co_2O_3 + 2KI + 6HCl = 2CoCl_2 + 2KCl + 3H_2O + I_2$$

and was titrated with 0.10 *M* thiosulfate solution of which 20.0 cm^3 were required. An identical mass of the complex was dissolved in water and treated with excess silver nitrate when 0.860 g silver chloride precipitated. Deduce the formula of the complex.
(b) What conclusions can be drawn from the electronic spectrum of the complex which exhibited absorption maxima at: 472nm (ϵ_{max} = 5.5), 338nm (ϵ_{max} = 4.9) and 200nm (ϵ_{max} = 2000), all ϵ_{max} in m^2mol^{-1}.

4. A vanadium complex had a room temperature magnetic moment of 2.80 μ_B. 1.32 g of the complex, heated at 105°C, lost 0.145 g. 0.200g of the complex, dissolved in dilute sulphuric acid and titrated with 0.0203 *M* permanganate, required 32.7cm^3. An identical mass was similarly treated and titrated, after precipitating vanadium as hydroxide, required 24.2 cm^3 for oxidation by permanganate. The complex contained 24% of K. Work out the formula of the complex and state the type(s) of isomerism expected.

5(a). Two compounds A and B have the same molecular formula and contain 18.25% Ti, 40.54% Cl, 36.59% O and 4.61% H. When 0.6559g of the green compound A was dissolved and quickly treated with $AgNO_3$ solution, 0.3583 g AgCl was precipitated. An identical mass of the violet compound B was dissolved and titrated with 0.10 *M* $AgNO_3$ solution of which 75.0 cm^3 were required. The absorption spectra of the solutions had the following wavelengths of maximum absorption:

A:520 nm and a shoulder at 668 nm

B:476 nm and a shoulder at 526 nm

Both A and B gave an effective room temperature magnetic moment of $1.7\mu_B$. Show the structure of A and B and identify the type(s) of isomerism encountered. What conclusions can be drawn from the magnetic and spectral data?

5(b). 25 cm^3 of a solution of B (0.8745 g dm^{-3}) was titrated in an inert atmosphere with an Fe(III) solution (6.66×10^{-3} *M*) using NH_4CNS as an indicator. Calculate the volume of the iron solution required for complete reaction. Write an equation for the reaction. Why was the titration carried out in this atmosphere?

6(a). Calculate the lattice energy *L* of CuBr, using the Born-Landé equation, taking *M* as 1.638, *n* as 9.5 and r_{eq} as 246 pm. Then calculate *L* using the thermochemical data: $\Delta_a H^\circ(\text{Cu}) = 339$; $I_1(\text{Cu}) = 745$; $\Delta_a H^\circ(\text{Br}_2,\text{l}) = 112$, $\Delta_{ea} H^\circ(\text{Br}) = -342$. $\Delta_f H^\circ(\text{CuBr}) = 105$ (all in kJ mol^{-1}). Work out the difference between the two lattice energies: δL and comment on the sum of $r^+ = 96$ pm and $r^- = 195$ pm compared to the measured distance r_{eq}.

6(b). Given the following values of *L*(CuCl)=117 (δL = 89) kJ mol^{-1} and *L*(CuI)=171 (δL = 89) kJ mol^{-1}, discuss the trend in the copper (I) halides and relate it to the polarizabilities of the halides below:

X^-	Cl^-	Br^-	I^-
polarizability 10^{-40} F m^2	4.10	5.34	7.95

6(c). Work out the radius ratio in the three halides and comment on the observation that they all crystallise in the zinc blende structure.

Ion	Cl^-	I^-
Radius/pm	180	218

7. Use the Latimer diagram in this chapter for V in acid solution to solve the following problems.

(a) 1.26 g of $VOSO_4.nH_2O$ were dissolved in an excess of dilute sulfuric acid and titrated with 0.02 mol dm^{-3} $KMnO_4$ solution, of which 50.00 cm^3 were required before the solution became permanently pink. Calculate the value of *n* and ΔG°(at 298 K) for the redox reaction which occurs.

(b) A 25 cm^3 aliquot of a solution containing NH_4VO_3 (12.81 g dm^{-3}) was acidified and treated with Zn/amalgam until a violet colour predominated. After filtration of the remaining amalgam, an excess of the NH_4VO_3 solution was added to the filtrate which then became blue. The blue solution was titrated with the $KMnO_4$ solution (as in (a)) and 81.36 cm^3 were required. Another 25 cm^3 aliquot of the NH_4VO_3 solution was acidified and treated with Mg ribbon, a green solution resulting. The decanted solution was titrated with the $KMnO_4$ solution of which 54.36 cm^3 were required. Calculate the percentage purity of the NH_4VO_3

and determine the oxidation states of V in the violet and green solutions. Write down properly balanced equations for the reactions taking place in the two procedures.

(For $MnO_4^- + 8H^+ + 5e^- \rightarrow Mn^{2+} + 4H_2O$ the E° value is 1.51 V at 298 K)

8. (a) Plot $-\Delta_{hyd}H^{\circ}$ (hydration enthalpy) for the +2 ions below against the atomic number of M.

Ion M^{2+}	Ca	Ti	V	Cr	Mn	Fe	Co	Ni	Cu	Zn
$-\Delta_{hyd}H^{\circ}$	1598	1782	1895	1925	1862	1958	2079	2121	2121	2059

(b) Given that the longest wavelength spin-allowed bands in the electronic spectra of $[V(H_2O)_6]^{2+}$ and $[Ni(H_2O)_6]^{2+}$ have maxima at 813 and 1176 nm respectively, calculate Δ_o for both ions and their L.F.S.E. values. Use these latter values and the following L.F.S.E. values below to correct $\Delta_{hyd}H^{\circ}$ and plot the corrected values again against atomic number.

Ion M^{2+}	Cr	Fe	Co	Cu
LFSE (kJ mol^{-1})	99	50	89	90

(c) Plot the average $M^{2+}...OH_2$ distance, d, in the double sulfates $[(NH_4)_2M(SO_4)_2]$ given below against the atomic number of M

M^{2+}ion	V	Cr	Mn	Fe	Co	Ni	Cu	Zn
d/pm	212.8	216.4	217.5	215.6	209.2	205.6	208.5	209.2

The average distances are not far from the extreme values except for Cr^{2+} and Cu^{2+} where the extreme values are 205.2 and 232.7, and 196.4 and 222.2 pm respectively.

(d) Comment on the shapes of each plot; correlate the different plots with ligand field theory and comment on the distances for Cr^{2+} and Cu^{2+}.

Appendix

I DERIVATION OF THE CHARACTER OF A GROUP OF ORBITALS

In Chapter 4 reference is made to the rule-of-thumb method of determining the character of a group of orbitals with respect to any particular symmetry operation. That is, write down the number of orbitals which are unaffected by the operation. The underlying logic of the rule is explained here in terms of the two 1s orbitals which participate in the formation of the covalent bonding of the dihydrogen molecule. There are two vectors v_A and v_B which can be considered to represent the displacement of the respective hydrogen atoms, A and B, from the centre of the molecule, the latter being the Cartesian zero on which the point group $D_{\infty h}$ is based. These vectors may or may not be altered by carrying out a symmetry operation on the molecule or the two orbitals concerned. If a C_∞ operation is carried out (rotation about the molecular axis by any angle) the two vectors are unchanged. This can be expressed mathematically by the equations:

$$v_A = v_A + 0v_B \tag{A.1}$$

$$v_B = 0v_A + v_B \tag{A.2}$$

The corresponding transformation matrix is: $\begin{pmatrix} 1 & 0 \\ 0 & 1 \end{pmatrix}$

which has a trace (i.e. the sum of the numbers along the diagonal from top-left to bottom-right) of 2. Thus, 2 is the character of the orbitals with respect to the C_∞ operation. The same number corresponds to the number of orbitals unaffected by the operation.

The matrix representing the effects of the *i* operation is similarly generated from the equations:

$$v_A = 0v_A + v_B \tag{A.3}$$

$$v_B = v_A + 0v_B \tag{A.4}$$

is: $\begin{pmatrix} 0 & 1 \\ 1 & 0 \end{pmatrix}$

which has a trace of zero, again coinciding with the number of orbitals that are unaffected by the operation, in this case both having swapped places with each other.

The short cut method of determining a character of a set of orbitals by deciding on the number of them unaffected by a symmetry operation is usually straightforward, although complication can arise with some point groups.

II CHARACTER TABLES

The character tables for the point groups which are relevant to most of the problems and discussions in this book are listed here. The transformation properties of the s, p and d orbitals are indicated by their usual algebraic descriptions appearing in the appropriate row of each table. The transformation properties of the rotations around each of the Cartesian axes for the point groups are indicated by the R terms in their appropriate rows.

C_{2v}	E	C_2	$\sigma_v(xz)$	$\sigma_v(yz)$		
A_1	1	1	1	1	z	x^2, y^2, z^2
A_2	1	1	-1	-1	R_z	xy
B_1	1	-1	1	-1	x, R_y	xz
B_2	1	-1	-1	1	y, R_x	yz

C_{3v}	E	$2C_3$	$3\sigma_v$		
A_1	1	1	1	z	$x^2 + y^2, z^2$
A_2	1	1	-1	R_z	
E	2	-1	0	$(x, y)(R_x, R_y)$	$(x^2 - y^2, xy)(xz, yz)$

D_{3h}	E	$2C_3$	$3C_2$	σ_h	$2S_3$	$3\sigma_v$		
A_1'	1	1	1	1	1	1		$x^2 + y^2, z^2$
A_2'	1	1	-1	1	1	-1	R_z	
E′	2	-1	0	2	-1	0	(x, y)	$(x^2 - y^2, xy)$
A_1''	1	1	1	-1	-1	1		
A_2''	1	1	-1	-1	-1	-1	z	
E″	2	-1	0	-2	1	0	(R_x, R_y)	(xz, yz)

D_{4h}	E	$2C_4$	C_2	$2C_2'$	$2C_2''$	i	$2S_4$	σ_h	$2\sigma_v$	$2\sigma_d$		
A_{1g}	1	1	1	1	1	1	1	1	1	1		$x^2 + y^2, z^2$
A_{2g}	1	1	1	-1	-1	1	1	1	-1	-1	R_z	
B_{1g}	1	-1	1	1	-1	1	-1	1	1	-1		$x^2 - y^2$
B_{2g}	1	-1	1	-1	1	1	-1	1	1	1		xy
E_g	2	0	-2	0	0	2	0	-2	0	0	(R_x, R_y)	(xz, yz)
A_{1u}	1	1	1	1	1	-1	-1	-1	-1	-1		
A_{2u}	1	1	1	-1	-1	-1	-1	-1	1	1	z	
B_{1u}	1	-1	1	1	-1	-1	1	-1	-1	1		
B_{2u}	1	-1	1	-1	1	-1	1	-1	1	-1		
E_u	2	2	-2	0	0	-2	0	2	0	0	(x, y)	

T_d	E	$8C_3$	$3C_2$	$6S_4$	$6\sigma_d$		
A_1	1	1	1	1	1		$x^2 + y^2 + z^2$
A_2	1	1	1	-1	-1		
E	2	-1	2	0	0		$(2z^2 - x^2 - y^2, x^2 - y^2)$
T_1	3	0	-1	1	-1	(R_z, R_x, R_y)	
T_2	3	0	-1	-1	1	(x, y, z)	(xy, xz, yz)

$C_{\infty v}$	E	C_2	$2C_\infty^\phi$	...	$\infty\sigma_v$		
Σ^+	1	1	1	...	1	z	x^2+y^2, z^2
Σ^-	1	1	1	...	−1	R_z	
Π	2	2	$2\cos\phi$	...	0	$(x, y)(R_x, R_y)$	(xz, yz)
Δ	2	2	$2\cos 2\phi$	...	0		(x^2-y^2, xy)
Φ	2	2	$2\cos 3\phi$	...	0		
...	...	...	...	...	...		

$D_{\infty h}$	E	$2C_\infty^\phi$	...	$\infty\sigma_v$	i	$2S_\infty^\phi$	...	∞C_2		
Σ_g^+	1	1	...	1	1	1	...	1		x^2+y^2, z^2
Σ_g^-	1	1	...	−1	1	1	...	−1	R_z	
Π_g	2	$2\cos\phi$	...	0	2	$-2\cos\phi$	...	0	(R_x, R_y)	(xz, yz)
Δ_g	2	$2\cos 2\phi$	...	0	2	$2\cos 2\phi$	...	0		(x^2-y^2, xy)
...	...	...	...	...	...	...	...	...		
Σ_u^+	1	1	...	1	−1	−1	...	−1	z	
Σ_u^-	1	1	...	−1	−1	−1	...	1		
Π_u	1	$2\cos\phi$	...	0	−2	$2\cos\phi$	...	0	(x, y)	
Δ_u	1	$2\cos 2\phi$	...	0	−2	$-2\cos 2\phi$	...	0		
...	...	...	...	...	...	...	...	...		

O_h	E	$8C_3$	$6C_2$	$6C_4$	$3C_2$ $(=C_4^2)$	i	$6S_4$	$8S_6$	$3\sigma_h$	$6\sigma_d$		
A_{1g}	1	1	1	1	1	1	1	1	1	1		$x^2+y^2+z^2$
A_{2g}	1	1	−1	−1	1	1	−1	1	1	−1		
E_g	2	−1	0	0	2	2	0	−1	2	0		$(2z^2-x^2-y^2, x^2-y^2)$
T_{1g}	3	0	−1	1	−1	3	1	0	−1	−1	(R_z, R_x, R_y)	
T_{2g}	3	0	1	−1	−1	3	−1	0	−1	1		(xy, xz, yz)
A_{1u}	1	1	1	1	1	−1	−1	−1	−1	−1		
A_{2u}	1	1	−1	−1	1	−1	1	−1	−1	1		
E_u	2	−1	0	0	2	−2	0	1	−2	0		
T_{1u}	3	0	−1	1	−1	−3	−1	0	1	1	(x, y, z)	
T_{2u}	3	0	1	−1	−1	−3	1	0	1	−1		

III THE REDUCTION OF A REPRESENTATION TO A SUM OF IRREDUCIBLE REPRESENTATIONS

In the text, when the character of a set of orbitals is deduced to give a reducible representation, the reduction to a sum of irreducible representations has been carried out by inspection of the appropriate character table. In some instances this procedure can be lengthy and unreliable. The formal method can also be lengthy, but it is highly reliable, although not to be recommended for simple cases where inspection of the character table is usually sufficient. The formal method will be explained by doing an example.

The three 1s atomic orbitals of the trigonally pyramidal C_{3v} NH_3 molecule have the character given below:

C_{3v}	E	$2C_3$	$3\sigma_v$
$3 \times N(1s)$	3	0	1

The coefficients of the symmetry elements along the top of the above classification (the same as those across the top of the C_{3v} character table), i.e. 1, 2 and 3 give a total of six which is the *order* of the point group, denoted by h. The relationship used to test the hypothesis that the reducible representation contains a particular irreducible representation is:

$$a(\text{irreducible}) = 1/h\ \Sigma[g \times \chi(R) \times \chi(I)]$$

where g is the number of elements in the class (i.e. the coefficients of the symmetry operations across the top line of the character table, $\chi(R)$ is the character of reducible representation $\chi(I)$ is the character of the irreducible representation.

In the present case the test for the presence of an A_1 representation depends upon the sum:

$$a(A_1) = 1/6[1 \times 3 \times 1 + 2 \times 0 \times 1 + 3 \times 1 \times 1] = 6/6 = 1$$

implying that the reducible representation contains one A_1 representation.

The test for an A_2 representation is:

$$a(A_2) = 1/6[1 \times 3 \times 1 + 2 \times 0 \times 1 + 3 \times 1 \times -1] = 0/6 = 0$$

so confirming that the irreducible representation does not contain an A_2 representation.

That leaves the possibility that it would contain an E representation, for which the cquation is:

$$a(E) = 1/6[1 \times 3 \times 2 + 2 \times 0 \times -1 + 3 \times 1 \times 0] - 6/6 = 1$$

confirming that the reducible representation does contain an E representation. There being no further sums to do in this case it can now be concluded that the representation of the three 1s orbitals of the hydrogen atoms of the ammonia molecule transform as the sum: $A_1 + E$.

Further Reading

As is indicated in the preface, this book is to be regarded as core material for a university course of inorganic chemistry and is expected to be supplemented by other books of a more advanced nature or books which treat the same subject matter in a different way.

There are two major compilations of inorganic chemistry which are ideal as source-books for the original literature references and which contain a comprehensive surveys of the subject which should encompass any modern course.

Advanced Inorganic Chemistry Sixth Edition by F. Albert Cotton, Geoffrey Wilkinson, Manfred Bochmann and Carlos A. Murillo with a chapter on Boron by Russell Grimes, John Wiley and Sons, New York is published in 1998. This is a completely updated account of the chemistry of the elements and with new specialist chapters. With the third edition of *Basic Inorganic Chemistry* by F. Albert Cotton, Geoffrey Wilkinson and P. L. Gaus now containing the theoretical introduction to the subject as well as a brief survey of the inorganic chemistry of the elements, there is very little theoretical treatment in the 6th edition of Cotton & Wilkinson. Nevertheless it will retain its place as the primary source of information for advanced undergraduates and postgraduates.

Chemistry of the Elements Second Edition by N. N. Greenwood and A. Earnshaw, Butterworths, Ltd., 1998 is another welcome and updated source of information. The book contains some theory, but would still have to be supplemented by a theoretical text for a rounded course in the subject.

Two texts are recommended which contain less information than the two major works described above. They do, however contain extended discussions of the theoretical background of inorganic chemistry and are books that would cover the majority of requirements for current degree courses in inorganic chemistry. They would need to be supplemented by more specialist books for the later stages of the degree course.

Modern Inorganic Chemistry Second Edition by W. L. Jolly, McGraw-Hill, Inc., 1991 and *Inorganic Chemistry* Second Edition by D. F. Shriver, P. W. Atkins and C. H. Langford, Oxford University Press, 1997.

A major textbook with a new approach to the subject is *Essential Trends in Inorganic Chemistry* by D. M. P. Mingos, Oxford University Press, 1998 contains a breadth of chemistry discussion based upon periodic trends and which includes much factual material that has slipped down the hole between the current GCSE A Level curriculum and where university courses used to begin. During the discussion of the trends sufficient theoretical treatments are introduced to make the book suitable for the first two years of a degree course in the subject.

Any one textbook or a coupling of two suitable books will not suffice for the complete coverage of a degree course and there are now approaching one hundred of the Oxford University *Oxford Chemistry Primers* available. Any one of these contains sufficient material for an eight-lecture course and the student should be guided by the tutors as to which to buy. There are two which are of general application: *Essentials of Inorganic Chemistry* 1 (1995) and *Essentials of Inorganic Chemistry* 2 (1998), both by D. M. P. Mingos.

Solutions to Problems

Chapter 2

1. (a) $^{35}Cl + {}^{1}n \rightarrow {}^{32}P + {}^{4}He$, (b) $^{235}U + {}^{1}n \rightarrow {}^{97}Sr + {}^{137}Xe + 2{}^{1}n$, (c) $^{27}Al(\alpha, n)^{30}P$
(d) $^{35}Cl(n, p)^{35}S$, (e) $^{55}Mn(n, \gamma)^{56}Mn$

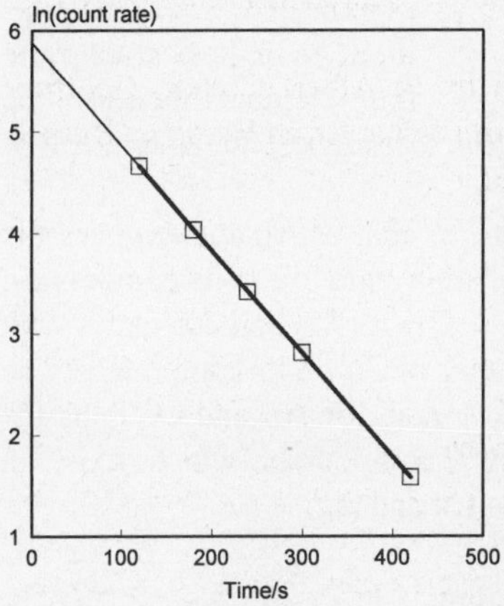

2. A plot of ln A_t versus t gives a straight line with a slope of -0.0102 leading to a value of $t_{1/2} = \ln 2/0.0102 = 68.0$ s.
If semi-log graph paper is used, extrapolation of the line back to time zero gives a count rate of 357.8 cps and a count rate of half this value occurs after a time of approximately 68 s.

3. Mass $= 3.7 \times 10^{10} \times 40 \times 60 \times 60 \times 140/(N_A \times \ln 2) = 1.79 \times 10^{-6}$ g.

4. Binding energies per nucleon for ^{12}C, ^{14}N, ^{16}O and ^{19}F are (C) $[(6 \times 1.00785) + (6 \times 1.008665) - 12] \times 931.5/12 = 7.69$ MeV,
(N) $[(7 \times 1.00785) + (7 \times 1.008665) - 14.003074] \times 931.5/14 = 7.49$ MeV,
(O) $[(8 \times 1.00785) + (8 \times 1.008665) - 15.994915] \times 931.5/16 = 7.99$ MeV,
(F) $[(9 \times 1.00785) + (10 \times 1.008665) - 18.998405] \times 931.5/19 = 7.79$ MeV.
Although there is a general increase in the values, the values for ^{12}C and ^{16}O are particularly high and their greater stability is because their mass numbers are multiples of 4, suggesting that α particles are constituents of their nuclei.

5. Ln $(25.5/20.5) = \ln 2/(5568 \times t)$, $t = 1750$ y. $\ln(25.5/A) = 0.693 \times 5568 \times 4000 = 0.4979$
$A = 15.5$ c.p.m.

6. For β^+ decay $E_{max} = [M_Z - M_{Z-1} - 2m_e]c^2$, for α decay $E_{max} = [M_{A,Z} - M_{A-4,Z-2} - M_{He}]c^2$

Chapter 3

1. $1/656.3 = R[1/2^2 - 1/3^2]$; λ(3rd Balmer) $= 656.3 \times [1/2^2 - 1/3^2]/[1/2^2 - 1/5^2] = 434.1$ nm
λ(2nd Paschen) $= 656.3 \times [1/2^2 - 1/3^2]/[1/3^2 - 1/5^2] = 1282$ nm

2. There are twenty electrons eventually removed, $Z = 20$, Ca is the element. Its electronic configuration is $1s^22s^22p^63s^23p^64s^2$. The first two points are due to the removal of the two 4s electrons. The increases in I as more electrons are removed are due to the increasing Z_{eff} and decreases in size of the ions formed. Sharp rises occur when the first 3p, the first 2p and the first 1s electrons are removed, indicative of the shell structure of the electronic arrangement.

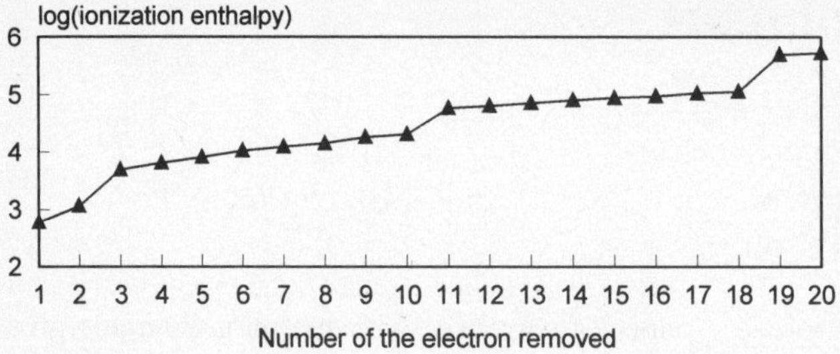

3. Set I belongs to Mg; below Be (and Be has lower I than B), also Mg is before Si in the period. Set II belongs to Si; It has the lowest I_3. Also $I_2 > I_2$(Mg) as expected. Set III belongs to B; I_2 highest among the four elements because the electron is removed from a 2s orbital. Set IV belongs to Be; I_3 is highest due to removal of an electron from a 1s orbital.

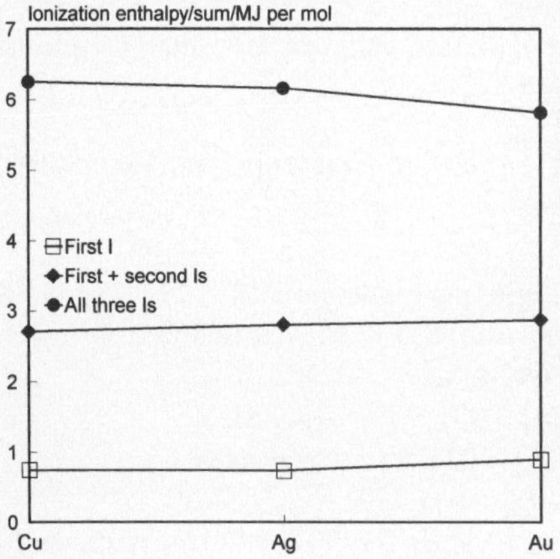

4. The lowest I_1 is that of Ag and explains the dominant oxidation state of I in its compounds. The lowest $I_1 + I_2$ is for Cu and correlates with the predominance of Cu^{II} in its compounds. The lowest $I_1 + I_2 + I_3$ is for Au and explains the numerous Au^{III} compounds which exist in comparison to Cu and Ag.

5. F configuration is $1s^2 2s^2 2p^5$, $Z_{eff} = 9 - 2 - (6 \times 0.35) = 4.9$

χ(Allred-Rochow) $= 0.744 + (3590 \times 4.9)/72^2 = 4.14$

Chapter 4

(a) $C_{\infty v}$, (b) $D_{\infty h}$, (c) $D_{\infty h}$, (d) $D_{\infty h}$, (e) C_{2v}, (f) C_{2v}, (g) C_{2v}, (h) C_s, (i) C_{2v}, (j) C_{2v}, (k) C_s, (l) D_{2h}, (m) D_3, (n) C_{3v}, (o) C_{2v}, (p) C_s, (q) C_2, (r) C_i, (s) C_{4v}, (t) D_{2d}, (u) D_{3h}, (v) D_{4h}, (w) D_{2d}

Chapter 5

1. As the atom size increases the overlap between their s orbitals to form σ bonds decreases and the less efficient overlap causes the dissociation energies to decrease.

2. N_2 From the vibrational frequencies the 1st signal is from a weakly bonding m.o. The 2nd is from a strongly bonding m.o. The 3rd is from a virtually non-bonding m.o. The electronic configurations and electronic states are (in order of increasing ionization energy and ignoring the core electrons) $(2\sigma_g^+)^2(2\sigma_u^+)^2 1\pi_u^4(3\sigma_g^+)^1$ $[^2\Sigma_g^+]$, $(2\sigma_g^+)^2(2\sigma_u^+)^2 1\pi_u^3(3\sigma_g^+)^2$ $[^2\Pi_u]$, $(2\sigma_g^+)^2(2\sigma_u^+)^1 1\pi_u^4(3\sigma_g^+)^2$ $[^2\Sigma_u^+]$. CO The 1st ionization is from a non-bonding orbital, the 2nd and 3rd ionizations are from strongly bonding m.os. The electronic configurations and electronic states are (in order of increasing ionization energy and ignoring the core electrons) $4\sigma^2 1\pi^4 5\sigma^1$ $[^2\Sigma]$, $4\sigma^2 1\pi^3 5\sigma^2$ $[^2\Pi]$, $4\sigma^1 1\pi^4 5\sigma^2$ $[^2\Sigma]$. The non-bonding 5σ orbital of CO is at a higher energy than the corresponding orbital in N_2 $(3\sigma_g^+)$ and is localized on the carbon atom

so that it is more easily donated to a metal atom or ion than the N_2 orbital. CO forms more stable and more numerous complexes than does N_2.

3. $N_3S_3F_3$ S to N bond order = 1; $S_4N_4^{2+}$ S to N bond order = 1 and 2; NSF S to N bond order = 3

4. The plot of the O–O bond dissociation energy of the species, O_2, O_2^+, O_2^-, O_2^{2-} and the O–O bond length against bond order for the same species correlates with the m.o. diagrams given below. The first three species are paramagnetic with 2, 1 and 1 unpaired electrons respectively. The peroxide ion has no unpaired electrons and is diamagnetic.

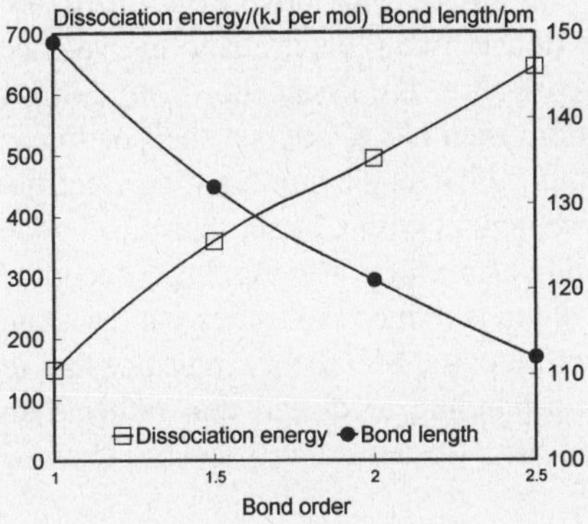

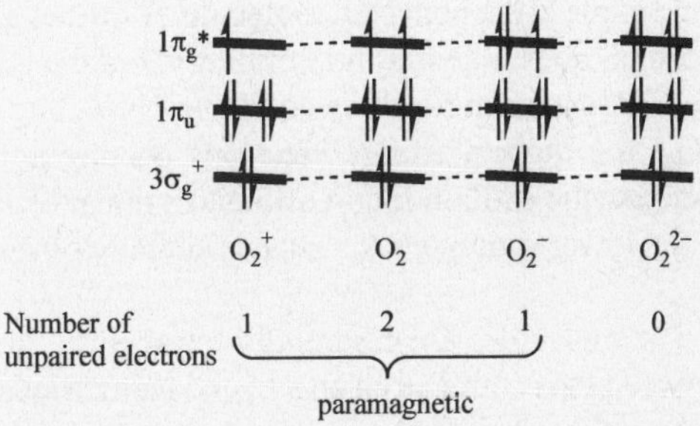

Chapter 6

1. (a) VSEPR theory: N has (5 + 3)/2 = 4 pairs of electrons, one of which is a lone pair. The geometry around N is expected to be a tetrahedral distribution of the four electron pairs, the shapes of the molecules being trigonally pyramidal. This is so for $N(CH_3)_3$, but the Si atom has 3d orbitals which can participate in the bonding. The Si can make use of its d_{yz} orbitals arranged to lie along the N–Si directions (in the σ_v planes) and which then transform as a_2'' + e', the a_2''combination having the same symmetry as the p_z orbital of the N atom in D_{3h} symmetry. This extra bonding interaction causes the NSi_3 group to have trigonally planar symmetry.
(b) VSEPR theory: N has (5 + 3)/2 = 4 pairs of electrons, one of which is a lone pair. The NH_3 and NF_3 molecules are based on a tetrahedral distribution of the four electron pairs, the shapes of the molecule being trigonally pyramidal. The same logic applies to $N(SiH_3)_3$, but the Si atoms form a trigonal plane with the N atom at the centre. Cl in ClF_3 has 7 + 3 = 10 valence electrons arranged in five pairs, three of which are bonding pairs. The electron pair distribution is based on a trigonal bipyramid with the two non-bonding pairs occupying two of the positions in the trigonal plane. The molecule is T-shaped, although in reality it is more like a blunt arrowhead.
MO theory: The ammonia molecule is most stable in C_{3v} symmetry with the stabilization of the non-bonding $3a_1$ orbital (an sp hybrid in the fourth tetrahedral position) making the largest contribution to the bonding. Because $\chi(N) > \chi(H)$ the bonding pairs are closer to N, but $\chi(N) < \chi(F)$ and the bonding pairs are closer to the F atom. In NF_3 the lone pair moment

opposes the N–F moments giving the molecule a small dipole moment. In NH_3 the bond moments are of the opposite polarity to those in NF_3 and, added to the lone pair moment give a large dipole moment of opposite sign to that of NF_3.
(c) PCl_5 is D_{3h} trigonally pyramidal in the gaseous state. In the solid it is composed of tetrahedral PCl_4^+ and octahedral PCl_6^- ions. The extra stabilization from a lattice energy as opposed to van der Waals interactions between neutral molecules is the dominant factor responsible for the difference in structures.
(d) NO_2^- (O–N–O = 115°) is C_{2v} as is NO_2 (O–N–O = 134°), but NO_2^+ is linear $D_{\infty h}$. VSEPR theory: The N^+ atom at the centre of NO_2^+ has s^2p^2 of which two electrons p^2 are reserved for π-bonding to the two O atoms. These donate two σ electrons to the valence shell of the N^+ so there are only two σ pairs which lead to the linear ion. If an extra σ electron is added to neutralize the charge, the N atom then has a valence shell of five σ electrons which would arrange into two bonding pairs with a single non-bonding electron. The three regions of σ charge gives a triangular arrangement with a larger than 120° wide bond angle in the bent NO_2 molecule. A further addition of an electron to the singly occupied σ orbital does not alter the symmetry, but the bond angle is reduced to somewhat less than the 120° expected for three equivalent repelling electron pairs. MO theory indicates that in NO_2^+ the highest energy electrons occupy a π non-bonding level and that rather than populate the antibonding π orbitals in linear NO_2 and NO_2^- greater stability can be achieved in the lower symmetry C_{2v} shapes actually observed.

2. The molecules are essentially tetrahedral, but contain single as well as multiple bonds between the central atom an a ligand atom. Since multiple bond–single bond repulsions are greater than those between single bonding pairs or non-bonding pairs, the angles in the second column are larger than the corresponding angles in the first column. In the series of S^{VI} compounds the size of the non-oxygen ligands appear to have an effect on the bond angles. (b) VSEPR theory: N has (5 + 3)/2 = 4 pairs of electrons, one of which is a lone pair. The NH_3 and NF_3 molecules are based on a tetrahedral distribution of the four electron pairs, the shapes of the molecule being trigonally pyramidal.

Chapter 7

1. For Cl^- $Z_{eff} = 17 - (7 \times 0.35 + 8 \times 0.85 + 2) = 5.75$ and for K^+ $Z_{eff} = 19 - (7 \times 0.35 + 8 \times 0.85 + 2) = 7.75$ $r(K^+) = 314 \times 5.75/(5.75 + 7.75) = 134$ pm and $r(Cl^-) = 314 - 134 = 180$ pm.
2. For I^- $Z_{eff} = 53 - (7 \times 0.35 + 10 \times 0.85 + 36) = 6.05$ and for Cs^+ $Z_{eff} = 55 - (7 \times 0.35 + 10 \times 0.85 + 36) = 8.05$ $r(Cs^+) = 395 \times 6.05/(6.05 + 8.05) = 169$ pm and $r(I^-) = 395 - 169 = 226$ pm.
3. The radius ratios, r^+/r^-, for the Group 2 oxides and sulfides are given in the table, together with the expected structures and the actual ones adopted by the compounds. The radius ratio rules are only approximate, even for predominantly ionic substances such as SrO and BaO and would not be expected to be very accurate for less ionic substances such as sulfides.

	O^{2-}	S^{2-}
Mg^{2+}	0.46, NaCl, correct	0.35, 4:4, NaCl
Ca^{2+}	0.71, NaCl, correct	0.54, NaCl, correct
Sr^{2+}	0.81, CsCl, NaCl	0.61, NaCl, correct
Ba^{2+}	0.96, CsCl, NaCl	0.73, NaCl, correct

4. $L(MO) = -\Delta_f H^\circ(MO) + \Delta_a H^\circ(M) + I_{1+2} + 248 + 704$
$L(MO_2) = -\Delta_f H^\circ(MO_2) + \Delta_a H^\circ(M) + I_{1+2} + I_{3+4} + 496 + 2 \times 704$

Oxide	SnO	SnO_2	PbO	PbO_2
L/kJ mol^{-1}	3656	11773	3532	11703

The much more negative $\Delta_f H^\circ(SnO_2)$ compared to those for PbO_2 or SnO shows the grester stability of the +4 oxidation state for Sn. $L(SnO_2) \approx L(PbO_2)$. The greater stability of SnO_2 is due to the lower values of $I_{1+2} + I_{3+4}$ compared to Pb. In the case of the Pb oxides, the slightly more negative $\Delta_f H^\circ(PbO_2)$ compared to that for PbO is mainly due to the much higher lattice energy of the former compound.

5. $L(AuCl) = 35 + 369 + 891 + 121 - 364 = 1052$ kJ mol^{-1}
$L(AuCl_3) = 118 + 369 + 5811 + 363 - 3 \times 364 - 1092 = 5569$ kJ mol^{-1}
$\Delta H^\circ = -118 - (3 \times -35) = -13$ kJ mol^{-1}
The greater stability of $AuCl_3$ is due to the much higher L. The disproportionation reaction indicates the relatively higher stability of the III state.

6. (a) BaO: $E_{ea}(2) = -560 - 157 - 1468 - 248 + 140 + 3130 = 837$ kJ mol^{-1}
BaS: $E_{ea}(2) = -444 - 157 - 1468 - 248 + 200 + 2624 = 532$ kJ mol^{-1}
$E_{ea}(1)$ is exothermic for both O and S, but less for O than S because of greater interelectronic repulsion in the smaller atom. The $E_{ea}(2)$ values are endothermic because of the difficulty of overcoming the repulsion between the incoming electron and the already uni-negative ions. Again the difference between the two values is some measure of the difference in interelectronic repulsion. (b) L values for Ba chalcogenides are smaller then the corresponding Mg compounds because Ba^{2+} is larger than Mg^{2+}.

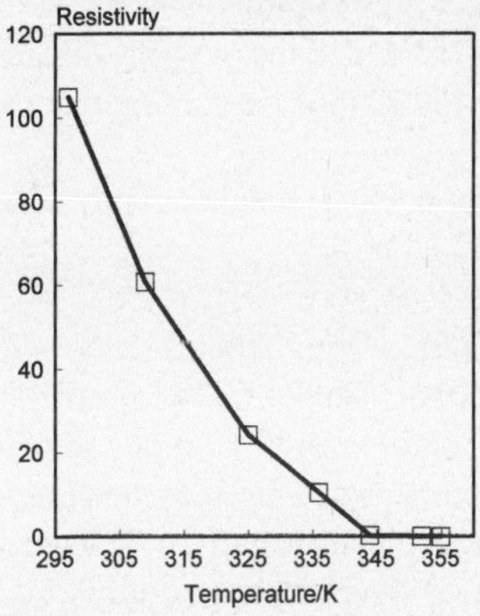

7. Below 340 K the V oxide is a semiconductor. At 340 K a transition to an electrical conducting phase occurs. The equal V...V distances indicate a metallic conducting phase.

8. $\%Cu^{III} = (20.85 - 16.1) \times 0.032 \times 63.55 \times 100/(0.15 \times 1000) = 6.4\%$
$\%Cu^{II} = \%Cu(total) - \%Cu^{III} = 27.90 - 6.44 = 21.46\%$
Formula = $YBa_2Cu^{III}_{0.69}Cu^{II}_{2.31}O_{6.84}$
The solid becomes superconducting on cooling. Solids with less O than a critical value are semiconducting.

Chapter 8

1. The following thermodynamic data (all in kJ mol^{-1} at 298K) are related to the decomposition of s block nitrates to their oxides. The higher $\Delta_{decomp}H^\circ$ for $LiNO_3$ indicates its greater stability. The trends can be related to the polarizing powers of the cations.

	$Ca(NO_3)_2$	$Ba(NO_3)_2$	$LiNO_3$
$\Delta_{decomp}H^⦵$	371.1	506.4	396.8
$\Delta_{decomp}G^⦵$	241.8	374.2	
$\Delta_{decomp}S^⦵$	0.434	0.443	
$T_{decomp} = \Delta_{decomp}H^⦵/\Delta_{decomp}S^⦵$	583°	869°	

2. $\Delta G^⦵ = -RT \ln K_{s.p.}$ $19950 = -8.314 \times 298 \times \ln K_{s.p.}$ $K_{s.p.} = 3.18 \times 10^{-4}$ At pH = 13, i.e. pOH = 1, $[OH^-] = 0.1$ *M*, $[Sr^{2+}] = 3.18 \times 10^{-4}/10^{-2} = 3.18 \times 10^{-2}$; Solubility of $Sr(OH)_2 = 3.18 \times 10^{-2}$ mol dm^{-1}. The solubility of the hydroxides increases down the group as their lattice energy decreases. The solubility of the sulfates decreases down the group as the cation hydration enthalpy decreases, with increased r^+.

Chapter 9

1. (a) the average Si–H intrinsic bond enthalpy = (452 + 399 + 872 – 34)/4 = 407 kJ mol^{-1}
(b) the average Si–H thermochemical bond enthalpy = (452 + 872 – 34)/4 = 323 kJ mol^{-1}
(c) $\Delta_f H^⦵$(hypothetical SiH_2,g) = 452 + 436 – (2 × 323) = 242 kJ mol^{-1}
(d) $\Delta H^⦵$ for: $2SiH_2(g) \rightarrow SiH_4(g) + Si(s)$ = 34 – (2 × 242) = –450 kJ mol^{-1}
The intrinsic Si–H bond enthalpy > thermochemical bond enthalpy because of the inclusion of the s → p promotion energy. The highly endothermic $\Delta_f H^⦵$ for SiH_2 indicates that it is unlikely to exist, confirmed by the highly negative $\Delta H^⦵$ for its disproportionation.

2. (i) The average O–H bond enthalpy in H_2O(g) = (242 + 249 + 436)/2 = 464 kJ mol^{-1}; the average S–H bond enthalpy in H_2S(g) = (21 + 272 + 436)/2 = 365 kJ mol^{-1}.
(ii) The oxygen–oxygen bond enthalpy in O_3(g) = 602/2 = 301kJ mol^{-1}; the oxygen–oxygen bond enthalpy in H_2O_2(g) = (133 + 436 + 496) – (2 × 464) = 137 kJ mol^{-1} assuming that the O-H bond energy in the latter is the same as that in H_2O(g).
(iii) The sulfur-sulfur bond enthalpy in H_2S_2(g), assuming that the S–H bond enthalpy is the same as that in H_2S(g) = (13 + 544 + 436) – (2 × 365) = 263 kJ mol^{-1}.

3. (a) *E*(B–F) = (1111 + 237 + 590)/3 = 646 kJ mol^{-1}; $\Delta_f H^⦵$(BF,g hypothetical) = 590 + 79 – 646 = 23 kJ mol^{-1}. (b) $L(AlF_3,s)$ = 314 + 5137 + 237 – 1044 + 695 = 5339 kJ mol^{-1}. (c) An ionic $B^{3+}(F^-)_3$ requires $\Delta_a H^⦵(B) + I_{1\rightarrow 3}(B)$ = 590 + 6879 = 7469 kJ mol^{-1} for the formation of the B^{3+}(g) cation which is too high to be compensated by the attachment enthalpy of the fluorine atoms and the lattice enthalpy. The larger electronegativity difference in the Al–F case means that ionic bonding is preferred. (d) As the electronegativity of the halogens decrease the structures of Al halides vary from the ionic AlF_3 to layer lattices and to molecular lattices containing Al_2X_6 dimers. Solid Al_2Cl_6 has a layer lattice with six-coordinate Al, but at the melting point there is a change to the Al_2Cl_6 dimeric molecular structure and the electrical conductivity falls to almost zero. The bromides and iodides have molecular lattices containing Al_2X_6 dimers.

Chapter 10

1. $\Delta_o = N_A hc/\lambda = 6.023 \times 10^{23} \times 6.626 \times 10^{-34} \times 3 \times 10^8/[\lambda(\text{nm}) \times 10^{-9}]$ = 151 kJ mol^{-1} for $[Ni(bipy)_3]^{2+}$ and 122 kJ mol^{-1} for $[Ni(bipy)(DMF)_4]^{2+}$. For an $[MA_nB_{6-n}]$ complex $\Delta_o = [n\,\Delta_o(A) + (6 - n)\Delta_o(B)]/6$ where $\Delta_o(A)$ and $\Delta_o(B)$ are the values for the complexes MA_6 and

MB_6 respectively. For $[Ni(bipy)(DMF)_4]^{2+}$ $\Delta_o = 122 = [(2 \times 151) + 4\Delta_o(DMF)]/6$ which gives $\Delta_o(DMF) = 108$ kJ mol^{-1}, $\lambda_{max} = 1106$ nm.

2. (a) $[Fe(MeCN)_6]^{2+}$ is d^6 (high spin) in O_h symmetry, the band at 917 nm is the Laporte forbidden $t_{2g} \rightarrow e_g$ ($^5T_{2g} \rightarrow {}^5E_g$) transition which is of low intensity. $[Fe(CN)_6]^{4-}$ is d^6 (low spin) in O_h symmetry. The d–d bands at 270 nm and 320 nm correspond to the $^1A_{1g} \rightarrow {}^1T_{2g}$ and $^1A_{1g} \rightarrow {}^1T_{1g}$ transitions and the much weaker band at 420 nm is a spin-forbidden transition (singlet to triplet, not shown on the Tanabe-Sugano diagram). (b) The bands of *cis*-$[Co(en)_2F_2]^+$ at 360 nm and 510 nm are the $^1A_{1g} \rightarrow {}^1T_{2g}$ and $^1A_{1g} \rightarrow {}^1T_{1g}$ transitions. The *trans*-isomer has the T_{1g} split by the lower symmetry which is more pronounced in the *trans*-form due to the difference in the positions of F^- and en in the spectrochemical series. The spectrum of $[CoF_6]^{3-}$ has two bands at 690 nm and 877 nm which are the $^4T_{1g} \rightarrow {}^1T_{2g}$ and $^4T_{1g} \rightarrow {}^4T_{1g}(P)$ transitions which occur at lower energies because of the low field exerted by the $6F^-$ environment. The complex is high spin and paramagnetic. The ground state configuration is $t_{2g}^4e_g^2$ and is subject to Jahn-Teller splitting. (c) The metals are in fairly high oxidation states (+3 or +4) and the bands are due to LMCT from the ligand to metal t_{2g} orbitals. As the reducing power of L increases from Cl to I, the energy of the transition decreases. The transfer is easier for the transition to the +4 state M rather than the +3 state. The lower metal Os is more stable in higher oxidation states, therefore, there are higher energy CT bands. (d) $[Mn(H_2O)_6]^{2+}$ is d^5 high spin and has a ground state $^6A_{1g}$ with no other sextet states. All transitions are Laporte and spin forbidden and have low intensities. $[Mn(H_2O)_6]^{3+}$ is d^4 high spin and has a ground state configuration $t_{2g}^3e_g^1$ which is Jahn-Teller distorted contributing to the broad band which is observed for the $^5E_g \rightarrow {}^5T_{2g}$ transition. (e) $[Fe(phen)_3]^{2+}$ and $[Fe(phen)_3]^{3+}$ contain the strong field ligand phen so both complexes are low spin. $[Fe(phen)_3]^{2+}$ has an intense absorption band at 510 nm which is MLCT (Mπ to Lπ*). $[Fe(phen)_3]^{3+}$ has an absorption band at 590 nm which is LMCT (Lπ* to Mπ). The bands in the UV region appear to be common to both complexes and are probably ligand $\pi \rightarrow \pi^*$ transitions. (f) $[Co(H_2O)_6]^{2+}$ is d^7 high spin and the band at 513 nm is $^4T_{1g}(F) \rightarrow {}^4T_{1g}(P)$. On the addition of of HCl (conc.) the solution turns blue and contains the $[CoCl_4]^{2-}$ complex. This is more intensely coloured with absorption maxima at 625, 670 and 700 nm. The 700 nm band is $^4A_2 \rightarrow {}^4T_1(P)$ and there is an envelope around it which is due to transitions to doublet excited states which acquire extra intensity be spin-orbit coupling. The intensity of the main peak is due to a relaxing of the Laporte rule since tetrahedral species do not have an inversion centre. (g) In the series $[Cr(NH_3)_6]^{3+}$, $[Cr(NH_3)_5CN]^{2+}$ and $[Cr(NH_3)_4(CN)_2]^+$ the bands at 464, 451 and 437 nm respectively are due to $^4A_{2g} \rightarrow {}^4T_{2g}$ transitions with energies equal to Δ_o. As ammonia is replaced with the strong-field ligand, cyanide ion Δ_o increases and the wavelength of the absorption maximum decreases. The rule of average environment can be checked. $\Delta_o = N_Ahc/\lambda = 6.023 \times 10^{23} \times 6.626 \times 10^{-34} \times 3 \times 10^8/[\lambda(\text{nm}) \times 10^{-9}]$ so Δ_o for $[Cr(NH_3)_6]^{3+}$ is 258 kJ mol^{-1}, that for $[Cr(NH_3)_5CN]^{2+}$ is 266 kJ mol^{-1} and so that for $[Cr(NH_3)_4(CN)_2]^+$ is expected to be 266 + (266 − 258) = 274 kJ mol^{-1}. The latter value gives 437 nm for the maximum wavelength absorption for the complex which is as observed. In the cases of the Cu^{II} complexes the d^9 configuration leads to a strong Jahn-Teller effect and causes the rule to break down. In $[Cu(NH_3)_4(H_2O)_2]^{2+}$ the two water ligands occupy *trans*-positions with long bonds. The replacement of the water molecules with one or two extra ammonia ligands does not produce the possibly expected decreases in wavelength of maximum absorption. The Jahn-Teller effect produces a broadening and a

distortion of the single band and in $[Cu(NH_3)_5(H_2O)]^{2+}$ and $[Cu(NH_3)_6]^{2+}$ the extra ammonia ligands are at greater distances away from the Cu^{II} centres than the four ligands in the square planes.

3. (a) $Cr_2O_7^{2-}$ can oxidize Fe^{2+} to Fe^{3+}, Ti^{3+} or Sn^{2+} can reduce Fe^{3+} to Fe^{2+}. (b) The aliquot of solution contains 0.348 g of the iron oxide. Moles of $Ti^{3+} = 31.34 \times 0.096/1000 \equiv Fe^{III}$ in oxide. Mass of $Fe^{III} = 31.34 \times 0.096 \times 55.85/1000 = 0.168$ g. Moles $Cr_2O_7^{2-} = 45.10/(60\times 1000)$. $Cr_2O_7^{2-} \equiv 6Fe^{II}$ so moles Fe $= 45.10 \times 6/(60\times 1000)$. Mass of total Fe $= 55.85 \times 45.10 \times 6/(60\times 1000) = 0.252$ g. Mass of O $= 0.348 - 0.252 = 0.096$ g. Atom ratio: O:Fe $= 0.096/16 : 0.252/55.85 = 6 \times 10^{-3} : 4.5 \times 10^{-3} = 4:3$, therefore the iron oxide has the formula Fe_3O_4.

4. (a) $[(CH_3)_4N]_2[MnCl_4]$ has T_d Mn^{2+} $e^2t_2^3$ with $n = 5$, orbital contributions are quenched, $\mu_e = \mu_{s.o.} = \sqrt{35} = 5.9\mu_B$, $K_2[Mn(IO_3)_6]$ is Mn^{4+} octahedral t_{2g}^3 $n = 3$ $\mu_{s.o.} = \sqrt{15} = 3.87\mu_B$ so there is some orbital contribution to μ_e by spin-coupling of the ground term and the next higher term ($^4A_{2g}$ and $^4T_{2g}$) because λ' is positive $\mu_e < \mu_{s.o.}$ $K_4[Mn(CN)_6]$ is O_h Mn^{2+} low spin t_{2g}^5 $n = 1$ and $\mu_{s.o} = \sqrt{3} = 1.73\mu_B$. $\mu_e > \mu_{s.o.}$ by an orbital contribution to the magnetic moment. $K_3[Mn(C_2O_4)_3]$ is octahedral Mn^{3+} high spin $t_{2g}^3e_g^1$ $\mu_{s.o.} = \sqrt{24} = 4.9\mu_B$ Orbital contribution by spin -orbit coupling of ^5E_g and $^5T_{2g}$ because λ' is positive $\mu_e < \mu_{s.o.}$ $K_3[Mn(CN)_5OH]$is Mn^{3+} distorted octahedral low spin t_{2g}^4 $n = 2$ $\mu_{s.o.} = \sqrt{8} = 2.83\mu_B$ An orbital contribution to the magnetic moment gives higher μ_e. $[Fe(CN)_6]^{3-}$ is Fe^{3+} low spin octahedral t_{2g}^5 $n = 1$ $\mu_{s.o.} = \sqrt{3} = 1.73\mu_B$ with higher μ_e because of an orbital contribution. $[Fe(H_2O)_6]^{2+}$ is Fe^{2+} high spin octahedral $t_{2g}^4e_g^2$ $n = 4$ $\mu_{s.o.} = \sqrt{24} = 4.9\mu_B$ augmented by an orbital contribution. $[Fe(H_2O)_6]^{3+}$ is Fe^{3+} high spin octahedral $t_{2g}^3e_g^2$ with the orbital contribution quenched, $n = 5$ $\mu_{s.o.} = \sqrt{35} = 5.9\mu_B$. $[FeCl_4]^{2-}$ is Fe^{2+} tetrahedral $e^3t_2^3$ with $n = 4$ $\mu_{s.o.} = \sqrt{24} = 4.9\mu_B$. Spin-orbit coupling introduces a contribution, with $\mu_e > \mu_{s.o.}$ and λ' negative.

(b) $[Cr(H_2O)_6]^{3+}$ has a ground term $^4A_{2g}$ orbital contribution to the magnetic moment but this is quenched and there is an orbital contribution from spin-orbit coupling, $\mu_e = \mu_{s.o.}(1 - \alpha\lambda'/\Delta_o)$ because the subshell < half-filled and λ' is positive. $\mu_{s.o.} = \sqrt{15} = 3.87\mu_B > \mu_e$. Small difference because Δ_o is large. $[Ni(H_2O)_6]^{2+}$ has a ground term $^3A_{2g}$ therefore as above $\mu_e = \mu_{s.o.}(1 - \alpha\lambda'/\Delta_o)$ $\mu_{s.o.} = \sqrt{8} = 2.83\mu_B$ λ' negative, therefore $\mu_e > \mu_{s.o.}$ Because Δ_o is smaller than for $[Cr(H_2O)_6]^{3+}$ the difference between μ_e and $\mu_{s.o.}$ is larger. $[Co(H_2O)_6]^{2+}$ is $t_{2g}^5e_g^2$ high spin (H_2O weak ligand field) $\mu_{s.o.} = \sqrt{15} = 3.87\mu_B$ μ_e is much larger because the orbital contribution is not quenched. Ground term is $^4T_{1g}(F)$.

Chapter 11

1. $\Delta S^\ominus$ for $2C(s) + O_2(g) = 2CO(g) = (2 \times 197.9) - (2 \times 5.7) - 205 = 179.4$ J $K^{-1}mol^{-1}$, 1 mole gas gives 2 moles gas. $\Delta S^\ominus$ for $C(s) + O_2(g) = CO_2(g) = 214 - 205 - 5.7 = 3.3$ J $K^{-1}mol^{-1}$, no change in moles of gas. $\Delta S^\ominus$ for $Ca(s) + Cl_2(g) = CaCl_2(s) = 114 - 41.6 - 223 = -150.6$ J $K^{-1}mol^{-1}$, loss of one mole of gas. $\Delta S^\ominus$ for $^4/_3Al(s) + O_2(g) = {^2/_3}Al_2O_3(s) = (^2/_3 \times 51) - (^4/_3 \times 28.3) - 205 = -208.7$ J $K^{-1}mol^{-1}$, loss of one mole of gas.

2. $\Delta S^\ominus$ for HgO(s, red) = Hg(1) + $\frac{1}{2}O_2(g)$ at 298 K $= 77.4 + \frac{1}{2} \times 205 - 72 = 107.9$ J $K^{-1}mol^{-1}$

ΔG° for HgO(s, red) = Hg(1) + ½O_2(g) at 298 K = $\Delta H^{\circ} - T\Delta S^{\circ}$ = 91000 − (298 × 107.9) = 59 kJ mol^{-1}. For ΔG° to be zero T = 91000/107.9 = 843 K. HgO is thermally unstable above 843 K

3(a). Moles of ammonia = 0.012, Moles of Co = 0.1 × 20/1000 = 0.002, Moles of Cl^- = 0.86/(107.9 + 35.5) = 0.006, the complex has the formula $[Co(NH_3)_6]Cl_3$. (b) Low spin d^6 in an octahedral field. d−d bands at 472 nm and 338 nm are the transitions $^1A_{1g} \rightarrow {}^1T_{1g}$ and $^1A_{1g} \rightarrow {}^1T_{2g}$ The 200 nm band is LMCT.

4. The magnetic moment suggests 2 unpaired electrons, i.e. a V^{III} complex. The percentage of water = 0.145 × 100/1.32 = 11%, moles of water in 0.200 g = 0.2 × 0.11/18 = 1.22 × 10^{-3} moles. In the first titration both the oxalate ion and the V^{III} are oxidized, in the second only the oxalate ion is in solution to be oxidized. The relevant equations are:
$5V^{III} + 2\,Mn^{VII} \rightarrow 5V^{V} + 2Mn^{II}$
$5C_2O_4^{2-} + 2Mn^{VII} \rightarrow 2Mn^{II} + 2CO_2$
Moles of V^{III} = 0.0203 × (32.7 − 24.2)/1000 × 5/2 = 0.43 × 10^{-3} moles
Moles of oxalate ion = 0.0203 ×24.2/1000 × 5/2 = 1.23 × 10^{-3} moles
Moles K = 0.2 × 0.24/39.1 = 1.23 × 10^{-3} moles
The complex has the formula $K_3[V(C_2O_4)_3]\cdot 3H_2O$ (1.23/0.43 ≈ 3)
Optical isomers are possible.

5(a) Both A and B gave an effective room temperature magnetic moment of 1.7μ_B consistent with the one unpaired electron of $Ti^{III}t_{2g}^{1}$. The single band spectra are consistent with the one d electron, the shoulders caused mainly by the Jahn-Teller effect on the e_g^{1} excited configuration. The atomic ratios are: Ti:Cl:O:H = 1:3:6:12 with a RMM value of 262.37. 0.6559g of the green compound A contain 0.6559/262.37 = 2.5 × 10^{-3} moles. 0.3583 g AgCl = 2.5 × 10^{-3} moles so A has one free Cl^- and a cation with a charge of +1. Complex B releases free chloride ions equivalent to 0.10 × 75/1000 = 7.5 × 10^{-3} moles and has 3Cl^- with a cationic charge of +3. Complex A is $[Ti(H_2O)_4Cl_2]Cl\cdot 2H_2O$ and complex B is $[Ti(H_2O)_6]Cl_3$. This is an example of hydration isomerism. Complex A can have *cis*- and *trans*-isomers. Complex A is the *trans*-isomer because the *cis*-isomer would be expected to have two bands consistent with its lower symmetry. Complex B absorbs at lower wavelengths than complex A because it has six water ligands while A has two weaker field chloride ion ligands.
(b) The solution of B is 0.8745/262.37 = 3.33 × 10^{-3} *M*. Ti^{III} is oxidized to Ti^{IV}:
$Ti^{III} + Fe^{III} \rightarrow Ti^{IV} + Fe^{II}$
and required 12.5 cm^3 of the Fe^{III} solution (6.66 x 10^{-3} *M*). The titration has to be carried out in an inert atmosphere to avoid dioxygen oxidizing the Ti^{III}.

6. (a) The lattice energy CuBr = 827 kJ mol^{-1} using the Born-Landé equation and 959 from the thermochemical data The difference between the two lattice energies and the lower than expected interionic distance cannot be due to any errors in the data. They are caused by CuBr being considerably covalent. (b) The departure from ionic behaviour leads to pronounced δL values. As the polarizability of the halide ion increases from Cl to I, the departure from ionic behaviour increases as does the magnitude of δL. (c) The radius ratios in the three halides are:

Halide	CuCl	CuBr	CuI
r^+/r^-	0.533	0.492	0.440

Although the NaCl structure is expected from these values the compounds all adopt the zinc blende structure, due to the departure from true ionic behaviour.

7. (a) Moles of $VO^{2+} = 5 \times 0.02 \times 50/1000 = 5 \times 10^{-3}$ and the mass of $VOSO_4 = 5 \times 10^{-3} \times 163 = 0.815$ g. Mass of water = 1.26 − 0.815 = 0.445 g = 0.445/18 = 2.47×10^{-2} moles. Therefore, $n = 5$. The standard potential for the reaction:
$5VO^{2+} + MnO_4^- + H_2O \rightarrow 2H^+ + 5VO_2^+ + Mn^{2+}$ is given by 1.51 − 1.0 = 0.51 V, which gives $\Delta G^{\circ} = -nFE^{\circ} = -5 \times 96.5 \times 0.51 = -246$ kJ mol^{-1}. (b) V is violet in its II state, green in its III state and blue in its IV state. In the first experiment the V^V is reduced to the III state, the excess of V^V allowing the reaction: $V^{II} + 2V^V \rightarrow 3V^{IV}$ Mg reduces the V state to the III state. These observations are consistent with the two titres of 81.36 cm^3 and 54.36 cm^3 which are in a 3:2 ratio. 81.36 cm^3 of the Mn^{VII} solution are equivalent to 81.36/3 = 27.12 cm^3 of the V^V solution = $0.02 \times 5 \times 27.12/1000$ moles of V or $0.02 \times 5 \times 27.12/1000 \times 116.95$ (RMM of NH_4VO_3) = 0.317 g of the compound. The compound has a purity of $0.317/(12.81/40) \times 100 = 98.99\%$.

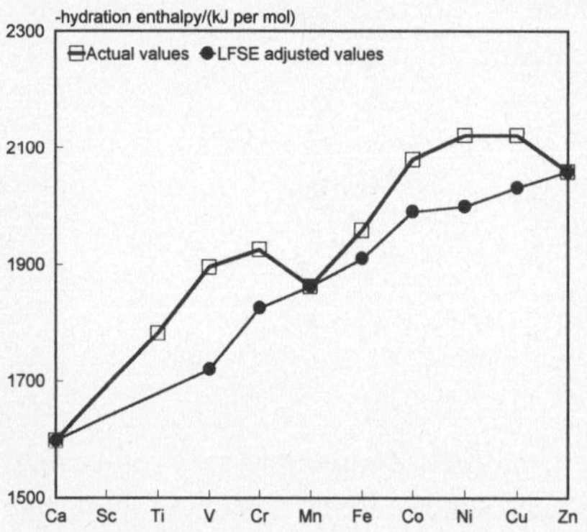

8. (b) Δ_o for $[V(H_2O)_6]^{2+} = N_A hc/\lambda = 147$ kJ mol^{-1} and for $[Ni(H_2O)_6]^{2+}$ is 102 kJ mol^{-1}. LFSE for $[V(H_2O)_6]^{2+} = -(^6/_5 \times 147) = -176$ kJ mol^{-1} and for $[Ni(H_2O)_6]^{2+}$ is $-(^6/_5 \times 102) = -122$ kJ mol^{-1}. The corrected $-\Delta_{hyd}H^{\circ}$ values for $[V(H_2O)_6]^{2+}$ and $[Ni(H_2O)_6]^{2+}$ are 1895 − 147 = 1748 and 2121 − 122 = 1999 kJ mol^{-1} respectively. LFSEs for Ca^{2+}, Mn^{2+} and Zn^{2+} are zero. The double hump shape of the unadjusted hydration enthalpies is the result of the variations in LFSE values which increase to a maxima V^{2+} d^3 and Ni^{2+} d^8. The plot which has had the ligand field effects removed is a more regular variation as the Z_{eff} increases. The variations in the M−OH_2 distance demonstrates the effect of LFSE changes. The quoted extreme values for Cr^{2+} and Cu^{2+} are due to the operation of the Jahn-Teller effect.

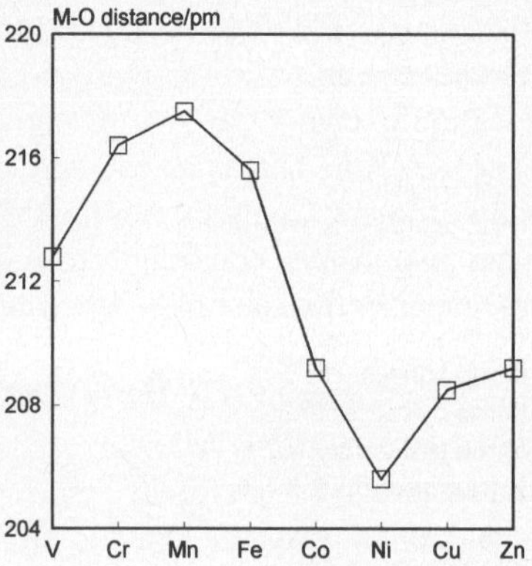

Index

Horwood Books of Chemical Relevance

REACTIONS MECHANISMS OF METAL COMPLEXES

R.W. HAY, School of Chemistry, University of St. Andrews

ISBN: 1-898563-41-1 200 pages *ca.* £15.00 1998

This text provides a general background as a course module in the area of inorganic reaction mechanisms, suitable for advanced undergraduate and post-graduate study and/or research. It tends to be an area which many students find quite difficult, possibly because mechanistic aspects are sometimes introduced rather late in inorganic chemistry courses. The topic has important research applications in the metallurgical industry, and is of interest in the sciences of biochemistry, biology, organic, inorganic and bioinorganic chemistry.

In addition to coverage of substitution reactions in four-, five- and six-coordinate complexes, the book contains further chapters devoted to isomerisation and racemisation reactions, to the general field of redox reactions, and to the reactions of coordinated ligands. This last area is not normally covered in undergraduate texts. It is strongly relevant in other fields such as organic, bioinorganic and biological chemistry, providing a bridge to organic reaction mechanisms. The book also contains a chapter on the kinetic background to the subject with many illustrative examples which will prove very useful to those beginning research.

Contents: Introduction to the field; The kinetic background; Substitution reactions of octahedral complexes; Substitution reactions in four- and five-coordinate complexex; Isomerisation and racemisation reactions; Reactions of coordinated ligands.

MÖSSBAUER SPECTROSCOPY: Principles and Applications

A. G. MADDOCK, Department of Chemistry, University of Cambridge *and* Fellow of St. Catherine's College

ISBN: 1-898563-16-0 272 pages £39.50 1997

Mössbauer spectroscopy has proved itself a most versatile technique, finding applications in extraordinary diverse areas of science and industry. Staring from physics and chemistry, it quickly spread into biochemistry, mineralogy, corrosion science, geochemistry and archaeology, with applications in industrial and scientific research. Dr Maddock aims to help advanced university students, professionals and research workers who ask the question "What's in it for us?". After a concise account of experimental techniques, he emphasises those applications in which there are few, if any, alternative ways of obtaining the same information about electron fields and the nuclei.

He explains areas of industrial interest, including the currently important applications related to tin and iron on which there is much activity in research and development, and interprets the extension of Mössbauer techniques to main group, transitional and other suitable elements. Great attention is paid throughout the book to factors which may lead to misinterpretation of spectra. An important chapter covers the complexities of interpreting emission spectra.

Contents: Basis of Mössbauer spectroscopy; Practical aspects of the technique; Further principles; Applications with tin; Applications with iron chemistry; Digression on magnetic effects; Further studies if iron compounds; The CEMS technique: applications in mineralogy and biochemistry; Applications to other elements; Mössbauer emission spectroscopy.

"A lot of theory in his book, presented in unusual and refreshing style. The first three chapters cover a large amount of background theory, practical aspects such as sources, cryostats and texture, and thickness effects are presented. It makes comparisons between tin and iron before discussing chemical systems involving iron. A useful discussion of magnetic effects, including ordering, spin-flop and magnetic changes, before more complicated Fe-57 spectra are dealt with. Treatment of iron-containing materials is thorough and up-to-date, and mineralogical and biological systems are included." - *Chemistry in Britain*

SYMMETRY AND GROUP THEORY IN CHEMISTRY

MARK LADD, Department of Chemistry, University of Surrey
Foreword by Professor The Lord Lewis, FRS, Warden, Robinson College, Cambridge

ISBN 1-898563-39-X 320 pages £20.00 1998

This clear and up-to-date introduction treats the subject matter thoroughly, in a logical sequence and in depth. Group theory has the power to draw together molecular and crystal symmetry, and the book will be helpful also to those reading any of the sciences wherein chemistry forms a significant part.

Copiously illustrated, including many stereoviews of molecules. Hints on stereoviewing are given, and reference is made in the text to programs developed by the author. The programs aid the derivation, study and recognition of point groups and provide for other germane procedures, and are available on Internet *www.horwood.net/publish*

Rigorous mathematics is avoided but more advanced mathematical topics are developed in the appendices. A generous supply of worked examples and a wealth of appropriate programs and exercises, with solutions to problems are given in extended tutorial form.

Contents: Symmetry everywhere; Symmetry operations and symmetry elements; Group theory and point groups; Representations and character tables; Group theory and wave functions; Group theory and chemical bonding; Group theory, molecular vibrations and electron transitions; Group theory and crystal symmetry.

"This book provides the enabling background to rationalize and synthesize the use of symmetry to problems in a wide range of chemical applications, and is a necessary part of any modern course of chemistry."
Professor The Lord Lewis, Warden, Robinson College, Cambridge

"The book treats the subject matter of its title in a logical sequence, thoroughly, and in depth. It would form an excellent text for a major core second or third year course."
Dr John Burgess, Reader in Inorganic Chemistry, University of Leicester

CHEMISTRY IN YOUR ENVIRONMENT

User-friendly, Simplified Science

J. BARRETT, Department of Chemistry, Imperial College, University of London

ISBN 1-898563-01-2 Library Hardback 250 pages 90 diagrams 1994
ISBN 1-898563-03-9 Paperback

Introduces chemical "mysteries" and shows the importance of chemistry to life quality, and how physics, metallurgy, geology and engineering, and the sciences of chemistry. biology, biochemistry and microbiology, have contributed to our use of the Earth's resources. Highlights the beneficial and harmful uses of chemicals, and the benefits chemists of made to industry, agriculture, medicine and other human activity.

"An admirable simple course on chemistry. It demands concentration, but it will be rewarded" - *New Scientist*

"A super book which I thoroughly recommend ... pundits often get things wrong because so many of them are seriously chemically challenged. I suggest you take a large dose of *Chemistry in Your Environment.* Dr Barrett's book is so welcome" - Dr John Emsley in *Imperial College Reporter*

ANTIOXIDANTS in Science, Technology, Medicine and Nutrition

GERALD SCOTT, Department of Chemistry, Aston University, Birmingham

ISBN 1-898563-31-4 350 pages Hardback 1997

"This distillation of a lifetime of work by one of the world's leading experts. It should find an essential place on the desk of anyone interested in the oxidative deterioration of organic materials and the ways of stopping it. A must have for anyone in the field." *Polymer Degradation and Stability* (Dr N. Billingham, University of Sussex)

"Professor Scott's considerable expertise comes through clearly in chapters on peroxidation in chemistry and chemistry technology, chain-breaking antioxidants, preventive antioxidants, synergism and technical performance. A welcome addition to the fields of free radicals, will hopefully generate healthy debate between the classical oxidation chemists and those in fields of biology, nutrition and medicine." - *Chemistry and Industry* (Garry Duthie, Rowett Research Institute, Aberdeen)

"The idea of collectively presenting information about antioxidants in technology and biology is certainly useful. The extensive chapter references provide a historical overview of the literature dating back to early in this century. Many figures, tables and schemes explain chemical and biological reactions." - *Choice, American Library Association* (Dr N. Duran, Illinois State University, USA)

EXPERIMENTAL DESIGN TECHNIQUES IN STATISTICAL PRACTICE

W.P. GARDINER, Department of Mathematics, Glasgow Caledonian University, *and* G. GETTINBY, Department of Statistics and Modelling Science, University of Strathclyde, Glasgow

ISBN 1-898563-35-7 256 pages £24.95 1997

This book describes classical and modern design structures together with new developments which now play a significant role for quality improvement and product development, principally for interpretation of data from industrial and scientific research. The material is reinforced with software output for analysis purposes, and covers much detail of value to users, such as model building and derivation of expected mean squares.

Advanced undergraduates and graduates will find it enjoyable to read and comprehend. The book lends itself for use as a reference source for advanced work in statistics, pharmaceutical chemistry, medical science, chemistry, biology, marketing, management, and production engineering. Users might be nutritionists developing diets for patient care, industrialists looking for quality production, economists assessing company performance, marketing analysts developing product sales strategy, and engineers devising constructional models.

The authors' approach is based on computer software (SAS and Minitab), and contains many practical and applicable exercises which will help readers understand, present and analyse data using the principles and techniques described in the book. The pragmatic approach offers technical training for use of designs, and teaches statistical and non-statistical skills in design and analysis of project studies throughout science and industry. For example, the Taguchi methods have been used in Japan for improvement and product development and have contributed significantly to that country's recent industrial success.

Contents: Introduction; Inferential data analysis for simple experiments; One-factor designs; One-factor blocking designs; Factorial experimental designs; Hierarchical designs; Two-level fractional factorial designs; Two-level orthogonal arrays; Taguchi methods; Response-surface methods; Appendices: Statistical tables; Glossary; Problems and answers.